# Analytical Lens Design
# (Second Edition)

Online at: https://doi.org/10.1088/978-0-7503-5774-6

## Series Editor

R Barry Johnson, a Senior Research Professor at Alabama A&M University, has been involved for over 50 years in lens design, optical systems design, electro-optical systems engineering, and photonics. He has been a faculty member at three academic institutions engaged in optics education and research, has been employed by a number of companies, and has provided consulting services.

Dr Johnson is an IOP Fellow, an SPIE Fellow and Life Member, an OSA Fellow, and was the 1987 President of SPIE. He serves on the editorial board of *Infrared Physics & Technology* and *Advances in Optical Technologies*. Dr Johnson has been awarded many patents, has published numerous papers and several books and book chapters, and was awarded the 2012 OSA/SPIE Joseph W Goodman Book Writing Award for Lens Design Fundamentals (second edition). He is a perennial co-chair of the annual SPIE Current Developments in Lens Design and Optical Engineering Conference.

## Foreword

Until the 1960s the field of optics was primarily concentrated in the classical areas of photography, cameras, binoculars, telescopes, spectrometers, colorimeters, radiometers, etc. In the late 1960s optics began to blossom with the advent of new types of infrared detector, liquid crystal display (LCDs), light emitting diode (LEDs), charge coupled device (CCDs), laser, holography, and fiber optics along with new optical materials, advances in optical and mechanical fabrication, new optical design programs, and many more technologies. With the development of the LED, LCD, CCD, and other electro-optical devices, the term 'photonics' came into vogue in the 1980s to describe the science of using light in the development of new technologies and the operation of a myriad of applications. Today optics and photonics are truly pervasive throughout society and new technologies are continuing to emerge. The objective of this series is to provide students, researchers, and those who enjoy self-education with a wide-ranging collection of books, each of which focuses on a topic relevant to the technologies and applications of optics and photonics. These books will provide knowledge to prepare the reader to be better able to participate in these exciting areas now and in the future. The title of this series is *Emerging Technologies in Optics and Photonics*, in which 'emerging' is taken to mean 'coming into existence', 'coming into maturity', and 'coming into prominence'. IOP Publishing and I hope that you will find this series of significant value to you and your career.

A full list of titles published in this series can be found here: https://iopscience.iop.org/bookListInfo/emerging-technologies-in-optics-and-photonics.

# Analytical Lens Design (Second Edition)

**Rafael G González-Acuña**
*Tecnológico de Monterrey, Monterrey, Mexico*

**Héctor A Chaparro-Romo**
*Oxford Immune algorithmics, Monterrey, Mexico*

**Julio C Gutiérrez-Vega**
*Tecnológico de Monterrey, Monterrey, Mexico*

**IOP** Publishing, Bristol, UK

ISBN 978-0-7503-5774-6 (ebook)
ISBN 978-0-7503-5772-2 (print)
ISBN 978-0-7503-5775-3 (myPrint)
ISBN 978-0-7503-5773-9 (mobi)

DOI 10.1088/978-0-7503-5774-6

Version: 20230501

IOP ebooks

British Library Cataloguing-in-Publication Data: A catalogue record for this book is available from the British Library.

Published by IOP Publishing, wholly owned by The Institute of Physics, London

IOP Publishing, No.2 The Distillery, Glassfields, Avon Street, Bristol, BS2 0GR, UK

US Office: IOP Publishing, Inc., 190 North Independence Mall West, Suite 601, Philadelphia, PA 19106, USA

## Dedication of the first edition

*…To our families.*

*…To independent and unsupported scientists.*

*…To free, critical and independent thinking.*

*…At the reason of doubt and why of the things.*

*…To the inexorable set of light, time and matter.*

*…and above all,*

*To the adverse and dignifying work that maintains the sense of existence.*

## Dedication of the second edition

*…To God*

# Contents

# Preface

## Preface of the first edition

Why this book? Because it is simply unique, it tells a story like no other book of optical design or geometrical optics; it tells the story of stigmatic lenses with a rigor that only a closed general analytical equation can tell.

The predominant paradigm in optical design is based almost entirely on optimization methods and concepts exported from the paraxial optics. The problem with this paradigm is not the many particularly useful solutions in engineering, the problem is that the mathematical rigor is lost.

This book is for people who are only here for the love of mathematics. We consider ourselves researchers who focus on mathematical rigor and its intrinsic beauty, naturally we are irritated with all kinds of hybrid concepts, such as approximate rays, principal planes, chief ray, etc.

Our objective here is to develop a lens design theory only using two postulates, the Fermat principle and Snell's law. The Fermat principle and Snell's law will be our axioms; the results of this work will be our theorems, lemmas and corollaries. In other words, are we looking for how far can we go if Fermat and Snell are true? What else is true? Defining the truth, as the implications that we have if Fermat and Snell are true.

Without fear of reaching anything, it is preferable not to strain that truth with a truncated step and not achieve anything; it is valid to say I do not know. Let's see how far we get!

Naturally, the results found in this book have the potential to be applied in all the areas of optical design. But with the beautiful characteristic that they are general and unique at the same time; since they came by drastically ignoring any optimization method and truncated theory.

With this book, we want to highlight the beautiful and exciting story behind the stigmatic lens! The reader should read all the chapters in their definite order in the book.

**Rafael G González-Acuña**
**Héctor A Chaparro-Romo**
**Julio C Gutiérrez-Vega**

## Preface of the second edition

Why a second edition? Well, because the original edition is good, but optics still advances, and new ideas, new results have come in the last years. These ideas come from the bases presented in the first edition.

The new edition presents the original fifteen chapters, that are the core of the theory behind the stigmatic lens. These chapters, without fear of reaching anything, develop a lens design theory only using two postulates, the Fermat principle and Snell's law. These chapters where made for people who are only here for the love of mathematics, optics and mostly refraction phenomena.

The new edition also adds six new chapters following the philosophy of the previous fifteen chapters. It is not only limited to stigmatic lenses, but also aplanatic lenses, stigmatic mirror systems and aplanatic mirrors. The new chapters lead the reader to extend the ideas of the first edition to areas that were not explored in that edition. Like freeform mirrors or aplanatic optical system with an arbitrary number of lenses.

The second edition has added two more chapters to the third part of this treatise, which addresses the ideas of an aplanatic lens and an aplanatic optical system with an arbitrary number of lenses.

Also, the second edition has added a fourth part which studies stigmatism and aplanatism in systems made purely with mirrors. This new part also takes the reader hand in hand, from the most basic ideas to the stigmatism in mirror to systems generated by an arbitrary number of freeform mirrors. Potential applications are discussed with illustrative examples.

I recommend the reader to read this book in order. I hope you can enjoy reading this book.

**Rafael G González-Acuña**

# Acknowledgements


## Acknowledgements of the first edition

### Acknowledgements of Rafael G González-Acuña

I want to thank so many people; I don't know where to start; this work personally means a lot to me. I need to tell you that I really didn't even think that I was going to graduate as a physicist, much less that I would write a book about optics.

First of all, I want to thank God, all-powerful, for enlightening me on my way despite my sins. Lord, forgive me …. This book is to honor thy glory.

I want to thank my family.

To my mother Carmen Leticia Acuña Medellín and to my father Rogelio González Cantú for encouraging me to follow my dreams, to teach me the habit of reading and to believe in me, I would be nothing without you ….

To my brother Rolando Enrique González Acuña for supporting my parents and me paying three semesters of the tuition of my bachelor degree, you did not have to do it, and I hope this treaty makes you feel that you did not lose money on me ….

Héctor Alejandro Chaparro Romo, Comrade, we did it! We had everything against, practically without support but we did it, comrade, we did it!

To Professor Barry Johnson, thank you very much for your interests in my research and for impelling me in this adventure of writing a book, I thank you very much, you are part of this dream!

To Julio C Gutiérrez-Vega for his support and patience!

To Israel Meléndez for his support and grammar check!

I would also like to thank Yoshio Catillejos, Adrian Lozano, Ileana Paulette, Max, Eliel, Mike, Joel, Homero, Mawa, Rojo, Feri, Tamayo, Chapa, Yepiz, Erick, Balderas, Rohan, Mateusz, Julian, Luis, Dor, Rulo, Benjas, Ferrer, Arturo, Vera, Vivi, Juan Lara, Bernardino, Cuevas, Abundio, Reinhard Klette, Alois Herkommer, Andrea, Michelle, Ianilli, Berumen, la familia de Chaparro, to IOP, to Conacyt, to Yachay Tech University, to Wolfram Research, to ITO, to ITESM, to UNDAM, AUT, CIO and many more! Thank you very much!

### Acknowledgements of Héctor A Chaparro-Romo

I would like to thank all those people who have made this project possible, mainly I want Professor Barry Jonhson and the IOP family.

I thank my parents, José Eleazar Chaparro Villegas and María del Refugio Romo González for their time, support, advice and understanding.

I would also like to thank my brother, Eleazar Enrique Chaparro Romo for all the feedback and observations throughout this assignment, to Professor José Raúl Miranda Tello for his friendship, advice and for all the time he has dedicated to forming high quality capital human.

Rafael, when we found the solution to this problem, you took a heavy load off my shoulders, and the joy from it is perpetual, again thanks for supporting me and continuing the step in company. I will always be in debt to you for not letting me walk alone among so many numbers and equations, you are a good comrade.

Finally, I would like to thank my professors Ricardo Benjamin Flores Hernandez, Juan Camilo Valencia Estrada, Zacarias Malacara Hernandez and Enrique Castro Camus of the Research Center in Optics, it is mainly for them who woke up the passion for research and for optical design. To Professor Julio Gutierrez Vega for his support for Rafael during his doctorate.

**Acknowledgements of Julio C Gutiérrez-Vega**

I should like to thank the many people that made this work possible. Discussions with Rafael G González-Acuña were quite useful in understanding the subtlety of geometrical optics and optical design. This book would not have been possible without his perseverance and constant effort. My thanks go to multiple anonymous reviewers for their critical comments and numerous suggestions.

## Acknowledgements of the second edition

**Acknowledgements of Rafael G González-Acuña**

I want to thank God for giving me the opportunity to have a second edition of *Analytical Lens Design*. Doing science is not only a pleasure for me, but I also know that God likes us doing science. The Glory is all yours!

I want to thank my mother Carmen Leticia Acuña Medellín and my father Rogelio González Cantú for their support and patience! To my family as well.

I want to thank IOP Publishing for their interest in having a second edition of *Analytical Lens Design*, and the readers of the first edition of *Analytical Lens Design*.

I want to thank the authors of the first edition of *Analytical Lens Design*. Héctor Alejandro Chaparro Romo, Comrade, this time we have separate paths, but with optics still on the way, thank you so much for everything!

I want to thank Julio C Gutiérrez-Vega, Professor thank you so much for all your help these years, I go back in time and remember the PhD days and the endless discussions about optics we use to have and also about football!

I would also like to thank Yoshio Catillejos, Adrian Lozano, Ghulam Murtaza, Eeero Tuulos, Eero Salmelin, Jianglei, Qiuyuan Zhang, Mikko Juhola, Paulius Vijeikis, Simon Thibault, Marko Eromaki, Pekka Ayras, Zane Sibley, Michelle Rocha, Aki Vannien, Markus Virta, Professor Barry Johnson, Paula Sola La Serna, Jorge Sánchez-Capuchino, and many more!

Also, I want to thank several institutions from Huawei Technologies, to Conacyt, to Yachay Tech University, to Wolfram Research, to Instituto Tecnológico y de Estudios Superiores de Monterrey, to Universidad Abierta y a Distancia de México, Centro de Investigaciones en Óptica A.C, Instituto de Astrofísica de Canarias, and Quadoa Optical CAD!

# Author biographies

## About the authors of the first edition

### Rafael G González-Acuña

**Rafael G González-Acuña** studied industrial physics engineering at the Tecnológico de Monterrey and studied a master's degree in optomechatronics at Centro de investigaciones en Óptica, A.C. He is currently studying his PhD at the Tecnológico de Monterrey. His doctoral thesis focuses on the design of free spherical aberration lenses. He is co-author of the solution to the problem of designing bi-aspheric singlet lenses free of spherical aberration.

### Héctor A Chaparro-Romo

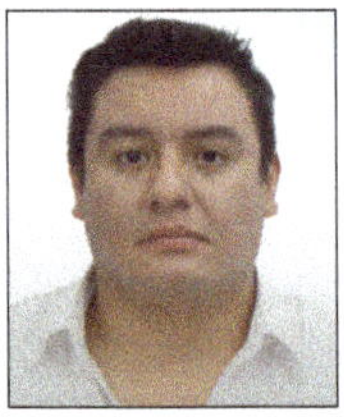

**Héctor A Chaparro-Romo**: Electronic Engineer specialized in scientific computation and years of experience in optics research and applications, he is co-author of the solution to the problem of designing bi-aspheric singlet lenses free of spherical aberration. He is an independent and multidisciplinary researcher, currently focused on the study of computer networks and advanced design of economic optical systems in his self-employed home office.

### Julio C Gutiérrez-Vega

**Julio C Gutiérrez-Vega** is a full professor in the Physics Department and heads the Photonics and Mathematical Optics Group at the Tecnológico de Monterrey, México. He has authored and co-authored about 101 journal papers and 83 conference proceedings, reporting more than 3500 citations to date with an h-index of 30. He served on the editorial committee of the journals *Optics and Photonics News* and *Optics Express* and was a member for 12 years of the scientific committee of the SPIE conference Laser Beam Shaping. Gutiérrez-Vega is a fellow member of OSA.

## About the author of the second edition

### Rafael G González-Acuña

**Rafael G González-Acuña** studied industrial physics engineering at the Tecnológico de Monterrey and studied a master's degree in optomechatronics at Centro de investigaciones en Óptica, A.C. He has a bachelor degree in mathematics from Universidad Abierta y a Distancia de México. He has a in PhD optics from Tecnológico de Monterrey. His doctoral thesis focuses on the design of free spherical aberration lenses. He is co-author of the solution to the problem of designing bi-aspheric singlet lenses free of spherical aberration. He has won several international award and scholarships. He has coauthor 32 paper is lens design. All of them as first and corresponding author. He works now a lens design for Huawei Technologies.

# Part I

## A historical, mathematical and optical introduction for beginners

**IOP** Publishing

Analytical Lens Design (Second Edition)

**Rafael G González-Acuña, Héctor A Chaparro-Romo and Julio C Gutiérrez-Vega**

# Chapter 1

# A brief history of stigmatic lens desing

This chapter gives a summary about the story of the problem of spherical aberration-free lens design. The study begins from the time of Euclid until the era of optical design programs. Contributions of renowned scientists behind the problem are reviewed.

## 1.1 The rise of geometrical optics

One of the oldest sciences, along with geometry, is geometrical optics. Geometrical optics can be seen as the extension of geometry through optics. It is so evident that what is true in geometry is true in geometrical optics.

All the phenomena studied by the paradigm under geometrical optics are based on Euclidean geometry. The magic of Euclidean geometry is that we are so related to it in everyday life activities.

With the knowledge of geometry, humankind understood the concepts of distance between objects. Humanity defined the space where objects are placed. These very abstract concepts were the key to all sciences. Geometry, in our personal point of view, is the playground where humankind understood mathematical rigor, proofs, and its very useful consequences and majesty.

The idea of measuring a distance related to a number in the real line $\mathbb{R}$ is so powerful that this notion implies that we can generate geometric bodies with precise distances between the points that compose them. And this is applied in every artifact developed by humankind from clothes to computers, from burning glasses to the most advanced telescopes.

With geometrical optics, humankind understood how to develop and design countless optical devices. These artifacts help us to see the world, the Universe and creation with different eyes. They change our point of view. All these apply if and only if the optical device has a specific geometry such that it can be associated with the concept of stigmatism. A gallery of optical devices can be seen in figure 1.1

doi:10.1088/978-0-7503-5774-6ch1

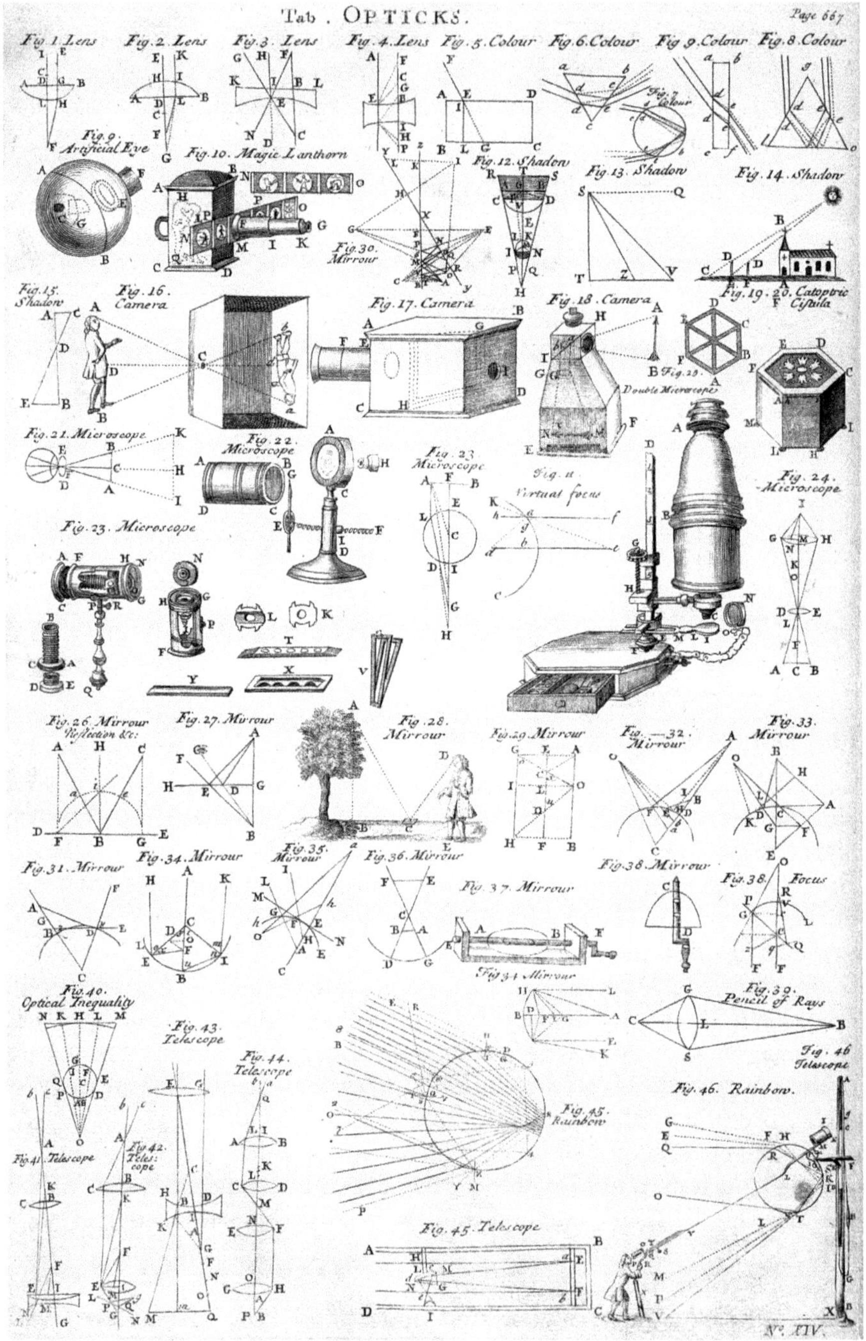


**Figure 1.1.** Illustrations of various optical instruments from Cyclopaedia. Cyclopaedia was an encyclopedia published by Ephraim Chambers in London in 1728. This image under the name of *Table of Opticks, Cyclopaedia, Volume 2* has been obtained by the authors from the Wikimedia website, where it is stated to have been released into the public domain. It is included within this book on that basis.

Let's define stigmatism as the property of an optical system to be similar to a geometric system as a whole. Therefore, a stigmatic system has a direct correspondence to a geometric arrangement. For example, the geometric foci of an ellipse are precisely the foci for an elliptical mirror. In the following chapter, we will give a more proper definition of stigmatism.

This chapter is a brief history of geometrical optics, but strategically, here we focus only on the most critical aspects of the history of stigmatic imaging. This chapter is the first step in our journey to get the general equation of lenses free of spherical aberration. The lenses free of spherical aberration are stigmatic.

## 1.2 Optics of the ancient Greeks and Arab world

In our personal point of view, geometry and geometrical optics are legitimate sons of a single father, Euclid. While we are writing these words, we have not quite found how to describe Euclid's majestic intelligence. A portrait of Euclid is in figure 1.2.

Euclid was a Greek mathematician and geometer. He probably lived from 325 BC to 265 BC. He is known as *the father of geometry*. However, the fact is that we do not know much about him. There are several theories about him. The first one is that he is the mathematician behind the magnum opus *Elements* and *Optics*, the other argument is that Euclid was the leader of a group of mathematicians and the term Euclid refers to a team of mathematicians. But the most accepted is that Euclid was the name of a flesh and blood citizen of Alexandria under the reign of Ptolemy I. Pappus of Alexandria (c. 320 AD) briefly gave evidence of the existence of Euclid, mentioning that Apollonius spent time with the pupils of Euclid. Paraphrasing, Pappus of Alexandria *spent a very long time with the pupils of Euclid at Alexandria, and it was thus that he acquired such a scientific habit of thought*.

Euclid in his prolific career wrote the following treaties: *Elements*, *Data*, *On Divisions of figures*, *Catoptrics*, *Phenomena* and *Optics* . In this chapter we are going to review the three most important for our purposes, which are *Elements*, *Catoptrics* and *Optics*.

*Elements* was the most prominent treatise of his career. The *Elements* of Euclid is a mathematical and geometric treatise that consists of thirteen books. In the first book, Euclid develops 48 propositions from 23 definitions (such as point, line and surface), five postulates and five conventional notions (axioms). Among these propositions is the first known general proof of the Pythagorean theorem. The other twelve books are implications of the five postulates and conventional notions. A fragment of *Elements* is shown in figure 1.3.

The common notions are:

1. Things equal to the same things are equal to each other.
2. If equals are added to equals, all are equal.
3. If equals are subtracted from equals, the remains are equal.
4. Things that match one another are equal to each other.
5. The whole is greater than the part.

The postulates are:

1. A straight line can be drawn by joining any two points.
2. A straight line segment can extend indefinitely in a straight line.

3. Given a straight line segment, a circle with any center and distance can be drawn.
4. All right angles are equal to each other.
5. By a point outside a line, a single parallel can be drawn

**Figure 1.2.** A portrait of Euclid. This image under the name of *Artgate Fondazione Cariplo - Cifrondi Antonio, Euclide* has been obtained by the author(s) from the Wikimedia website where it was made available Mcasanova under a CC BY-SA [2.0/3.0/4.0] licence. It is included within this book on that basis. It is attributed to Artgate Fondazione Cariplo.

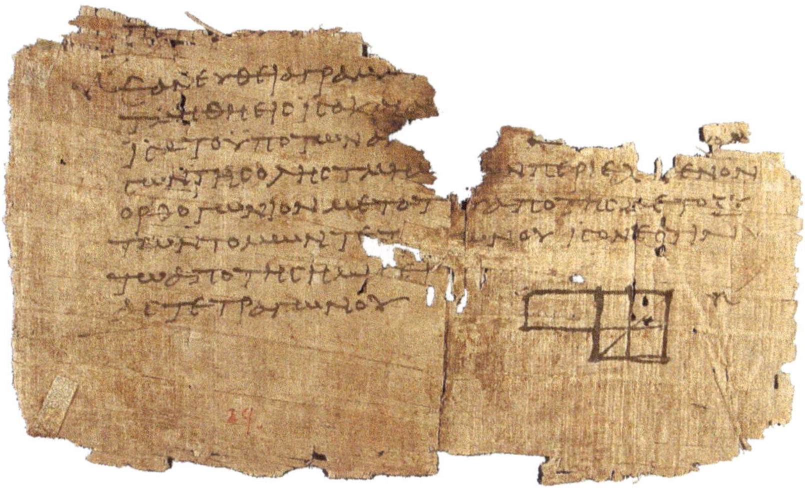

**Figure 1.3.** A conserved fragment of a copy of *Elements* of Euclid. This image under the name of P. Oxy. I 29. jpg has been obtained by the authors from the Wikimedia website, where it is stated to have been released into the public domain. It is included within this book on that basis.

With this, Euclid presented something more than geometry. He gave axiomatisation. An affirmation that is considered valid must either be contained within a base of initial statements, the so-called axioms, or it must be able to be demonstrated from them. The axioms are the fundamental pillars of every branch of mathematics, and from them, through mathematical demonstrations, the truth of any statement is deduced.

The axioms will, therefore, be statements that are accepted as truth and that their truth cannot be demonstrated from other axioms. A trivial or intuitive statement does not characterize an axiom; the axiom of choice being an example of an axiom that is not trivial.

The other type of statements referred to is theorems. These statements must be demonstrated using the axioms or other theorems already demonstrated. An immediate consequence of a theorem will be called a corollary.

Many parts of mathematics are axiomatized; thus, there is a set of axioms from we can deduce all the truths of that part of mathematics. For example, from Peano's axioms, it is possible to deduce all truths from arithmetics and by extension, from other parts of mathematics.

Axioms rule over all the mathematics behind physical models. Geometrical optics is not the exception. Geometrical optics is the study of the behavior of light through geometry. Once humanity understood the first axioms of geometry, it did not take too much time to relate it to optics. Euclid himself started to relate geometry with optics in his treatises *Catoptrics* and *Optics. Catoptrics* concerns the geometrical theory of plane and spherical concave mirrors. *Optics* is a work on vision geometry. The work focuses almost entirely on the geometry of vision, with few references to physical aspects. However, it gives the first premises of geometrical optics, that is, the rays of light travel in a straight line. These results of the research of Euclid are the first step to get stigmatic imaging.

Calculating the influence of Euclidean works on science is impossible, so we will only rate it as fundamental. One of the people influenced by Euclid is Apollonius of Perge. Apollonius of Perge (Perge, c. 262—Alexandria, c. 190 BC) was a Greek geometer famous for his work on conic sections. He was the one who gave the name of ellipse, parabola and hyperbola, to the figures we know. He managed to solve the general equation of the second degree employing conical geometry.

Only two works of Apollonius survive to this day: *Sections* in a given reason (the original is not preserved but there is an Arabic translation) and *The Conics* (only the original of half of the work is maintained, the rest is a translation into Arab). The second is the most critical work of Apollonius, moreover, together with the *Elements* of Euclid, it is one of the most essential math books. *The Conics* is made up of eight books.

1. Book I deals with the fundamental properties of these curves.
2. Book II deals with the conjugate diameters and the tangents of these curves.
3. Book III deals with the types of cones.
4. Book IV deals with the ways in which the sections of cones can be cut.
5. Book V studies maximum and minimum segments drawn with respect to a conic.
6. Book VI is about similar conics.
7. Book VII it is about conjugated diameters.
8. Book VIII has been lost, it is believed to be an appendix.

The importance of conic sections in optics is based on mirrors with conical shapes. These mirrors have the peculiarity of being stigmatic, which means that their geometric foci also turn out to be their optical foci.

The first person to notice these properties of the conic mirrors was Diocles. Although little is known about the life of Diocles, it is known that he was a contemporary of Apollonius of Perge.

Diocles studied the shape of a concave mirror, such as the spherical and parabolic mirror. He found that the parabolic mirror was stigmatic, for a distant object. Which means that if an object is very far from the mentioned mirror but is alienated with the axis of symmetry of the mirror, the mirror forms a stigmatic image of said object in its geometric focus. All this thanks to its predefined geometrical properties and also thanks to Euclid's fifth postulate. These results were published in his magnum opus *Burning Mirrors*.

The general objective of an optical system is to be stigmatic, which means that all the rays of a point object that pass over an optical system converge into a single image object. That is why we say that it has a stigmatic system that has a geometric corresponding, because of the conical figures.

For example, when designing a lens, it is intended that all the rays that emerge from a point object converge in a single image point and not a region.

Anthemius of Tralles like Diocles discovered another stigmatic mirror in his treatise *On Burning Glasses*. Anthemius found that the ellipse reflects light from a point source into a single point.

*Burning Mirrors* and *On Burning Glasses* had a significant influence on Arab mathematicians, particularly to Ibn al-Haytham Alhazen, a Muslim mathematician, physicist and astronomer of the eleventh century considered the creator of the scientific method.

The great thinker Alhazen was born around the year 965, from an Arab family, in Basora, present-day Iraq.

Alhazen arrived in Cairo under the reign of the Fatimid caliph Al-Hakim, a patron of science who was particularly interested in astronomy. A hydraulic project was recommended to the caliph to improve the regulation of Nile floods, a hard task. Still, later his fieldwork convinced him of the technical impossibility of this task. Alhazen continued to live in Cairo, in the neighborhood of the famous University of al-Azhar, until his death in 1040. Legend says that after deciding that the dam was not realizable and fearing the wrath of the caliph, Alhazen pretended madness and remained under house arrest since 1011 until the death of Al-Hakim in 1021. During this time, he wrote his influential *Book of Optics* in seven volumes and continued to write new treatises on astronomy, geometry, number theory, optics and natural philosophy. He is considered *the father of optics* for his work and experiments with lenses, mirrors, reflection and refraction. He wrote a comprehensive treatise on lenses, where he describes the image formed in the human retina due to the lens. A portrait of Alhazen is in figure 1.4.

Many of the advances presented by Alhazen were thanks to Ibn Sahl's studies. Ibn Sahl treated the optical properties of curved mirrors and lenses and has been described as the discoverer of the law of refraction, currently known as Snell's law. These results were presented in his magnum opus *On the Burning Instruments.*

Ibn Sahl knew very well the optics of ancient Greeks, but he went much further to the unexplored region of refraction. His treatise, *On the Burning Instruments,* is so unique that it makes Ibn Sahl the first mathematician known to have studied stigmatic lens design. In his time, the tenth-century, the main study was based on catoptrics. Ibn Sahl studied burning mirrors, both parabolic and ellipsoidal; he considered hyperbolic plano-convex lenses and hyperbolic biconvex lenses. All presented in his treatise, *On the Burning Instruments.* But his most notable achievement is the refraction law (Snellius's law) long before Snellius himself.

Sahl was the first to find the stigmatic refractive surface, today it is known as the Cartesian oval. Sahl proposed a stigmatic lens composed of two Cartesian ovals where the rays refracted in are collimated along the optical axis. Sahl could not find a general equation of stigmatic lenses. A fragment of *On the Burning Instruments* is shown in figure 1.5.

## 1.3 Snell, Descartes, Huygens, Newton and Fermat

It is a pity that the Europeans ignored much of the knowledge that the Arabs had in optics and had to reach the same results as Sahl independently.

For example, Copernicus, Kepler, and Galileo made terrific contributions to astronomy but they ignored the specific law of refraction obtained by Sahl. So they

**Figure 1.4.** A portrait of Alhazen. This image under the name of Hazan has been obtained by the authors from the Wikimedia website, where it is stated to have been released into the public domain. It is included within this book on that basis.

could not design a stigmatic lens properly. Sahl's law or the law of refraction was rediscovered by Willebrord Snell van Royen, also known as Snellius or Snell. In 1621, he enunciated the law of refraction of light, advancing, according to Christiaan Huygens in *Dioptrika* in 1703, to Descartes to whom the discovery was attributed by publishing it in *De la nature des lignes courbes* in 1637. The cover of the *La Géométrie* is shown in figure 1.6. *De la nature des lignes courbes* is the second book of *La Géométrie*.

Descartes, in his book *De la nature des lignes courbes*, independently rediscovered much of the knowledge provided by Shal in *On the Burning Instruments*. Descartes studied the conic mirrors and the Cartesian oval, but more importantly he introduced the problem of design of the stigmatic lens, when one of the refractive surfaces is given. René Descartes does not solve the problem; he leaves it to his successors. A portrait of René Descartes is shown in figure 1.7.

It seems that the problem was sufficiently impressive around the community because Christian Huygens mentioned it in the preface of his magnum opus, *Traité de la lumière* of 1690, that the fathers of calculus Sir Isaac Newton and Gottfried Wilhelm Leibniz were behind it.

Huygens mentions the problem several times in *Traité de la lumière*, he cites Descartes as the initial reference of the problem. The cover of the *Traité de la lumière* is shown in figure 1.8.

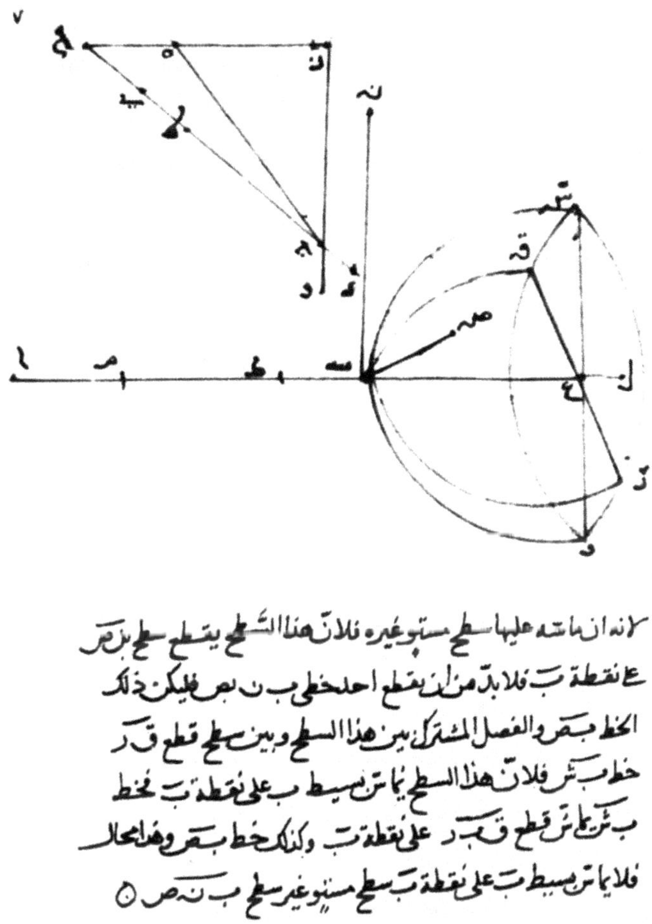

**Figure 1.5.** Part of the manuscript of Ibn Sahl, *On the Burning Instruments*. This image under the name of Ibn Sahl manuscript has been obtained by the authors from the Wikimedia website, where it is stated to have been released into the public domain. It is included within this book on that basis.

Finally, Huygens tried to solve it, but he does not present any equation. He just put several numerical calculations about the problem. A portrait of Huygens is shown in figure 1.9.

It is essential to mention that *Traité de la lumière* is one of the most critical books in optics, but not for this problem. It is because in the mentioned treatise Huygens presents his principle, which has his name. Huygens' principle expresses the idea that light is a wave. We are not going to cover this important event in the history of optics because here in this chapter we are going only to focus on an account of the stigmatic lens. The history of wave optics is an exciting subject that can take hundreds of pages to tell.

The other major book in optics of that generation is Newton's *Opticks*. In this book, Newton expresses his ideas about how light is composed of small particles.

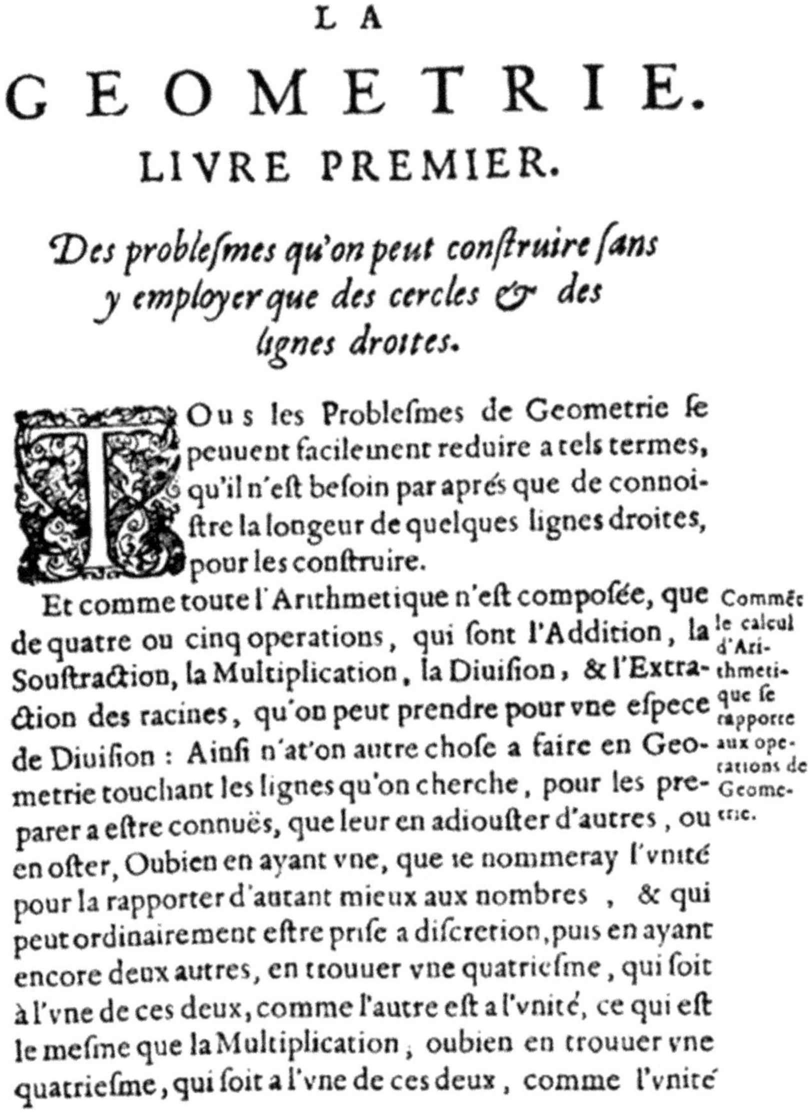

LA

GEOMETRIE.

LIVRE PREMIER.

*Des probleſmes qu'on peut conſtruire ſans y employer que des cercles & des lignes droites.*

Ous les Probleſmes de Geometrie ſe peuuent facilement reduire a tels termes, qu'il n'eſt beſoin par aprés que de connoiſtre la longeur de quelques lignes droites, pour les conſtruire.

Et comme toute l'Arithmetique n'eſt compoſée, que de quatre ou cinq operations, qui ſont l'Addition, la Souſtraction, la Multiplication, la Diuiſion, & l'Extraction des racines, qu'on peut prendre pour vne eſpece de Diuiſion : Ainſi n'at'on autre choſe a faire en Geometrie touchant les lignes qu'on cherche, pour les preparer a eſtre connuës, que leur en adiouſter d'autres, ou en oſter, Oubien en ayant vne, que ie nommeray l'vnité pour la rapporter d'autant mieux aux nombres, & qui peut ordinairement eſtre priſe a diſcretion, puis en ayant encore deux autres, en trouuer vne quatrieſme, qui ſoit à l'vne de ces deux, comme l'autre eſt a l'vnité, ce qui eſt le meſme que la Multiplication; oubien en trouuer vne quatrieſme, qui ſoit a l'vne de ces deux, comme l'vnité

Commẽt le calcul d'Arithmetique ſe rapporte aux operations de Geometrie.

**Figure 1.6.** Cover of the treatise *La Géométrie* of the Descartes. One of the book of this treatise is *De la nature des lignes courbes*. In this book Descartes independently rediscovered much of the knowledge provided by Shal in *On the Burning Instruments*. For example the Cartesian Oval. This image under the name of GeometryDescartes has been obtained by the authors from the Wikimedia website, where it is stated to have been released into the public domain. It is included within this book on that basis.

This belief created a big debate over who was right Newton or Huygens. Finally, time has given the verdict to Huygens.

However, *Opticks* is still a big step in lens design and stigmatic lens design. Newton presented chromatic aberration. Chromatic aberration is a type of optical

**Figure 1.7.** A portrait of René Descartes. This image under the name of Frans Hals - Portret van René Descartes has been obtained by the authors from the Wikimedia website, where it is stated to have been released into the public domain. It is included within this book on that basis.

distortion caused by the inability of a lens to focus all colors at a single point of convergence. The cover of *Opticks* is shown in figure 1.10.

Chromatic aberration is something that should be considered in lens design but can be ignored if we use conical mirrors since the reflection of light does not occur.

Newton in *Opticks* does not give a direct reference to the problem of the design of a stigmatic lens, but he draws in figure 3, part 1 of the first book of the mentioned treatise the idea of a stigmatic glass for three-point objects and three-point images. A portrait of Newton is shown in figure 1.11.

*Traité de la lumière* and *Opticks* serve as the basis for the next generation of optical designers that also were interested in designing a stigmatic lens.

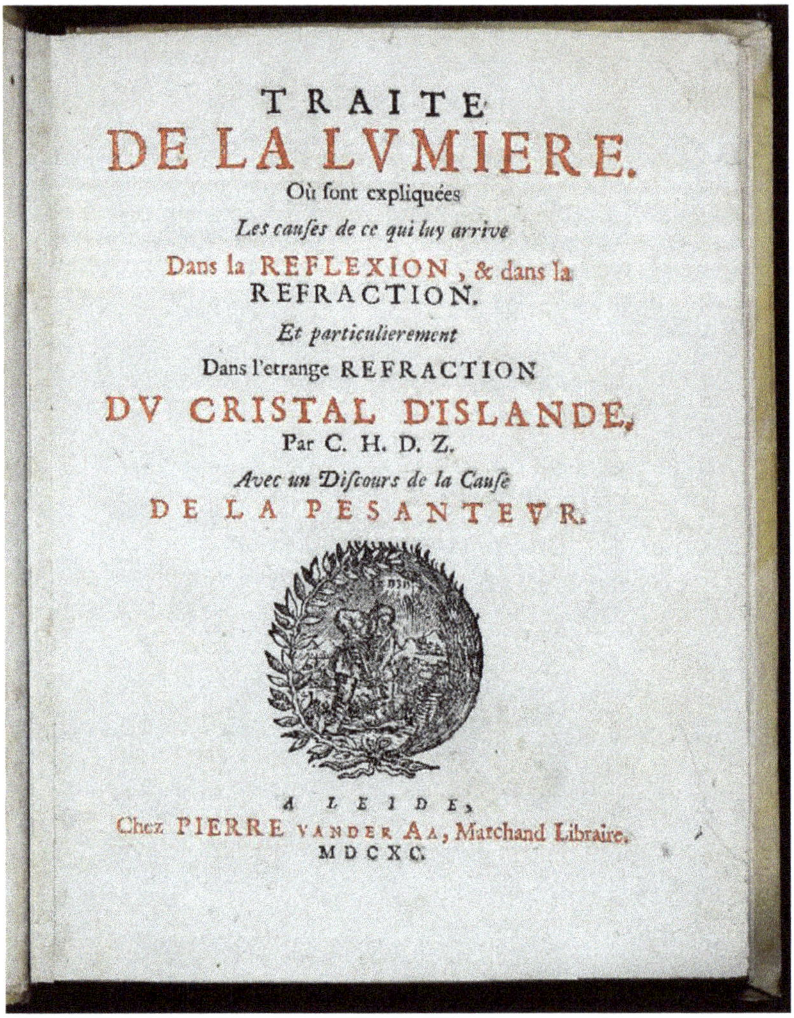


**Figure 1.8.** *Traité de la lumière* the magnum opus of Christiaan Huygens. In this treatise Huygens mentioned that the fathers of calculus Newton and Leibniz were behind the problem of the design a stigmatic lens. This image under the name of Web Huygens690 has been obtained by the authors from the Wikimedia website, where it is stated to have been released into the public domain. It is included within this book on that basis.

The result provided by Huygens was based on a numerical approach and he does not show a general equation.

Also, we want to mention that in this period was formulated the next principle that light follows,

*The path followed by the light when propagating from one point to another is such that the time taken to travel it is a minimum.*

Pierre de Fermat declared this principle in its modern form in a letter from 1662, hence it bears his name. The stigmatic lenses must obey this principle, which predicts

**Figure 1.9.** A portrait of Christiaan Huygens. This image under the name of Christian Huygens-painting has been obtained by the authors from the Wikimedia website, where it is stated to have been released into the public domain. It is included within this book on that basis.

that the time propagation of any ray that crosses a stigmatic lens is the same. In chapter 3, we are going to study it in detail.

## 1.4 19th and 20th Century

In the period of last two centuries, a significant advance in optics was made, with the introduction of the theory of waves and photonics. In terms of optical design, significant advances have been made in the design of lenses and optical systems. In this section, we are limited only to showing the works that we consider that

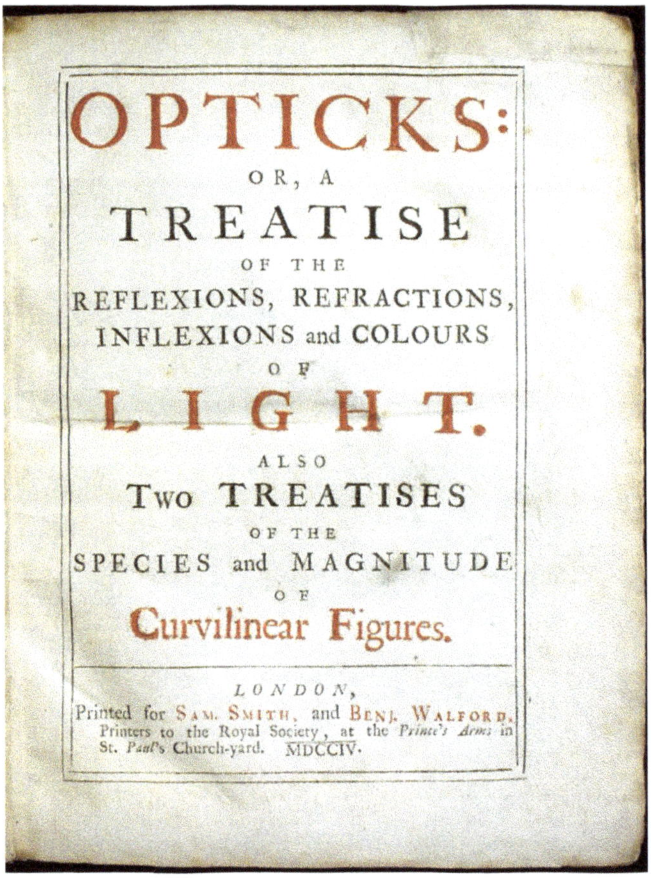


**Figure 1.10.** The cover of Newton's *Opticks*. This image under the name of Opticks has been obtained by the authors from the Wikimedia website, where it is stated to have been released into the public domain. It is included within this book on that basis.

contributed directly or indirectly but significantly to the problem of the design of a stigmatic lens.

According to Stravrudis, two significant figures in science were interested in the problem in this period. We are talking about Johann Carl Friedrich Gauss (1777–1855) and James Clerk Maxwell (1831–1879), but they failed to give a closed-form formula of a stigmatic lens.

Gauss proposed the design of the lens when the angle of incidence of the rays is tiny. When the angle is minimal, the law of refraction tends to have a linear response. This approximation is called paraxial optics or Gaussian optics.

On the other hand, James Clerk Maxwell treats the ideal lens as an optical device that maps a plain object into a plain image. For this, he applied a fractional linear transform; the problem of his approach was that he did not consider wavefront and refraction index in his ideal lens.

**Figure 1.11.** A portrait of Isaac Newton. This image under the name of GodfreyKneller IsaacNewton1680 has been obtained by the authors from the Wikimedia website, where it is stated to have been released into the public domain. It is included within this book on that basis.

David Brewster (1781–1868) was a Scottish scientist. He researched the field of optics directly with Snell's law. David Brewster made a comprehensive book about the optics known in his time. The book is called *Treatise on dioptriks*. In the mentioned book, Brewster referred to the stigmatic lens as non-existent in general.

Tullio Levi-Civita (1873–1941) was an Italian mathematician, famous for his work on tensor calculus, but who also made changes in other areas of mathematics and optics. Tullio Levi-Civita in 1900 published a paper under the name *Complementi al teorema di Malus-Dupin* were he gave several ideas on how the stigmatic lens or a stigmatic system of lenses should be. But he did not provide closed-form results.

Alexander Eugen Conrady (1866–1944) was a prominent optical designer and author of the book *Applied Optics and Optical Design*. In *Applied Optics and Optical Design* Conrady presented several approximated circumstances that spherical lenses are stigmatic, but he did not report a general solution.

Emil Wolf (1922–2018) was a Czech-American physicist. Wolf published several exciting works on lenses. One of the best known approximated solutions to the problem was proposed by Wolf and Max Born (Nobel Prize in Physics 1954) in their magnum Opus *Principles of Optics* in section 14.10.1. This attempt is partially based on the works of 1949 of Wasserman and Wolf called *On the Theory of Aplanatic Aspheric Systems* and the works *On the Determination of Aspheric Profiles* by Wolf and Pendry and *On the Designing of Aspheric Profiles* by Wolf. All the results of Wolf mentioned are not closed-form and partially deal with numerical approaches.

## 1.5 The computer era and the closure of a conjecture

The implementation of the computer in optical design completely changed the initial and rigorous paradigm implemented by the ancient Greeks. Computing capacity facilitated optical design based on trial and error and optimization algorithms. From a harsh and mathematical treatment, it became an almost purely numerical treatment. Nevertheless, the numerical results should not be less precious since they have given exciting results in all areas of optical design.

Thanks to this approximated paradigm, there is an infinite number of numerical solutions to the problem of the design of a spherical aberration-free lens. **But with this paradigm it is impossible to get a general formula to describe stigmatic lenses and solve some of the oldest myths in geometrical optics**. It is of general observance that no approximate solution mentioned closes the problem in question.

In 2018 we presented a closed analytical solution that closes the conjecture, and we show that the solution is unique.

This treatise is about how we come to the solution of said conjecture, showing step-by-step the derivation and extending it to its main implications. In this process there is no need of a computer, paraxial approximation, numerical nor optimization method.

## Further reading

Archibald R C 1950 The first translation of Euclid's Elements into English and its source *Am. Math. Mon.* **57** 443–52

Bell A E 2012 *Christian Huygens* (Redditch: Read Books Ltd)

Bellosta H 2002 Burning Instruments: From Diocles to Ibn Sahl *Arab. Sci. Philos.* **12** 285–303

Born M and Wolf E 2013 Principles of Optics: Electromagnetic Theory of Propagation *Interference and Diffraction of Light* (Amsterdam: Elsevier)

Boyer C B 1959 Descartes and the geometrization of algebra *Am. Math. Mon.* **66** 390–3

Brownson C D 1981 Euclid's optics and its compatibility with linear perspective *Arch. Hist. Exact Sci.* **24** 165–94

Conrady A E 2013 *Applied Optics and Optical Design, Part One* (North Chelmsford, MA: Courier Corporation)

Daumas M 1972 Scientific instruments of the seventeenth and eighteenth centuries and their makers *Scientific Instruments of the Seventeenth and Eighteenth Centuries and Their Makers* ed M Daumas (London: Batsford) 361 pp +142 plates

De Young G 1984 The Arabic textual traditions of Euclid's Elements *Hist. Math.* **11** 147–60

Descartes R 1637a *De la nature des lignes courbes*

Descartes R 1637b *La Géométrie*

Dijksterhuis F J 2004 *Lenses and waves: Christiaan Huygens and the Mathematical Science of Optics in the Seventeenth Century* vol 9 (Berlin: Springer Science)

Hartshorne R 2013 *Geometry: Euclid and beyond* (Berlin: Springer Science)

Heath T L *et al* 1956 *The Thirteen Books of Euclid's Elements* (North Chelmsford, MA: Courier Corporation)

Hogendijk J P 2002 The burning mirrors of Diocles: reflections on the methodology and purpose of the history of pre-modern science *Early Sci. Med.* **7** 181–97

Huygens C 1690 *Traité de la lumière*

Levi-Civita T 1900 *Complementi al teorema di Malus-Dupin: nota* (Rome: Tipografia della R. Accademia dei Lincei)

Luneburg R K and Herzberger M 1964 *Mathematical Theory of Optics* (Berkeley, CA: University of California Press)

McCarthy J P 1957 The cissoid of Diocles *Math. Gaz.* **41** 102–5

Mueller I 1981 *Philosophy of Mathematics and Deductive Structure of Euclid's 'Elements'* (New York: Dover)

Newton I 1704 *Opticks, or, a Treatise of the Reflections, Refractions, Inflections and Colours of Light* (New York: Dover)

Rashed R 1990 A pioneer in anaclastics: Ibn Sahl on burning mirrors and lenses *ISIS* **81** 464–91

Rashed R 2002 *Encyclopedia of the History of Arabic Science* (Milton Park: Routledge)

Scott J F 2016 *The Scientific Work of R Descartes: 1596-1650* (Milton Park: Routledge)

Seidenberg A 1975 Did Euclid's Elements, Book I, develop geometry axiomatically? *Arch. Hist. Exact Sci.* **14** 263–95

Serfati M 2005 R Descartes, Géométrie, Latin edition (1649) French edition (1637) *Landmark Writings in Western Mathematics 1640-1940* (Amsterdam: Elsevier) pp 1–22

Thomas-Stanford C 1926 *Early editions of Euclid's Elements* (London: Bibliographical Society) p 20

Tobin R 1990 Ancient perspective and Euclid's Optics *J. Warburg Courtauld Inst.* **53** 14–41

Toomer G J 2012a *Apollonius: Conics Books V to VII: The Arabic Translation of the Lost Greek Original in the Version of the Banū Mūsā* vol 9 (Berlin: Springer Science)

Toomer G J 2012b *Diocles, On Burning Mirrors: The Arabic Translation of the Lost Greek Original* vol 1 (Berlin: Springer Science)

Vaskas E M 1957 Note on the Wasserman-Wolf method for designing aspheric surfaces *J. Opt. Soc. Am.* **47** 669–70

Vitrac B 2013 Euclid *The Encyclopedia of Ancient History* (New York: Wiley)

Wassermann G D and Wolf E 1949 On the theory of aplanatic aspheric systems *Proc. Phys. Soc. Sect.* B **62** 2

Wolf E 1948 On the designing of aspheric surfaces *Proc. Phys. Soc.* **61** 494

Wolf E and Preddy W S 1947 On the determination of aspheric profiles *Proc. Phys. Soc.* **59** 704

**IOP** Publishing

Analytical Lens Design (Second Edition)

**Rafael G González-Acuña, Héctor A Chaparro-Romo and Julio C Gutiérrez-Vega**

# Chapter 2

# A mathematical toolkit for stigmatic imaging

In this chapter, we show the mathematical tools used to design lenses free of spherical aberrations. We start with set theory, topological spaces, metric spaces, conical curves and, finally, we review geometric algebra.

## 2.1 A mathematical toolkit

Once we have reviewed the history of geometrical optics for convenience, it is essential to set the mathematical machinery behind it. In this chapter, we present a toolkit only for mathematical concepts which are useful for us in stigmatic imaging[1]. This brief introduction points out the concepts of mathematics that are not usually studied by engineers in a formal mathematical language. Concepts like axioms of set theory, mapping, relations, homeomorphism are explored, while other mathematical tools often dominated by engineers like calculus are not presented.

This treatise is intended for students of optical or physical engineering in their last year, or graduate students in these disciplines. Therefore it assumes that the students know linear algebra and vector calculus.

## 2.2 Set theory

The notorious mathematician Paul Halmos, in his book *Naive Set Theory*, begins with the idea that every mathematician agrees with other mathematicians that a mathematician should know set theory. The disagreement is on how much. This treatise is not about mathematics, it's about optics, and we agree to start with some basic knowledge on set theory. The purpose is to present the required concepts of set theory, such that the reader is in a position that he/she can fully understand the mathematics behind stigmatic lenses without reading other sources. For a more advanced reading, the reader has to study the further reading for this chapter.

[1] Although we have not formally defined what stigmatic imaging is, this chapter is entirely mathematical and does not present any physical concept in general.

doi:10.1088/978-0-7503-5774-6ch2

An incredible fact in Nature that can be translated into the language of mathematics is that there are associations between the identities that surround us. A cluster of rocks, a bouquet of fruit, a bunch of leaves are all examples of sets in Nature. Sets are generated by a number of elements or members that conform to the set. An element of a set may be a rock, a fruit, or a leaf. It is essential to know that a set may also be an element of some other set. A bunch of leaves from a particular branch is part of the set of leaves from a whole tree.

The concept of a set is the foundation for all known mathematics. One of the main goals of mathematics is to study the nature of sets. Mathematics has countless examples of sets; for example a line, a set of points, a plane, a surface etc. Specific physical models can be precisely expressed in the language of mathematics. Therefore, set theory is exciting for us because we want to show the lens design theory in set theory, thus a lens, or a surface of a lens can be expressed as a set.

### 2.2.1 Axiom of extension

One of the fundamental ideas of set theory is the one that relates the concept of belonging. If $x$ belongs to $X$ ($x$ is a member of $X$, $x$ is contained in $X$, $x$ is an element of $X$), then in set theory we write this fact as,

$$x \in X. \tag{2.1}$$

Please see figure 2.1. Usually, in mathematical books, lower cases are for single elements, and upper cases are for sets. But that's not a general rule. Also if set $A$ is contained/included in $X$, then,

$$A \subset X, \tag{2.2}$$

also written as,

$$X \supset A. \tag{2.3}$$

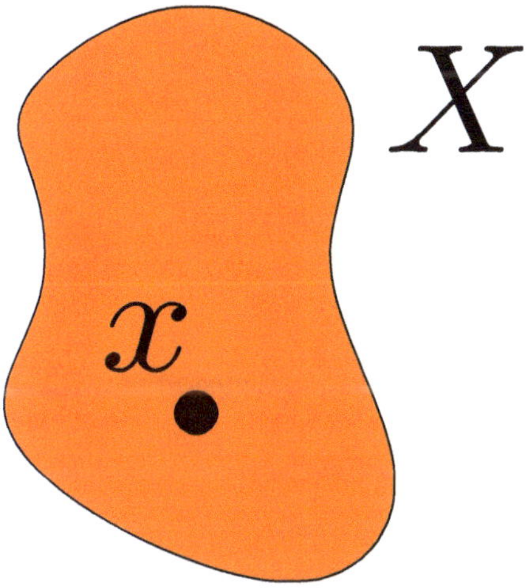


**Figure 2.1.** $x$ belongs to $X$, this means $x$ is a member of the set $X$.

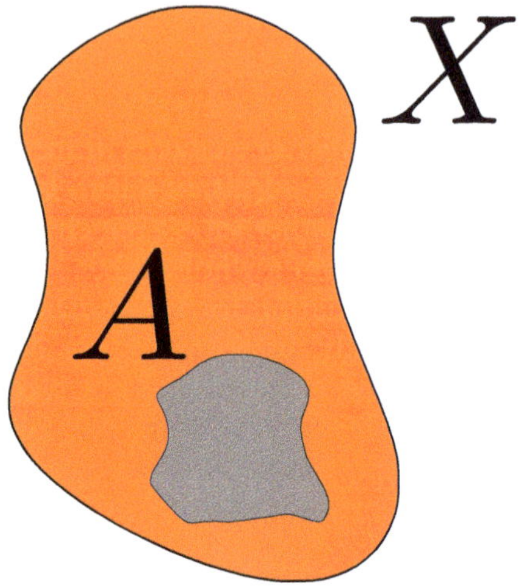


**Figure 2.2.** $A \subset X$, since $A$ is included in $X$, $A$ is a subset of $X$.

Please see figure 2.2. Observe that belonging $\in$ and inclusion $\subset$ are conceptually very different things indeed. Belonging is for elements, inclusion for sets. For every set $A$ is contained in itself, thus

$$A \supset A, \quad A \subset A. \tag{2.4}$$

This fact is expressed by saying that set inclusion is reflexive. If $A$ and $X$ are sets such that $A \subset X$ and $A \not\supset X$, $A$ is the proper subset $X$. If $A$, $B$, and $C$ are sets, with the particularity that they hold, $A \subset B$ and $B \subset C$, then $A \subset C$; this fact is called transitive.

A fundamental relation that two sets can have is if they are equal. The axiom of extension expresses the following statement on equality of sets.

**Axiom of extension.** *Two sets are equal if and only if they have the same elements*

When sets $A$ and $X$ are equal, then we can express this fact in set theory as,

$$A = X. \tag{2.5}$$

If the sets $A$ and $X$ are not equal, then this fact is expressed by writing,

$$A \neq X. \tag{2.6}$$

Therefore, the two sets are equal if and only if,

$$A \subset X, \quad A \supset X. \tag{2.7}$$

This fact is described by saying that set inclusion is anti-symmetric. Thus, if $A = X$, then necessarily $X = A$.

### 2.2.2 Axioms of specification and pairing

The next steps are to filter elements from a set and generate new sets. Make new sets from old ones. We can make sets from other sets with a specification that describes the membership of the new set. For example, we can generate a set of football players $A$ from the set of sports players $X$,

$$A = \{x \in X|S(x)\}, \tag{2.8}$$

where $S(x) = x$ is a football player. Let's write a numerical example, let $X = \{4, 5, 7, 8, 2, 4, 16\}$ and let $A$ be the set of elements of $X$ which are bigger than 6, so,

$$A = \{7, 8, 16\} = \{x \in X|x > 6\}, \tag{2.9}$$

where $\{x \in X|x > 6\}$ is the condition $S(x)$ of the membership of $A$. At this point it is appropriate to define the axiom of the specification.

**Axiom of specification.** *To every set A and to every condition S(x) there corresponds a set B whose elements are exactly those elements x of A for which S (x) holds*

With the axiom of specification, we can generate countless new sets, from older sets. But one of these *new* sets is so unique that it is crucial, the empty set. The empty set is defined as the result of,

$$\varnothing = \{x \in X|x \neq x\}. \tag{2.10}$$

Where the $X$ can be any set, the symbol $\varnothing$ is universally related to the empty set. The axiom of extension implies that there can be only one set with no elements, that set is the empty set. The empty set is a subset of every set, $\varnothing \subset X$ for every set $X$. Since there are no elements in the $\varnothing$, every element in the $\varnothing$ belongs to $X$.

With the empty set in mind, a fundamental set in set theory, and the axiom of specification we can extend these ideas to the Axiom of pairing, which is stated as,

**Axiom of pairing.** *For any two sets there exists a set that they both belong to.*

This means, that given a set $A$ and a set $X$, there exists a set $G$, such that $A$ is included in $G$ and $X$ is included in $G$. In other means $A \subset G$ and $X \subset G$, actually there is an infinite number of sets like $G$. We can quickly generate a set that includes $A$ and $X$ with the operation between sets.

### 2.2.3 Operations between sets

There are several operations between sets which give as a result other sets.

The first operation between sets that we are going to study is the union of sets. Given the sets $A$, $X$ the union of them is given by

$$A \cup X = \{x|x \in A \text{ or } x \in X\}. \tag{2.11}$$

Please see figure 2.3. The union between sets has the following features,

$$A \cup \varnothing = A, \tag{2.12}$$

$$A \cup X = X \cup A \text{ (commutativity)}, \tag{2.13}$$

$$A \cup (X \cup G) = (A \cup X) \cup G \text{ (associativity)}, \tag{2.14}$$

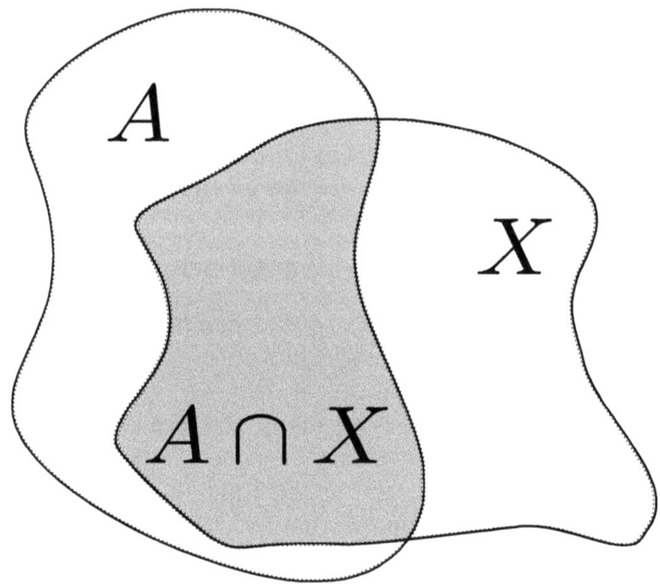


**Figure 2.3.** The colored area is the intersection between set $A$ and $X$ $A \cap X$.

$$A \bigcup A = A \text{ (idempontence)}, \tag{2.15}$$

$$A \bigcup X = X \text{ if and only if } A \subset X. \tag{2.16}$$

A suggestive example is the following,

$$\{x\} \bigcup \{y\} = \{x, y\}. \tag{2.17}$$

Actually, every set is the union of its elements, for example,

$$\{4, x, y, 9\} = \{4\} \bigcup \{x\} \bigcup \{y\} \bigcup \{9\}, \tag{2.18}$$

more general,

$$\bigcup_i^m \{x_i\} = \{x_1, x_2, x_3, \dots, x_m\}, \tag{2.19}$$

where $\bigcup_i^m$ denotes the union, of each of the $m$, $x_i$.

It is natural to think that after many unions between sets, their elements from one set are equal to elements from other sets. Let $A$ and $X$ be sets, then their *intersection* is

$$A \bigcap X, \tag{2.20}$$

which has the following definition,

$$A \bigcap X = \{x \in A | x \in X\} \tag{2.21}$$

by its symmetry it also can be defined as,

$$A \bigcap X = \{x \in X | x \in A\}. \tag{2.22}$$

For example,

$$\{4, x\} = \{4, 7, p, k, x\} \bigcap \{4, x, 9, j\}. \tag{2.23}$$

Some basic features of intersections between sets are,

$$A \bigcap \varnothing = \varnothing, \tag{2.24}$$

$$A \bigcap X = X \bigcap A, \tag{2.25}$$

$$X \bigcap (Y \bigcap Z) = (X \bigcap Y) \bigcap Z, \tag{2.26}$$

$$X \bigcap X = X, \tag{2.27}$$

$$Z \bigcap Y = Z, \text{ if and only if } Z \subset Y. \tag{2.28}$$

When two sets have no elements in common, just the empty set, they are called *disjoint*. Therefore, if $A$ and $X$ are disjoint sets we have,

$$A \bigcap X = \varnothing. \tag{2.29}$$

The *distributive laws* are two useful facts about unions and intersection between sets,

$$X \bigcap (Y \bigcup Z) = (X \bigcap Y) \bigcup (X \bigcap Z), \tag{2.30}$$

$$X \bigcup (Y \bigcap Z) = (X \bigcup Y) \bigcap (X \bigcup Z). \tag{2.31}$$

The demonstrations of the *distributive laws* are left as a task for the reader. Let $A$ and $X$ be sets; the *difference* between $A$ and $X$ is the elements that are members of $A$ but not members of $X$, thus,

$$A - X = \{x \in X | x \notin X\}, \tag{2.32}$$

where $\notin$ means not a member, we also can use the notation as $\notin \equiv \in'$, where $'$ is the complement operation. Thus,

$$A - X = \{x \in A | x \notin X\} = \{x \in A | x / X\}. \tag{2.33}$$

For example, $A'$ is the complement of $A$, which means it has all the elements which are non-members of $A$. For derivation proposes we can name a set such as all the sets to be mentioned are subsets of the same set $E$ and that all complements (unless otherwise specified) are formed relative to that $E$. In terms of this symbol notation, some basic features of complements are the following, Their proofs are given as an exercise to the reader.

$$(X')' = X, \tag{2.34}$$

$$\varnothing' = E, \tag{2.35}$$

$$E' = \varnothing, \tag{2.36}$$

$$X \cap X' = \varnothing, \tag{2.37}$$

$$X \cup X' = E, \tag{2.38}$$

$$X \subset Y \text{ iff } Y' \subset X', \tag{2.39}$$

where iff means if and only if. Important implications about the previews facts on complements are the *De Morgan laws*, which are stated as,

$$(A \cup B)' = A' \cap B', \tag{2.40}$$

$$(A \cap B)' = A' \cup B'. \tag{2.41}$$

De Morgan's Laws relate the intersection and union of sets through complements. We are not going to dig in to sets relations implicated by De Morgan's laws since this chapter is just a reminder to fresh the memory of the reader before the optics begins.

In general in sets, the order does not matter, for example, the set $\{a, b\}$, $\{a, b\} = \{b, a\}$, but in ordered pairs the order matters. An ordered pair $(a, b)$ is a pair of objects, such that the order in which the objects appear in the pair is significant. Therefore, the ordered pair $(a, b)$ is different from the ordered pair $(b, a)$ with the exception when $a = b$. Ordered pairs are beneficial for our proposes because we can specify a surface of a lens with a set of ordered pairs, as we will see in the following chapters.

The Cartesian product is a mathematical operation that returns a set from multiple sets. Let $X$ and $Y$ be sets, the Cartesian product $X \times Y$ is the set of all ordered pairs $(x, y)$ where $x \in X$ and $y \in Y$. Cartesian product $X \times Y$ in set notation is given by,

$$X \times Y = \{(a, b) | x \in X \text{ and } y \in Y\}. \tag{2.42}$$

The primary historical example is the Cartesian plane in analytic geometry. To represent geometrical shapes numerically and extract numerical information from shapes' numerical representations, René Descartes assigned to each point in the plane a pair of real numbers called its coordinates. The Cartesian products can express the shapes of geometrical bodies; lines, for our propose can represent the shape of optical elements, rays.

### 2.2.4 Relations and functions

Given two sets $X$ and $Y$, their Cartesian product $X \times Y$ and its elements are called ordered pairs; we can propose relations between the ordered pairs and the mentioned sets. In this book, we shall have no occasion to treat the theory of relationships that are ternary, quaternary, or worse. We just focus on binary relations.

A binary relation $R$ on $X$ and $Y$ is a subset of $X \times Y$; thus a binary relationship is a set of ordered pairs $(x, y)$ consisting of elements $x \in X$ and $y \in Y$. The set $X$ is named departure set, and the set $Y$ is called the destination set.

If the binary relation $R$, has some special features it is called a function. More precisely, If $X$ and $Y$ are sets, a relation function from (or on) $X$ to (or into) $Y$ is a function $f$ if $\operatorname{dom} f = X$ and such that for each $x \in X$ there is a unique element $y \in Y$

with $(x, y) \in f$. The uniqueness condition can be formulated explicitly as follows: if $(x, y) \in f$ and $(x, z) \in f$, then $y = z$, for each $x \in X$. The unique $y \in Y$ such that $(x, y) \in f$ is denoted by $f(x)$. If $f$ is a function, we shall write $f(x) = y$. The element $y$ is named the value of the function $f$ when the argument of the function is $x$; equivalently we may say that $f$ sends or *maps* $x$ onto $y$. The symbol of a function, transformation or mapping is,

$$f: X \to Y. \tag{2.43}$$

It is said that a function $f: X \to Y$ is *injective* if you only assign identical images to identical elements. If $x_1, x_2 \in X$ and $f(x_1) = f(x_2)$, then $x_1 = x_2$. It is called image of $x_1$ of the application of the $f: X \to Y$ on $x_1$. Thus $f(x_1)$ is the image of $x_1$. See figure 2.4.

A function $f: X \to Y$ is called *surjective* if every $y \in Y$, $\exists x \in X$, such as $f(x) = y$. See figure 2.5.

A function $f: X \to Y$ is called *bijective* if it is *injective* and *surjective*. In other means, it has a one-to-one correspondence between elements of $X$ and elements of $Y$, for every $f(x) = y \in Y$, then $\exists$ a corresponding $x \in X$. See figure 2.6. In the following chapters, we are going to use the above notation of functions. Functions are fundamental in this treatise since the ultimate goal for us is to obtain functions that describe surfaces of stigmatic lenses.

### 2.2.5 Continuity

A continuous function is a function for which sufficiently small changes in the input result in arbitrarily small changes in the output. Otherwise, a function is said to be a

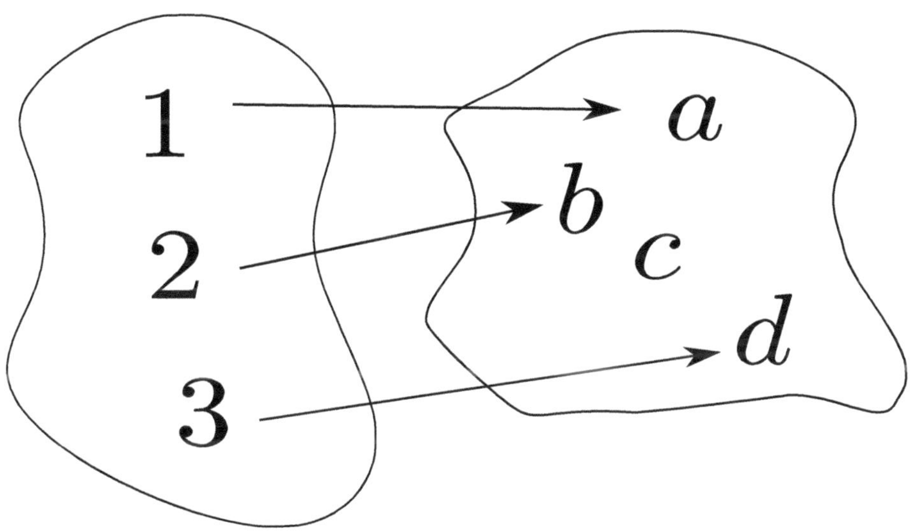


**Figure 2.4.** An injective function.

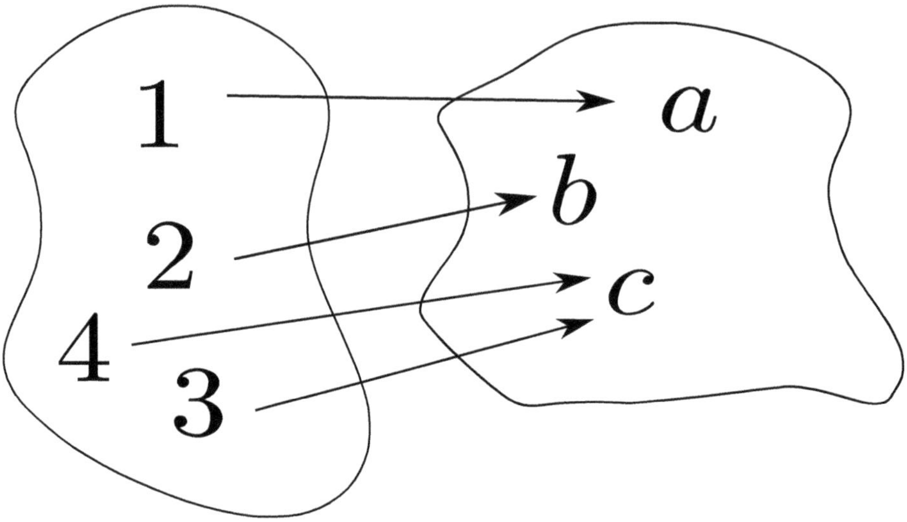


**Figure 2.5.** A surjective function.

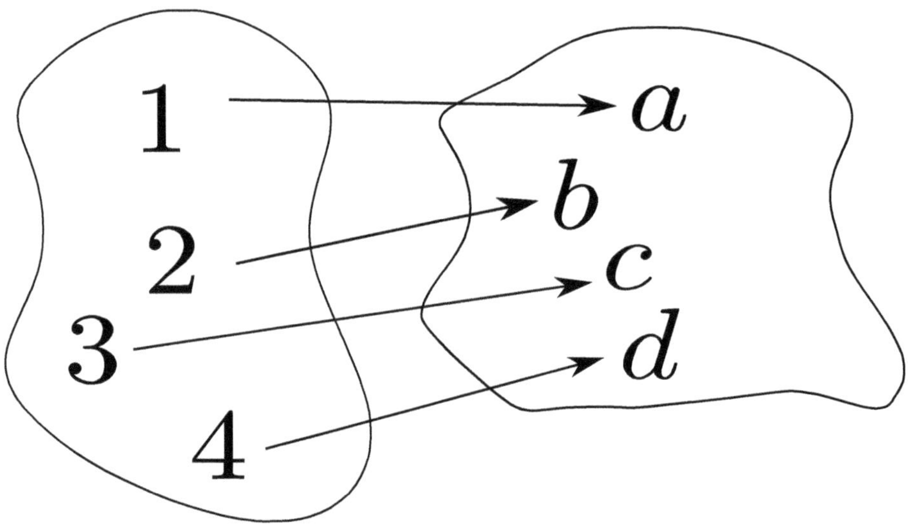


**Figure 2.6.** A bijective function.

discontinuous function. Formally speaking, a function $f$ is continuous on $x_0$ if $\forall x \in X$ and $\forall \epsilon > 0$, $\exists \delta > 0$[2],

$$|x - x_0| < \delta \Rightarrow |f(x) - f(x_0)| < \epsilon, \tag{2.44}$$

[2] The symbol $\forall$ means for all, and the symbol $\exists$ means exists. Therefore the statement $\forall x \in X$ and $\forall \epsilon > 0$, $\exists \delta > 0$, read as for all $x \in X$ and for all $\epsilon > 0$ there exists a $\delta > 0$.

where $|x - x_0|$ is the absolute value of the difference of the points $x$ and $x_0$, and $|f(x) - f(x_0)|$ is the absolute value of the difference between $f(x)$ and $f(x_0)$. These differences express how distant are $x$ and $x_0$, and $f(x)$ and $f(x_0)$. The differences between the elements in a set are the bases of the metric, a concept that we will see in detail in section 2.3.3.

We are especially interested in continuous functions since continuous functions describe many optical systems, but there are exceptional cases when the functions are discontinuous. We will explore the continuity conditions of stigmatic lenses in chapter 6. But, before properly studying the continuity of stigmatic lenses, we need to define the playground; topological and metric spaces.

## 2.3 Topological spaces

Topology is the branch of mathematics dedicated to the study of those properties of geometric bodies that remain unchanged by continuous transformations. It is a discipline that studies the properties of topological spaces and continuous functions. Topology is interested in concepts such as proximity connectivity, compactness, metrizability, among others.

In topology, two objects are equivalent in a much broader sense. They must have the same number of pieces, holes, intersections, etc. Topology allows one to bend, stretch, shrink, twist, etc, the objects, but only when it is done without breaking or separating what was attached, or sticking what was separated. For example, a triangle is topologically the same as a circle. We can transform a triangle into a circle, without breaking or pasting. But a circle is not the same as a segment since it must be split (or pasted) by some point.

A topological space is a mathematical structure that allows the formal definition of concepts such as convergence, connectivity, continuity, neighborhood, by subsets of a given set. The definition of a topological space is based only upon set theory. There are several equivalent definitions of this structure. The most common definition used is that in terms of open sets. But the reasonably more intuitive definition is that in terms of neighborhoods and so this one is given first. In figure 2.7 are examples of well-defined and not well-defined neighborhoods.

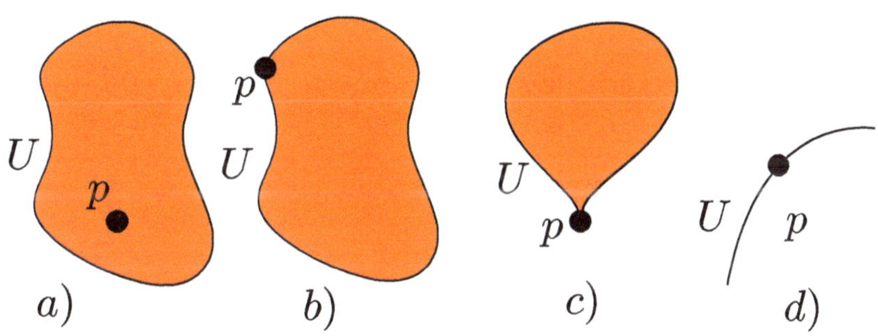


**Figure 2.7.** Example of neighborhoods in the Euclidean plane. The only valid neighborhood is (a), the other examples, (b), (c) and (d) are not well-defined neighborhoods.

### 2.3.1 Definition of a topological space via neighborhoods

Let's pick a point $p$ from the set $X$, (it can be Euclidean space). The neighborhood $U$ of $p$ is such that it should be a set of points near $p$, entirely surrounding $p$. The definition of the neighborhood is formulated in this way to be as free as possible. Taking the definition of neighborhood, $X$ is a topology if the following properties hold:

1. $p$ belongs to any neighborhood of $p$.
2. if $U$ is a neighborhood of $p$ and $V \subset U$, then $V$ is a neighborhood of $p$.
3. if $U$ and $V$ are neighborhoods of $p$, then $p \in U \cap V$.
4. If $U$ is a neighborhood of $p$, then there is a neighborhood of $p$ such as $V \subset U$, $V$ is a neighborhood of $p$.

**Definition 1.** A topological space is a set $X$, along with the assignment of each $p \in X$ of a collection of subsets of $X$, called *neighborhoods* of $p$ that satisfy the four properties mentioned above.

A classic example of such a system of neighborhoods is when $X$ is the Euclidean space, where a subset neighborhood is defined to be an open interval containing $p$. Examples of neighborhoods in the Euclidean plane are in figure 2.7.

### 2.3.2 Definition of a topological space via open sets

A topological space is an ordered pair $(X, T)$, where $X$ is a set and $T$ is a collection of subsets of $X$, satisfying the following axioms,

1. The empty set and $X$ itself belong to $T$. This means $\varnothing \in T$ and $X \in T$.
2. Any arbitrary (finite or infinite) union of members of $T$ still belongs to $T$.
3. The intersection of any finite number of members of $T$ still belongs to $T$.

It may seem strange that precise formulation of the concept of proximity is obtained from such a high formality and only upon set theory. The first thing observed is that different topologies can be defined in the same space $X$. The elements of $T$ are called open sets, and the collection $T$ is called a topology on $X$.

Once, we have formally defined what a topology is, we can officially define when topological spaces are equivalent.

### 2.3.3 Continuity and homeomorphism

Continuity is a concept that we have already studied in section 2.2.5. However, in that section, it was considered from the metric point of view.

Nevertheless, the concept of continuity is a topological concept. This may mean to define continuity, there is no lack of talking about distances, the use of the open set concept can be limited. Furthermore, if one has two sets, $X$ and $Y$, it is equivalent to say whatever functions between $X$ and $Y$ are continuous, to define a topology in $X$ and another in $Y$. This is why the concept of continuity is central in topology.

Topological definition of continuity is more general. It can be used for functions that go from one topological space to another, or instead, it is applied in spaces where the concept of distance may not be defined.

**Definition.** Let $X$ and $Y$ be topological spaces. A function $f$: $X \to Y$ it is continuous at $x \in X$ if $U$ is open at $X$ every time $f(U)$ is open at $Y$. $U$ is a neighborhood of $x \in X$ and $f(U)$ is a neighborhood of $f(x) \in Y$.

In simple words, they are continuous if the inverse image $f^{-1}(f(U))$ of any open set is an open set. With this definition, the idea of close points is expressed in terms of open sets: two points are close if they are in the same neighborhood, which is an open set.

The image of a set $U$ under a function $f$: $X \to Y$ is the function $f$ evaluated on $U$, $f(U)$. The inverse image is $f^{-1}(f(U))$, if $f$ is bijective then, $f^{-1}(f(U)) = U$.

Now, we have mentioned that in topology we can apply several transformations on a set, to end in another set that is different geometrically speaking—for example, implementing changes over a triangle to end with a circle.

We mentioned that the triangle and the circle are topologically equivalent because we can apply transformations to the triangle and end with a circle. When these transformations are continuous and reversible, we have a homeomorphism. Like the triangle and the circle that are topologically equivalent, a doughnut and a cup of coffee are also topologically equivalent, see figure 2.8.

Homeomorphisms are functions between topological spaces that provide us with the most fundamental notion of topological equivalence. These functions preserve all the properties between the topological spaces, since they are a direct correspondence between them.

The properties preserved by homeomorphisms are called topological properties. Two homeomorphic spaces share the same topological properties. If we find a topological feature that does not share two spaces, we can affirm that they are not equivalent; in other words, they are topologically different.

**Definition.** Let $X$ and $Y$ be topological spaces. A homeomorphism $f$: $X \to Y$ is a bijective function with inverse $f^{-1}$: $Y \to X$ such that $f$ and $f^{-1}$ are continuous. If there is a homeomorphism between $X$ and $Y$ we say that $X$ and $Y$ are homeomorphic.

**Figure 2.8.** Topologists cannot distinguish a doughnut from a cup of coffee, because the doughnut and cup of coffee are homeomorphic.

A homeomorphism is a bijection between topological spaces, such that it preserves the topological structure between them.

We have mentioned the topological properties preserved by homeomorphisms, but we haven't yet discussed them. In the next section, we present a brief review of the topological features, which are also intrinsically related to the shape of lenses and other optical systems.

### 2.3.4 Topological properties

The first one is **connectivity**. A connected set is a subset $C \subset X$ of a topological space $(X, T)$ that cannot be expressed as a disjoint union of two non-empty open topology sets. Intuitively, a connected set is the one that appears as a single piece, which cannot be *divided* or *partitioned.* If a set is not connected, then it is disconnected. See figure 2.9.

In the language of set theory we can express the connectivity as,

**Definition.** Let $A$, $B$, $C$ be open sets, thus $A, B, C \subset T$, such as $(A \cap B) \cap C = \varnothing$ and $C \subset A \cup B$. This implies that $C \subset A$ or $C \subset B$. Notice that if $C = X$ and meets the above, then we say that $(X, T)$ is a connected topological space.

**Compactness**, a compact space is a space that has properties similar to a finite set (a set that has a finite number of elements). The notion of compactness is a more general version of this property. More accurately, compactness is a property that generalizes the notion of a subset of Euclidean space being closed and bounded. Let's define what is a closed set.

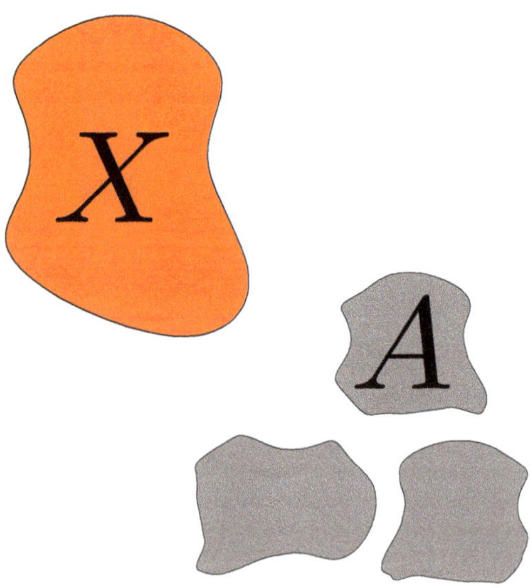


**Figure 2.9.** The set $X$ is connected and the set $A$ is disconnected.

**Definition.** A set is called closed if it is not open.

A more intuitive definition is that a closed set is a set which contains all its limit points. The definition of limit point is the following.

**Definition.** Let $S$ be a subset of a topological space $X$, $S \subset X$. A point $x \in X$ is a limit point of $S$ if every neighborhood $U$ of $x$ contains at least one point of $S$ different from $x$ itself, $S \cap (U - \{x\}) \neq \emptyset$.

For example, the interval (0,1) has as its limit points all the points of the interval [0,1].

A set is called bounded, if it is, in a certain sense, of finite size. This means that the set has all its points lying within some fixed distance of each other. The problem here is the word distance, to know the distance between two points in a topological space it is necessary to define a metric. The word bounded makes no sense in a topological space without a metric. In the following section, we are going to focus on the concept of the metric, once we fully understand it, the concepts of compactness and metrizability will be clear.

Some examples of bounded sets are the closed interval, the rectangle, or a finite set of points. These sets are called bounded, thus, if a set which is not bounded is called unbounded.

In our optical model, we want to know if in what circumstances stigmatic lenses are compact and in which cases the lenses are bound or not.

Finally, what are the implications of the intrinsic topological properties of stigmatic lenses in the forming imaging performance.

## 2.4 Metric spaces

In our study, we have reviewed the fundamental concepts of set theory; we have certain sets that form topological spaces. In this study, we have needed to mention the idea of the distance between two elements of a set. In this section, we will have a look at the concept of metrics and its implications in optical design.

A metric space is a set that has a distance function associated with it. This function is defined in the mentioned set, fulfilling properties attributed to a distance. So for any pair of points in the set, there is a distance assigned by this function.

**Definition.** A metric space is a set $X$, with a distance function $d$ assigned called metric $d$: $X \times X \to \mathbb{R}$, where $\mathbb{R}$ is the set of real numbers. The conditions satisfied by a metric are for $x_1, x_2, x_3 \in X$,

1. $d(x_1, x_2) > 0 \iff x_1 \neq x_2$,
2. $d(x_1, x_2) = 0 \iff x_1 = x_2$,
3. $d(x_1, x_2) = d(x_2, x_1)$ (symmetry),
4. $d(x_1, x_3) \leqslant d(x_1, x_2) + d(x_2, x_3)$ (triangle inequality).

Also, it is important to see that a metric can define a neighborhood. Metrically speaking, the neighborhood is the set of points that are distant from another equal to or less than a given distance, called radius.

**Definition** Let $(X, d)$ be a metric space. Let $a$ be real number and let $r > 0$. Let $a \in X$. The open neighborhood of center $a$ and radius $r$ is defined as the set $\{x \in X | d(a, x) < r\}$.

In particular, any metric space will also be a topological space because any function of distance defined on a given set induces a topology on that set. This is the topology induced by the open neighborhoods associated with the distance function of the metric space.

### 2.4.1 Euclidean metric

In this treatise there are three metrics that are fundamental: the Euclidean metric in the real line $\mathbb{R}$ for two points $x_1, x_2 \in \mathbb{R}$ is given by

$$\boxed{d(x_1, x_2) \equiv |x_1 - x_2|,} \tag{2.45}$$

where $|\cdot|$ is the absolute value.

The Euclidean metric in the plane $\mathbb{R}^2$ for $\vec{\mathbf{x}}, \vec{\mathbf{y}} \in \mathbb{R}^2$ is given by

$$\boxed{d(\vec{\mathbf{x}}, \vec{\mathbf{y}}) \equiv \sqrt{(x_1 - y_1)^2 + (x_2 - y_2)^2},} \tag{2.46}$$

where, $\vec{\mathbf{x}} = [x_1, x_2]$, $\vec{\mathbf{y}} = [y_1, y_2]$, and finally Euclidean metric for three dimensions,

$$\boxed{d(\vec{\mathbf{x}}, \vec{\mathbf{y}}) \equiv \sqrt{(x_1 - y_1)^2 + (x_2 - y_2)^2 + (x_3 - y_3)^2},} \tag{2.47}$$

where, $\vec{\mathbf{x}} = [x_1, x_2, x_2]$, $\vec{\mathbf{y}} = [y_1, y_2, y_3]$. Therefore, the Euclidean space is a metric space.

The Euclidean metric is fundamental for geometrical optics since the whole theory lies in the Euclidean space and is the only metric implemented in this treatise.

The Euclidean metric is the geodesic line between two points in the plane. The geodetic line is defined as the line of minimum length that joins two points on a given surface and it is contained in the surface.

The Euclidean metric has all the information of the Euclidean geometry, the five postulates[3].

Let's recall the postulates of *The Elements*:

1. Any two points determine a line segment.
2. A line segment can extend indefinitely in a straight line.
3. You can draw a circle given a center and any radius.

[3] Euclid's postulates refer to the treatise called *The Elements*, written by Euclid about 300 BC, exposing the geometric knowledge of classical Greece, deducing it from five postulates, the most prominent and straightforward.

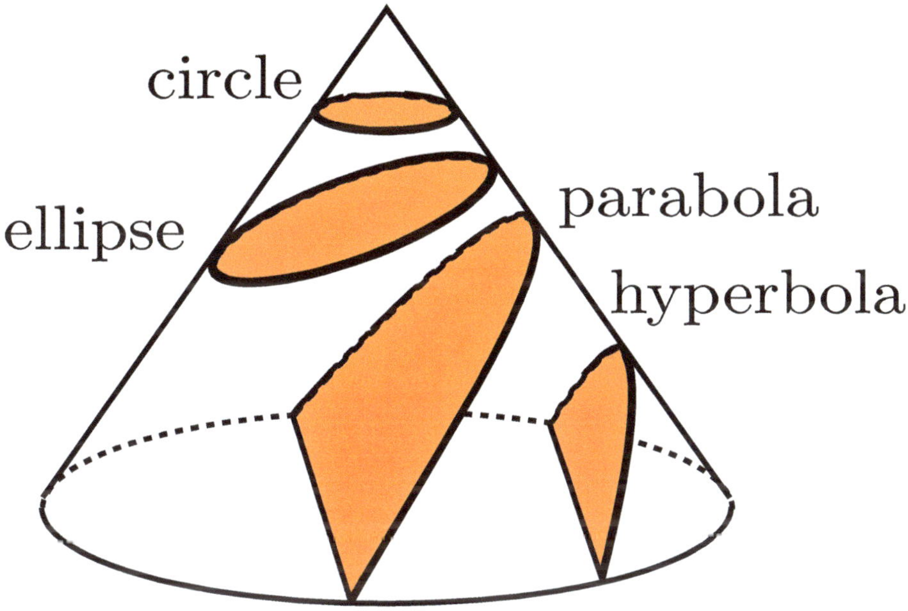


**Figure 2.10.** Conic sections.

4. All right angles are equal to each other.
5. Two parallel lines keep a finite distance between them.

All the postulates are self-evident except for the last one. The last one defines Euclidean geometry. Without it, we are left to absolute geometry[4].

Finally, we want to express that all optical systems lie on Euclidean space, so all the properties applied in the Euclidean space apply to optical systems. If the Euclidean space is a metric space, then the optical systems are metric spaces. If the Euclidean space is a topological space, then the optical systems are topological spaces.

These premises are essential for us to define stigmatic lenses as sets. This paradigm will help us recognize their geometric and topological intrinsic properties.

## 2.5 The conics

The conic sections are called to all curves resulting from the different intersections between a cone and a plane; if the mentioned plane does not pass through the vertex, then the conics are obtained. Conic sections are classified into four types: ellipse, parabola, hyperbola and circumference.

The conic sections are exceptional sets, which can be described as continuous functions. But what makes the conic sections so unique is because they are sets of all

[4] For a more profound lecture on non-Euclidean geometries see the citations in this treatise.

points that form a curved line such that each point satisfies a specific condition. In general, this type of set is called locus.

Figure 2.10 shows the conic sections as colored regions inside the cone.

Finally, before we begin, please note that all the distances implemented in the conic sections are given by Euclidean metrics.

### 2.5.1 The parabola

The parabola $\mathcal{P}$, the first conic section that we are going to study, is a locus (set of points) that lies in the Euclidean plane $\mathbb{R}^2$. See figure 2.11.

What makes interesting the parabola $\mathcal{P}$ is that for any point $P$ of the set the distance $|PF|$ to a fixed point $F$ is equal to the distance $|Pl|$ to a fixed-line $l$,

$$\boxed{\mathcal{P} = \{P \in \mathbb{R}^2 \mid |PF| = |Pl|\},} \tag{2.48}$$

where, $F$ is named the focus and $l$ the directrix. Please see figure 2.11.

Let's choose the coordinates in the Euclidean plane such that the parabola lies in the first and second quadrant, and its center is at the origin. Then, $F = (0, p)$, $p > 0$ and the directrix line is given by $y = -p$, so for ordered pair $P = (x, y)$ that holds $|PF| = |Pl|$ it holds for $|PF|^2 = |Pl|^2$, so we have,

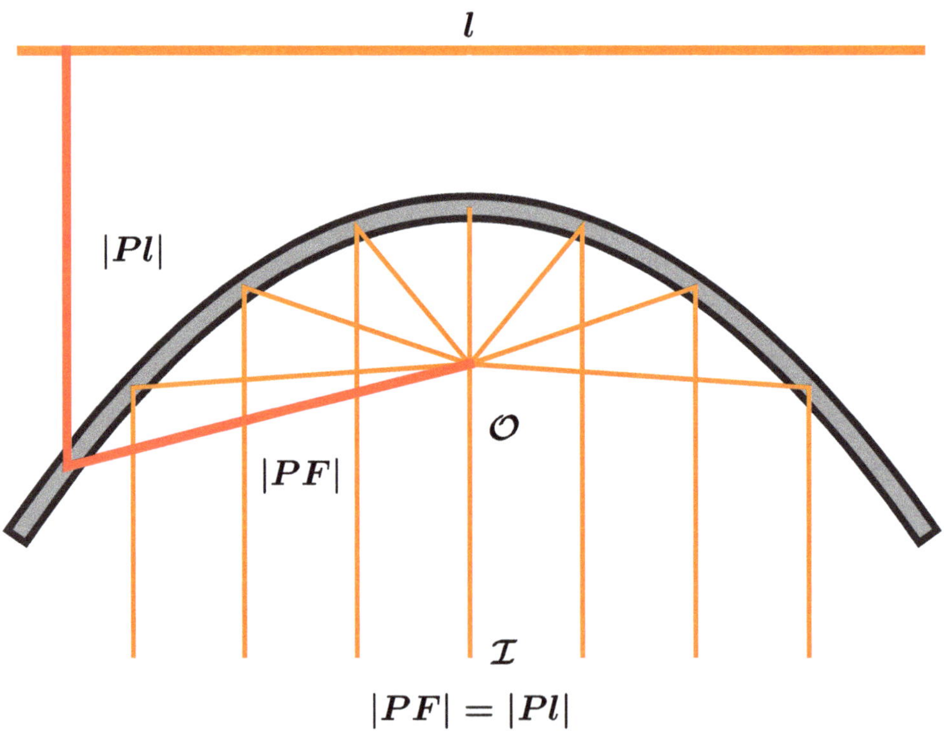


**Figure 2.11.** Diagram of a parabola.

$$x^2 + (y - p)^2 = (y + p)^2, \tag{2.49}$$

manipulating and solving for $y$,

$$y = \frac{x^2}{4p}, \tag{2.50}$$

where the above equation is a function of $x$, we can just call it $f(x)$,

$$f(x) = y = \frac{x^2}{4p}. \tag{2.51}$$

Every order pair $(x, y) = \left(x, \frac{x^2}{4p}\right)$ holds $|PF| = |Pl|$. Note that there is no other locus, or set of order pairs that hold for $|PF| = |Pl|$.

This unique geometrical property of the parabola gives an exclusive optical property for a mirror with the shape of the parabola. In the next chapter, we will see how attractive a mirror is with the shape of the parabola.

### 2.5.2 The ellipse

An ellipse $\mathcal{E}$ is another interesting locus, in the Euclidean plane $\mathbb{R}^2$, that comes from cutting a section of the cone. See figure 2.12.

Given two fixed points $F_1$, $F_2$ and a distance $2a$ which is greater than the distance between $F_1$, $F_2$, the ellipse is the set of ordered pairs $P$ such that $|PF_1| + |PF_2| = 2a$. Thus, the ellipse $\mathcal{E}$ is given by,

$$\boxed{\mathcal{E} = \{P \in \mathbb{R}^2 \mid |PF_1|+|PF_2|=2a\},} \tag{2.52}$$

where $F_1$, $F_2$ are the focuses of the ellipse. The distance $c$ of the focus to the center is called the focal distance or linear eccentricity. The eccentricity is $e = \frac{c}{a}$.

Let's choose the Cartesian coordinates such as the origin is the center of the ellipse. Therefore, we have $F_1 = (c, 0)$, $F_2 = (-c, 0)$. In this configuration the distance from an arbitrary ordered pair $P = (x, y)$ to the focus $(\pm c, 0)$ is $\sqrt{(x \mp c)^2 + y^2}$. Then the ordered pair $P = (x, y)$ must hold,

$$\sqrt{(x - c)^2 + y^2} + \sqrt{(x + c)^2 + y^2} = 2a. \tag{2.53}$$

We can remove the radicals by using $b^2 = a^2 - c^2$ and squaring,

$$\frac{x^2}{a^2} + \frac{y^2}{b^2} = 1. \tag{2.54}$$

Now we can solve for $y$ and we get,

$$y = \pm\frac{b}{a}\sqrt{a^2 - x^2}, \tag{2.55}$$

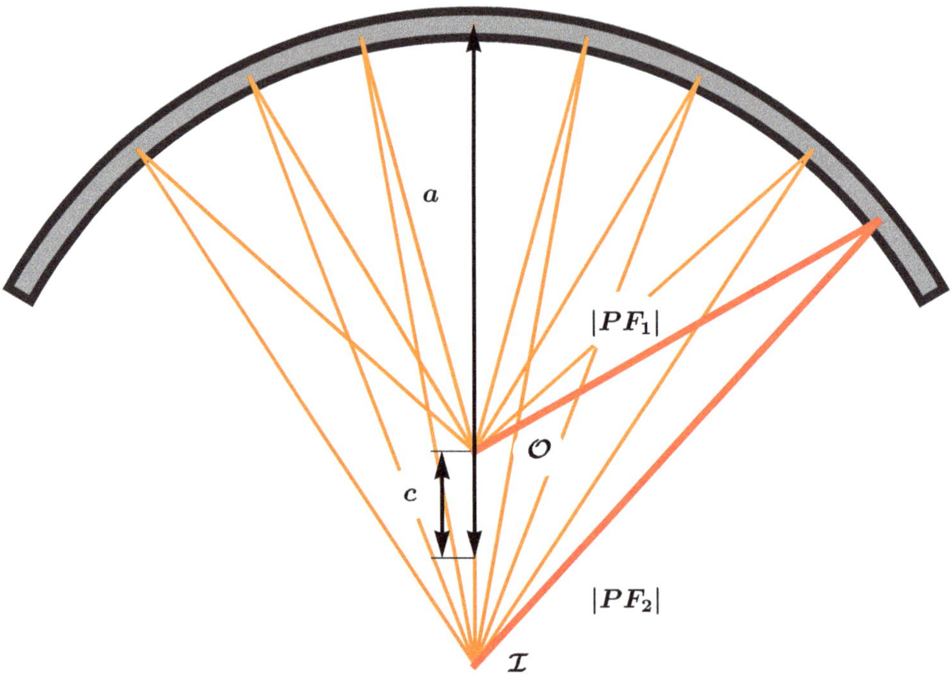


**Figure 2.12.** Diagram of a ellipse.

where $y$ is actually a function $f(x) = y$ that describes the shape of the ellipse $\mathcal{E}$. The elements of $\mathcal{E}$, the order pairs $(x, \pm\frac{b}{a}\sqrt{a^2 - x^2})$, are the only set of points that hold for $|PF_1| + |PF_2| = 2a$. This instinct geometrical property makes the elliptic mirror so attractive in optical design. In the next chapter we will see it in detail.

### 2.5.3 The hyperbola

The hyperbola $\mathcal{H}$ is a locus in Euclidean plane $\mathbb{R}^2$, with a similar definition to the ellipse. But this time, it is not a sum but a subtraction. See figure 2.13.

A hyperbola is a locus in which any ordered pair $P$ holds $|PF_1| - |PF_2| = 2a$. $F_1$, $F_2$ are called focuses of the hyperbola and $a > 0$. Thus,

$$\boxed{\mathcal{H} = \{P \in \mathbb{R}^2 \mid |PF_1| - |PF_2| = 2a\}.} \tag{2.56}$$

Like in the ellipse, $c$ is the distance from the focus to the center, and it is called the focal distance or linear eccentricity. The eccentricity is $e = \frac{c}{a}$.

Let's choose the Cartesian coordinates in such a way that the origin is the center of the hyperbola. Therefore, the focuses are $F_1 = (c, 0)$, $F_2 = (-c, 0)$. The distance from an arbitrary ordered pair $P = (x, y)$ to the focuses $(\pm c, 0)$ is given by $\sqrt{(x \mp c)^2 + y^2}$. Then calling the hyperbola condition for the ordered pair $P = (x, y)$,

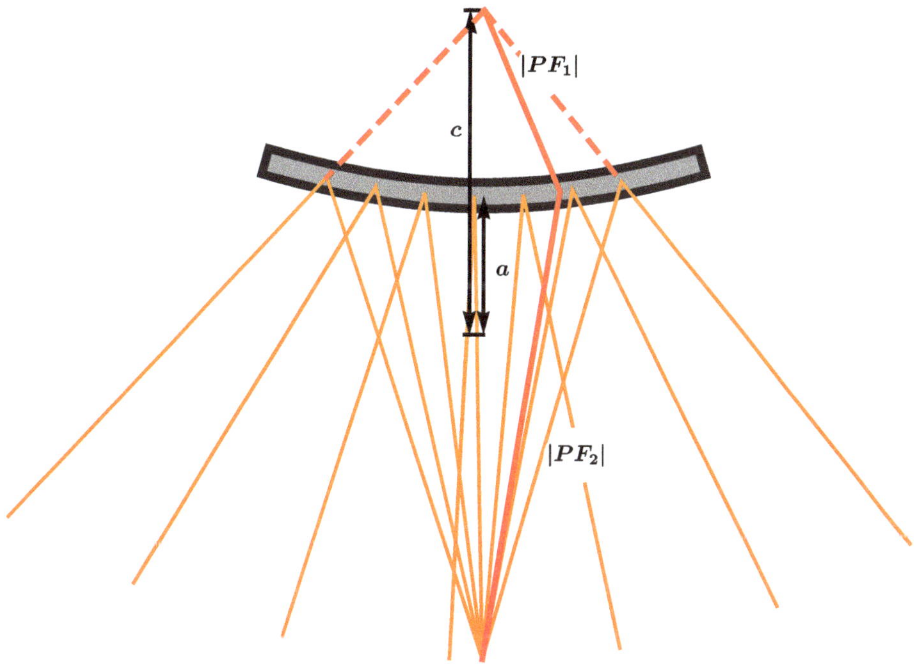


$||PF_1| - |PF_2|| = 2a$

**Figure 2.13.** Diagram of a hyperbola.

$$\sqrt{(x-c)^2+y^2} - \sqrt{(x+c)^2+y^2} = 2a. \tag{2.57}$$

Again, we can remove the radicals by using $b^2 = a^2 - c^2$ and squaring,

$$\frac{x^2}{a^2} - \frac{y^2}{b^2} = 1. \tag{2.58}$$

Solving for $y$ and we have function $f(x) = y$,

$$f(x) = y = \pm\frac{b}{a}\sqrt{x^2 - a^2}. \tag{2.59}$$

The ordered pairs $P \in \mathcal{H}$ are the only points that hold $|PF_1| - |PF_2| = 2a$. The hyperbolic mirror will be studied in the next chapter, it is relevant since by the definition its points hold $|PF_1| - |PF_2| = 2a$.

### 2.5.4 The circle

The circle $\mathcal{C}$ is a conic section, but some authors do not recognize it. Like the conic constant, it was studied for a long time by Euclid; paraphrasing the definition of Euclid:

*A circle is a plane figure bounded by one curved line, and such that all straight lines drawn from a certain point within it to the bounding line, are equal. The bounding line is called its circumference and the point, its center.*

The previous definition, given by Euclid in the first book of *The Elements*, remains the definition of the circle. Therefore, for an ordered pair $P = (x, y)$ and a center $C = (h, k)$ and $r > 0$, and ordered pair $P$ holds, $|CP| = r$. See figure 2.14. The set of the circle is given by,

$$\boxed{\mathcal{C} = \{P \in \mathbb{R}^2 \mid |CP|{=}r\},} \tag{2.60}$$

where $r$ is called the radius. $|CP| = r$ is expressed by the equation,

$$\sqrt{(x - h)^2 + (y - k)^2} = r. \tag{2.61}$$

If we choose the center of the circle at the origin, $C = (h = 0, k = 0)$, we get,

$$\sqrt{x^2 + y^2} = r, \tag{2.62}$$

solving for $y$ we have,

$$y = \pm\sqrt{r^2 - x^2}. \tag{2.63}$$

Please note that the solution is unique despite the $\pm$. There is only one set that its elements hold $|CP| = r$. The conics are just implications of the set theory and the Euclidean metric. They are unique sets by definition, they inherited the axioms of set theory. The properties that constructed the conic sections will give an exceptional kind of mirror, the conic mirror. Conic mirrors are stigmatic.

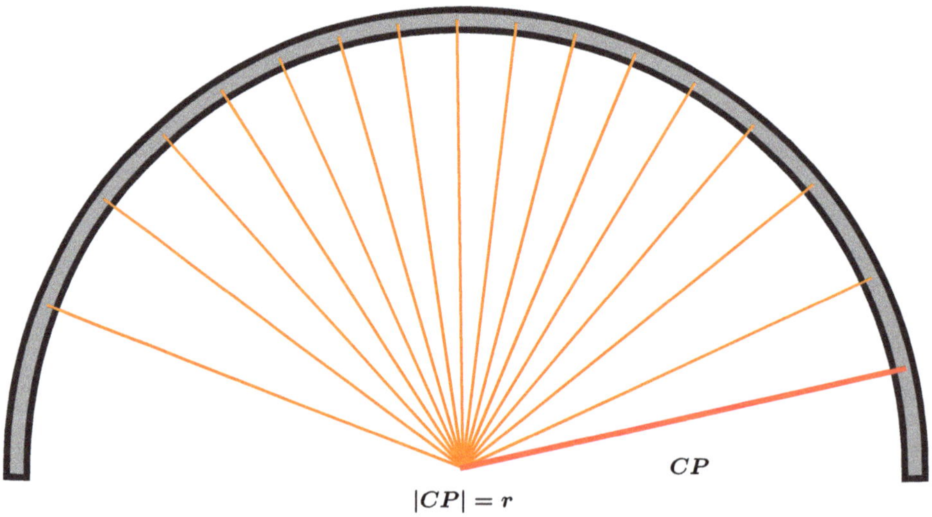


**Figure 2.14.** Diagram of a circle.

## 2.6 Geometric algebra

This is the last section of our mathematical toolkit. At this point, we have set the playground of optical devices, the Euclidean space, which is topological and metric space.

We have also defined some locus sets that have interesting geometrical and optical properties. These loci are the conic sections. The goal of the treatise is to define the locus of stigmatic lenses. To achieve it, we need to deal with vectors, which are mathematical identities with magnitude and direction.

David Hestenes, a theoretical physicist at Arizona State University, mentioned that the problem of vectors is that physicists had not properly learned how to multiply vectors and, as a result of the standard notation implemented using the cross product.

David Hestenes recovered the concepts of Clifford algebra, geometric algebras in his magnum opus *Space-Time Algebra.*

Geometric algebras are very relevant because in it can be defined the cross product for two dimensions without trouble. Something that is not defined in the standard vector notations learned by engineers and physicists.

### 2.6.1 Scalars, vectors, and vector spaces

Scalars are quantities with magnitude with no direction. The name scalars comes from the fact that they scale the magnitudes. Vectors, as mentioned before, have direction and magnitude. In this treatise scalars are written in lower case italic, for example, $a$, $b$, $c$ and vectors are written in bold lower case roman $\vec{\mathbf{u}}$, $\vec{\mathbf{v}}$, $\vec{\mathbf{w}}$.

Arrows usually represent vectors. The direction of the arrow represents the direction of the vector. The length of the arrow is the magnitude of the vector. In figure 2.15 are some vectors.

The norm is the magnitude of a vector and it is given by the metric. The norm computes the distance between the vector and the origin using the metric, in our case the metric is given by the Euclidean metric. For example the vector $\vec{\mathbf{w}} = [p, q]$, its norm is,

$$|\vec{\mathbf{w}}| = \sqrt{(p-0)^2 + (q-0)^2} = \sqrt{p^2 + q^2}, \tag{2.64}$$

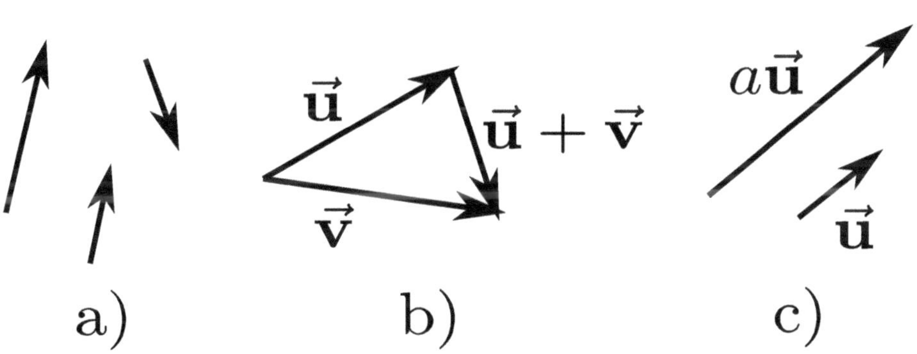


**Figure 2.15.** (a) some vectors (b) vector addition (c) scalar multiplication.

where $|\vec{\mathbf{w}}|$ denotes the norm of the vector $\vec{\mathbf{w}}$.

The multiplication of a vector $\vec{\mathbf{w}} = [p, q]$ with a scalar results in an escalation of the vector's magnitude,

$$a|\vec{\mathbf{w}}| = a\sqrt{p^2 + q^2}. \tag{2.65}$$

Vectors *live* in vector spaces, a vector space is the set of objects called vectors. The Euclidean space is a vector space. A formal definition of the vector space is the following.

**Definition.** A vector space $V$ is the set of objects called vectors. Two operations are defined on the vector space $V$, the vector addition and the scalar multiplication. Every vector space must have an element called zero vector $\vec{\mathbf{0}}$ and also must hold the following axioms for any vector, $\vec{\mathbf{u}}, \vec{\mathbf{v}}, \vec{\mathbf{w}} \in V$ and scalars $a, b \in \mathbb{R}$,

1. $\vec{\mathbf{v}} + \vec{\mathbf{w}} \in V$ and $a\vec{\mathbf{v}} \in V$ (closure),
2. $\vec{\mathbf{u}} + \vec{\mathbf{v}} = \vec{\mathbf{v}} + \vec{\mathbf{u}}$ (commutativity),
3. $(\vec{\mathbf{u}} + \vec{\mathbf{v}}) + \vec{\mathbf{w}} = \vec{\mathbf{u}} + (\vec{\mathbf{v}} + \vec{\mathbf{w}})$ (associativity),
4. $\vec{\mathbf{0}} + \vec{\mathbf{v}} = \vec{\mathbf{v}}$(additive identity),
5. $\forall \vec{\mathbf{v}} \in V, \ \exists \vec{\mathbf{u}} \in V$ such that $\vec{\mathbf{u}} + \vec{\mathbf{v}} = \vec{\mathbf{0}}$ (additive inverse),
6. $\forall \vec{\mathbf{v}} \in V$, $1\vec{\mathbf{v}} = \vec{\mathbf{v}}$(multiplicative identity),
7. $\forall \vec{\mathbf{v}} \in V$, $(ab)\vec{\mathbf{v}} = a(b\vec{\mathbf{v}})$ (associativity),
8. $\forall \vec{\mathbf{u}}, \vec{\mathbf{v}} \in V, \quad a(\vec{\mathbf{u}} + \vec{\mathbf{v}}) = a\vec{\mathbf{u}} + a\vec{\mathbf{v}}, \quad (a + b)\vec{\mathbf{u}} = a\vec{\mathbf{u}} + b\vec{\mathbf{u}}$ (distributive properties).

Examples of vector spaces are the Euclidean plane $\mathbb{R}^2$, the Euclidean space in three dimensions $\mathbb{R}^3$, or in more dimensions. The set of all polynomials of order less than or equal to $m \in \mathbb{Z}$, is a vector space. It perfectly holds the listed axioms above[5].

### 2.6.2 The inner product

One fundamental operation between vectors is the inner product.

**Definition.** The inner product of the nonzero vectors $\vec{\mathbf{u}}$ and $\vec{\mathbf{v}}$ is the scalar defined by

$$\boxed{\vec{\mathbf{u}} \cdot \vec{\mathbf{v}} = |\vec{\mathbf{u}}||\vec{\mathbf{v}}|\cos\theta, \quad 0 \leqslant \theta \leqslant \pi}, \tag{2.66}$$

where $\theta$ is the angle between the vectors. We define the case when one of the vectors is the zero vector as $\vec{\mathbf{u}} \cdot \vec{\mathbf{0}} \equiv 0$. Notice that inner product of vector $\vec{\mathbf{u}}$ with itself is its squared magnitude $\vec{\mathbf{u}} \cdot \vec{\mathbf{u}} = |\vec{\mathbf{u}}|^2$. See figure 2.16.

Let $\vec{\mathbf{u}}, \vec{\mathbf{v}}, \vec{\mathbf{w}} \in V$ and $a, b \in \mathbb{R}$, the next list presents features of inner product,

1. $\vec{\mathbf{u}} \cdot \vec{\mathbf{v}} = \vec{\mathbf{v}} \cdot \vec{\mathbf{u}}$ (commutativity),
2. $(a\vec{\mathbf{u}}) \cdot \vec{\mathbf{v}} = a(\vec{\mathbf{u}} \cdot \vec{\mathbf{v}}), \quad \vec{\mathbf{u}} \cdot (a\vec{\mathbf{v}}) = a(\vec{\mathbf{u}} \cdot \vec{\mathbf{v}})$ (homogeneity),

[5] Where $\mathbb{Z}$ is the set of natural numbers. $\mathbb{Z} = 1, 2, 3, 4, \ldots$. The positive integers.

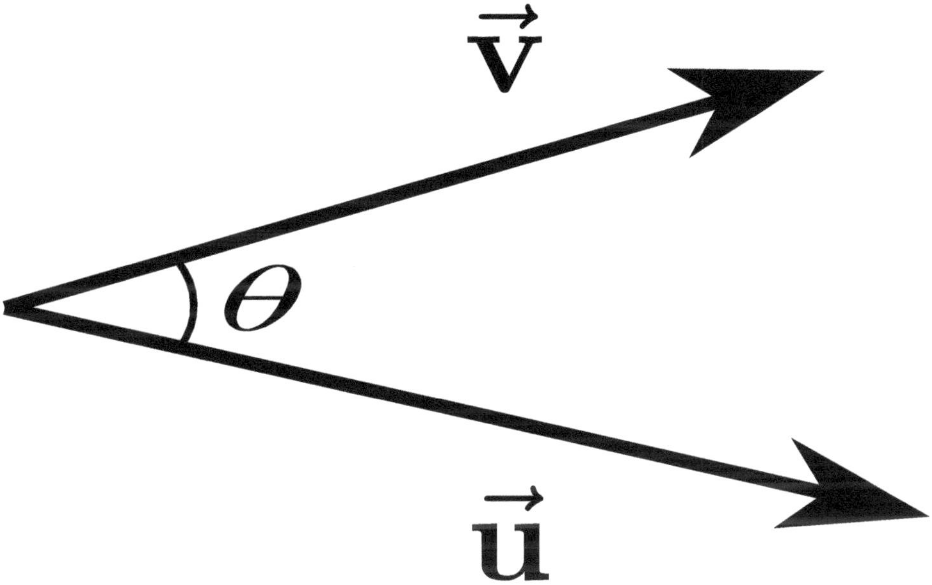


**Figure 2.16.** The inner product of $\vec{\mathbf{v}}$ and $\vec{\mathbf{u}}$.

3. $(\vec{\mathbf{u}} + \vec{\mathbf{v}}) \cdot \vec{\mathbf{w}} = \vec{\mathbf{u}} \cdot \vec{\mathbf{w}} + \vec{\mathbf{v}} \cdot \vec{\mathbf{w}}$, $\quad \vec{\mathbf{w}} \cdot (\vec{\mathbf{u}} + \vec{\mathbf{v}}) = \vec{\mathbf{u}} \cdot \vec{\mathbf{w}} + \vec{\mathbf{v}} \cdot \vec{\mathbf{w}}$ (distributive),
4. if $\vec{\mathbf{u}} \neq \vec{\mathbf{0}}$, then $\vec{\mathbf{u}} \cdot \vec{\mathbf{u}} > 0$. (positivity).

It is important to remark that when two vectors, let's say $\vec{\mathbf{u}}$ and $\vec{\mathbf{v}}$ are orthogonal to each other, thus the angle between them is $\theta = \pi/2$, then $\vec{\mathbf{u}} \cdot \vec{\mathbf{v}} = 0$.

At this point it is convenient to define the unit vectors. The unit vectors have the distinction that their magnitude is one.

For example in two dimensions in the Euclidean plane, the unit vectors are $\vec{\mathbf{e}}_1$ and $\vec{\mathbf{e}}_2$ for the horizontal and vertical direction respectively. Every vector can be expressed in terms of the unit vectors, for example,

$$\vec{\mathbf{u}} = [u_x, u_y] = u_x\vec{\mathbf{e}}_1 + u_y\vec{\mathbf{e}}_2, \tag{2.67}$$

where $u_x$, $u_y$ are the components of the vector $\vec{\mathbf{u}}$.

In three dimensions on the Euclidean space, we add another unitary vector $\mathbf{e}_3$, then,

$$\vec{\mathbf{u}} = [u_x, u_y, u_z] = u_x\vec{\mathbf{e}}_1 + u_y\vec{\mathbf{e}}_2 + u_z\vec{\mathbf{e}}_3. \tag{2.68}$$

In the case of unit vectors the inner product can be one or zero, it depends if they are in the same direction or not,

$$\vec{\mathbf{e}}_i \cdot \vec{\mathbf{e}}_j = \begin{matrix} 0 \text{ if } i \neq j \\ 1 \text{ if } i = j \end{matrix} \tag{2.69}$$

$\{\vec{\mathbf{e}}_1, \vec{\mathbf{e}}_2, \vec{\mathbf{e}}_3\}$ forms an orthonormal basis for the vector space in $\mathbb{R}^3$. This means by adding vectors every vector can be represented by a sum of them or a sum of them multiplied by a scalar.

### 2.6.3 The outer product

In this section, we define the outer product of two vectors. This operation is a fundamental concept of geometric algebra.

Given a vector $\vec{\mathbf{u}}$, someone can draw a line in direction of the direction and magnitude of the vector $\vec{\mathbf{u}}$. Then, if that line is extended in the direction and magnitude of vector $\vec{\mathbf{v}}$, it sweeps out a parallelogram. The area of the corresponding parallelogram is uniquely determined by this procedure. This area is a product of between $\vec{\mathbf{u}}$ and $\vec{\mathbf{v}}$. Notice that the area of this parallelogram is the same if we start with $\vec{\mathbf{u}}$ and end with $\vec{\mathbf{v}}$ or vice versa. But the orientation of the boundary lines is not commutative, in fact, they are opposites. Therefore, this parallelogram is an oriented area. Please see figure 2.17.

If one of the vectors is zero, then the area of the parallelogram is zero. If both vectors have the same direction, then, again the area of the parallelogram is zero.

We call this fancy product between vectors the outer product. The notation that we implement for this product, for the vectors $\vec{\mathbf{u}}$ and $\vec{\mathbf{v}}$, is given by,

$$\vec{\mathbf{u}} \wedge \vec{\mathbf{v}}, \tag{2.70}$$

where $\vec{\mathbf{u}} \wedge \vec{\mathbf{v}}$ is a bivector, a mathematical identity that has the magnitude of an area and orientation.

Some properties of the bivectors are listed,

1. The magnitude of a bivector $\vec{\mathbf{u}} \wedge \vec{\mathbf{v}}$ has the area of its respective parallelogram.
2. The orientation of a $\vec{\mathbf{u}} \wedge \vec{\mathbf{v}}$ is given the line that starts with $\vec{\mathbf{u}}$, then it turns and ends in $\vec{\mathbf{v}}$.

Observe from figure 2.17 that the orientation of $\vec{\mathbf{u}} \wedge \vec{\mathbf{v}}$ is the opposite of $\vec{\mathbf{v}} \wedge \vec{\mathbf{u}}$, therefore,

$$\vec{\mathbf{u}} \wedge \vec{\mathbf{v}} = -\vec{\mathbf{v}} \wedge \vec{\mathbf{u}}. \tag{2.71}$$

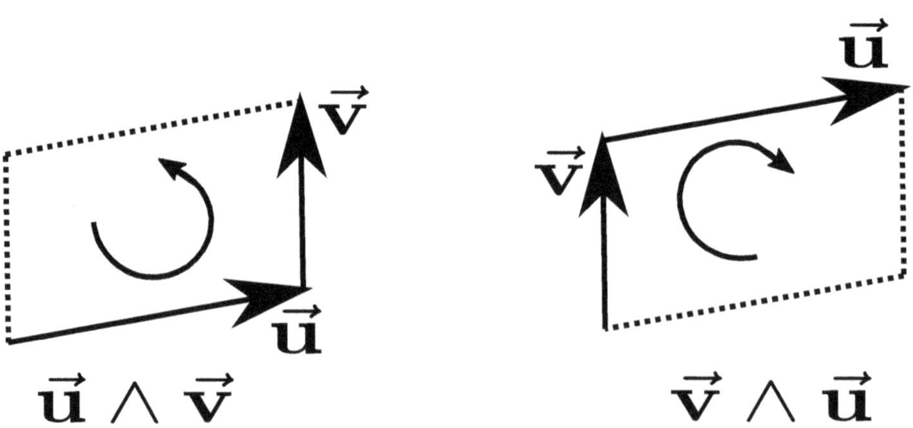


**Figure 2.17.** The outer product of $\vec{\mathbf{u}} \wedge \vec{\mathbf{v}}$ and $\vec{\mathbf{v}} \wedge \vec{\mathbf{u}}$. Notice that $\vec{\mathbf{u}} \wedge \vec{\mathbf{v}} = -\vec{\mathbf{v}} \wedge \vec{\mathbf{u}}$.

A formal definition of the outer product is the following,

**Definition.** Let $\{\vec{\mathbf{e}}_1, \vec{\mathbf{e}}_2\}$ be an orthonormal basis for the Euclidean plane $\mathbb{R}^2$. Let $\vec{\mathbf{u}}, \vec{\mathbf{v}} \in \mathbb{R}^2$. Let $\theta$ be the oriented angle between $\vec{\mathbf{u}}$ and $\vec{\mathbf{v}}$, then the outer product between $\vec{\mathbf{u}}$ and $\vec{\mathbf{v}}$ is given by,

$$\boxed{\vec{\mathbf{u}} \wedge \vec{\mathbf{v}} = |\vec{\mathbf{v}}||\vec{\mathbf{u}}|\sin\theta(\vec{\mathbf{e}}_1 \wedge \vec{\mathbf{e}}_2), \; -\pi < \theta \leqslant \pi.} \tag{2.72}$$

Please note that $\vec{\mathbf{u}} \wedge \vec{\mathbf{v}} \notin \mathbb{R}^2$.

Some properties of the outer product are,

1. $\vec{\mathbf{u}} \wedge \vec{\mathbf{u}} = 0$ (parallel),
2. $\vec{\mathbf{u}} \wedge \vec{\mathbf{v}} = -\vec{\mathbf{v}} \wedge \vec{\mathbf{u}}$ (antisymmetry),
3. $(a\vec{\mathbf{u}}) \wedge \vec{\mathbf{v}} = a(\vec{\mathbf{u}} \wedge \vec{\mathbf{v}})$, or, $\vec{\mathbf{u}} \wedge (a\vec{\mathbf{v}}) = a(\vec{\mathbf{u}} \wedge \vec{\mathbf{v}})$ (homogeneity),
4. $(\vec{\mathbf{u}} + \vec{\mathbf{v}}) \wedge \vec{\mathbf{w}} = \vec{\mathbf{u}} \wedge \vec{\mathbf{w}} + \vec{\mathbf{v}} \wedge \vec{\mathbf{w}}$, or, $\vec{\mathbf{w}} \wedge (\vec{\mathbf{u}} + \vec{\mathbf{v}}) = \vec{\mathbf{w}} \wedge \vec{\mathbf{u}} + \vec{\mathbf{w}} \wedge \vec{\mathbf{v}}$ (distributive).

Let's take an example of an outer product between two vectors. Let $\vec{\mathbf{u}} = a\vec{\mathbf{e}}_1 + b\vec{\mathbf{e}}_2$ and $\vec{\mathbf{v}} = c\vec{\mathbf{e}}_1 + d\vec{\mathbf{e}}_2$, then,

$$\begin{aligned}\vec{\mathbf{u}} \wedge \vec{\mathbf{v}} &= (a\vec{\mathbf{e}}_1 + b\vec{\mathbf{e}}_2) \wedge (c\vec{\mathbf{e}}_1 + d\vec{\mathbf{e}}_2)\\ &= ac(\vec{\mathbf{e}}_1 \wedge \vec{\mathbf{e}}_1) + ad(\vec{\mathbf{e}}_1 \wedge \vec{\mathbf{e}}_2) + bc(\vec{\mathbf{e}}_2 \wedge \vec{\mathbf{e}}_1) + bd(\vec{\mathbf{e}}_2 \wedge \vec{\mathbf{e}}_2)\\ &= ad(\vec{\mathbf{e}}_1 \wedge \vec{\mathbf{e}}_2) + bc(\vec{\mathbf{e}}_2 \wedge \vec{\mathbf{e}}_1)\\ &= (ad - bc)(\vec{\mathbf{e}}_1 \wedge \vec{\mathbf{e}}_2).\end{aligned} \tag{2.73}$$

The outer product between two vectors in $\mathbb{R}^2$ remembers the determinant of a matrix. For three dimensions vectors, $\vec{\mathbf{u}} = u_1\vec{\mathbf{e}}_1 + u_2\vec{\mathbf{e}}_2 + u_3\mathbf{e}_3$ and $\vec{\mathbf{v}} = v_1\vec{\mathbf{e}}_1 + v_2\vec{\mathbf{e}}_2 + v_3\mathbf{e}_3$, we have,

$$\begin{aligned}\vec{\mathbf{u}} \wedge \vec{\mathbf{v}} = {} & (u_2v_3 - u_3v_2)(\vec{\mathbf{e}}_2 \wedge \vec{\mathbf{e}}_3) + (u_3v_1 - u_1v_3)(\vec{\mathbf{e}}_3 \wedge \vec{\mathbf{e}}_1)\\ & + (u_1v_2 - u_2v_1)(\vec{\mathbf{e}}_1 \wedge \vec{\mathbf{e}}_2).\end{aligned} \tag{2.74}$$

It seems just like the cross product. The great advantage of the outer product is that it can be defined in two or more dimensions.

### 2.6.4 The geometric product

The geometric product is the fundamental concept of geometric algebra. It is what provides geometric algebra capability as a mathematical instrument.

**Definition.** The geometric product **uv** of vectors $\vec{\mathbf{u}}$ and $\vec{\mathbf{v}}$ is a scalar plus a bivector:

$$\boxed{\vec{\mathbf{u}}\vec{\mathbf{v}} = \vec{\mathbf{u}} \cdot \vec{\mathbf{v}} + \vec{\mathbf{u}} \wedge \vec{\mathbf{v}}.} \tag{2.75}$$

The following list has the properties of the geometric product.

1. $\vec{\mathbf{u}}\vec{\mathbf{v}} = \vec{\mathbf{u}} \cdot \vec{\mathbf{v}}$ if $\vec{\mathbf{u}}||\vec{\mathbf{v}}$ (inner product if parallel vectors),

2. $\vec{\mathbf{u}}\vec{\mathbf{v}} = \vec{\mathbf{u}} \wedge \vec{\mathbf{v}}$ if $\vec{\mathbf{u}} \perp \vec{\mathbf{v}}$ (outer product if perpendicular vectors),
3. $(a\vec{\mathbf{u}})\vec{\mathbf{v}} = a(\vec{\mathbf{u}})\vec{\mathbf{v}}$, or $\vec{\mathbf{u}}(a\vec{\mathbf{v}}) = a(\vec{\mathbf{u}}\vec{\mathbf{v}})$ (homogeneity),
4. $(\vec{\mathbf{u}} + \vec{\mathbf{v}})\vec{\mathbf{w}} = \vec{\mathbf{u}}\vec{\mathbf{w}} + \vec{\mathbf{v}}\vec{\mathbf{w}}$, or $\vec{\mathbf{w}}(\vec{\mathbf{u}} + \vec{\mathbf{v}}) = \vec{\mathbf{w}}\vec{\mathbf{u}} + \vec{\mathbf{w}}\vec{\mathbf{v}}$ (distributive),
5. $\vec{\mathbf{u}}(\vec{\mathbf{v}}\vec{\mathbf{w}}) = (\vec{\mathbf{u}}\vec{\mathbf{v}})\vec{\mathbf{w}}$ (homogeneity).

Notice that the geometric product can be also written with a reverse in the outer product,

$$\vec{\mathbf{u}}\vec{\mathbf{v}} = \vec{\mathbf{u}} \cdot \vec{\mathbf{v}} + \vec{\mathbf{u}} \wedge \vec{\mathbf{v}} = \vec{\mathbf{u}} \cdot \vec{\mathbf{v}} - \vec{\mathbf{v}} \wedge \vec{\mathbf{u}}. \tag{2.76}$$

Another way to express above equation is,

$$\boxed{\vec{\mathbf{u}} \cdot \vec{\mathbf{v}} = \frac{1}{2}(\vec{\mathbf{u}}\vec{\mathbf{v}} + \vec{\mathbf{v}}\vec{\mathbf{u}}),} \tag{2.77}$$

and,

$$\boxed{\vec{\mathbf{u}} \wedge \vec{\mathbf{v}} = \frac{1}{2}(\vec{\mathbf{u}}\vec{\mathbf{v}} - \vec{\mathbf{v}}\vec{\mathbf{u}}).} \tag{2.78}$$

The above elegant expressions can only be achieved through the notation of geometric algebra.

### 2.6.5 The imaginary number

We can get an astonishing result with geometric algebra, too good for the real plane $\mathbb{R}^2$. Let $\{\vec{\mathbf{e}}_1, \vec{\mathbf{e}}_2\}$ be an orthonormal basis for the Euclidean plane $\mathbb{R}^2$.

The geometric product of $\vec{\mathbf{e}}_1\vec{\mathbf{e}}_1$ is given by,

$$\begin{aligned}\vec{\mathbf{e}}_1\vec{\mathbf{e}}_1 &= \vec{\mathbf{e}}_1 \cdot \vec{\mathbf{e}}_1 + \vec{\mathbf{e}}_1 \wedge \vec{\mathbf{e}}_1 \\ &= |\vec{\mathbf{e}}_1|^2 + 0 \\ &= 1.\end{aligned} \tag{2.79}$$

For $\vec{\mathbf{e}}_2\vec{\mathbf{e}}_2$ we have,

$$\begin{aligned}\vec{\mathbf{e}}_2\vec{\mathbf{e}}_2 &= \vec{\mathbf{e}}_2 \cdot \vec{\mathbf{e}}_2 + \vec{\mathbf{e}}_2 \wedge \vec{\mathbf{e}}_2 \\ &= |\vec{\mathbf{e}}_2|^2 + 0 \\ &= 1.\end{aligned} \tag{2.80}$$

Let's define a convenient notation,

$$\vec{\mathbf{e}}_1\vec{\mathbf{e}}_1 \equiv \vec{\mathbf{e}}_1^2 = 1, \quad \vec{\mathbf{e}}_2\vec{\mathbf{e}}_2 \equiv \vec{\mathbf{e}}_2^2 = 1. \tag{2.81}$$

Now let's focus on $\vec{\mathbf{e}}_1\vec{\mathbf{e}}_2$.

$$\begin{aligned}\vec{\mathbf{e}}_1\vec{\mathbf{e}}_2 &= \vec{\mathbf{e}}_1 \cdot \vec{\mathbf{e}}_2 + \vec{\mathbf{e}}_1 \wedge \vec{\mathbf{e}}_2 \\ &= \vec{\mathbf{e}}_1 \wedge \vec{\mathbf{e}}_2 \\ &= -\vec{\mathbf{e}}_2 \wedge \vec{\mathbf{e}}_1.\end{aligned} \tag{2.82}$$

Then, if squared $\vec{\mathbf{e}}_1\vec{\mathbf{e}}_2$, thus $(\vec{\mathbf{e}}_1\vec{\mathbf{e}}_2)^2$,

$$\begin{aligned}(\vec{\mathbf{e}}_1\vec{\mathbf{e}}_2)^2 &= \vec{\mathbf{e}}_1\vec{\mathbf{e}}_2\vec{\mathbf{e}}_1\vec{\mathbf{e}}_2\\ &= -\,\vec{\mathbf{e}}_1\vec{\mathbf{e}}_1\vec{\mathbf{e}}_2\vec{\mathbf{e}}_2\\ &= -\,1\\ (\vec{\mathbf{e}}_1\vec{\mathbf{e}}_2) &= \sqrt{-\,1}\,.\end{aligned} \tag{2.83}$$

We have obtained the imaginary number, the bivector $\vec{\mathbf{e}}_1 \wedge \vec{\mathbf{e}}_2$ is the imaginary number,

$$\boxed{\vec{\mathbf{e}}_1\vec{\mathbf{e}}_2 = \vec{\mathbf{e}}_1 \wedge \vec{\mathbf{e}}_2 = i \equiv \sqrt{-1}\,.} \tag{2.84}$$

This elegant form of obtaining the legendary imaginary number $i \equiv \sqrt{-1}$ is so natural in geometric algebra. With this we can conclude that $\vec{\mathbf{e}}_1 \wedge \vec{\mathbf{e}}_2 \not\in \mathbb{R}^2$.

### 2.6.6 Multiplicative inverse of a vector

Now that we have defined the inner product, the outer product and the geometrical product, it is possible to determine the multiplicative inverse of a nonzero vector $\vec{\mathbf{u}}$ concerning the geometric product.

Assuming $\vec{\mathbf{u}}\vec{\mathbf{u}} \neq 0$, then $\vec{\mathbf{u}}\vec{\mathbf{u}} = \vec{\mathbf{u}} \cdot \vec{\mathbf{u}} + \vec{\mathbf{u}} \wedge \vec{\mathbf{u}} = |\vec{\mathbf{u}}|^2 + 0 = |\vec{\mathbf{u}}|^2$. Therefore,

$$1 = \frac{\vec{\mathbf{u}}\vec{\mathbf{u}}}{\vec{\mathbf{u}}\vec{\mathbf{u}}} = \frac{1}{|\vec{\mathbf{u}}|^2}\vec{\mathbf{u}}\vec{\mathbf{u}} = \left(\frac{1}{|\vec{\mathbf{u}}|^2}\vec{\mathbf{u}}\right)\vec{\mathbf{u}} = \vec{\mathbf{u}}\left(\frac{1}{|\vec{\mathbf{u}}|^2}\vec{\mathbf{u}}\right), \tag{2.85}$$

therefore, the multiplicative inverse of $\vec{\mathbf{u}}$ is given by

$$\boxed{\vec{\mathbf{u}}^{-1} = \frac{1}{|\vec{\mathbf{u}}|^2}\vec{\mathbf{u}}.} \tag{2.86}$$

The last equation is another elegant result obtained with geometric algebra.

All the physical models that involve vectors can be expressed with geometric algebra. Geometric algebra is more profound and elegant than the current mainstream notation implemented with vectors. William K Clifford understood this, he understood the work of Hermann Grassmann, and he gives us geometric algebra, which is also called Clifford algebra. It is a shame that geometric algebra is far from being the mainstream notation for vectors.

### 2.6.7 Application of Clifford algebra in the law of sines

In the last two subsections we will apply Clifford algebra to proof the law of sines and cosines. So, let's start with the law of sines. In figure 2.18 is the image of a triangle constructed by the vectors, $\vec{\mathbf{a}} = \vec{\mathbf{b}} + \vec{\mathbf{c}}$. Then,

$$\vec{\mathbf{b}} = \vec{\mathbf{a}} - \vec{\mathbf{c}}, \tag{2.87}$$

apply the outer product of $\vec{\mathbf{a}}$ on both sides,

$$\begin{aligned}\vec{\mathbf{b}} \wedge \vec{\mathbf{a}} &= \vec{\mathbf{a}} \wedge \vec{\mathbf{a}} - \vec{\mathbf{c}} \wedge \vec{\mathbf{a}},\\ \vec{\mathbf{b}} \wedge \vec{\mathbf{a}} &= -\,\vec{\mathbf{c}} \wedge \vec{\mathbf{a}},\end{aligned} \tag{2.88}$$

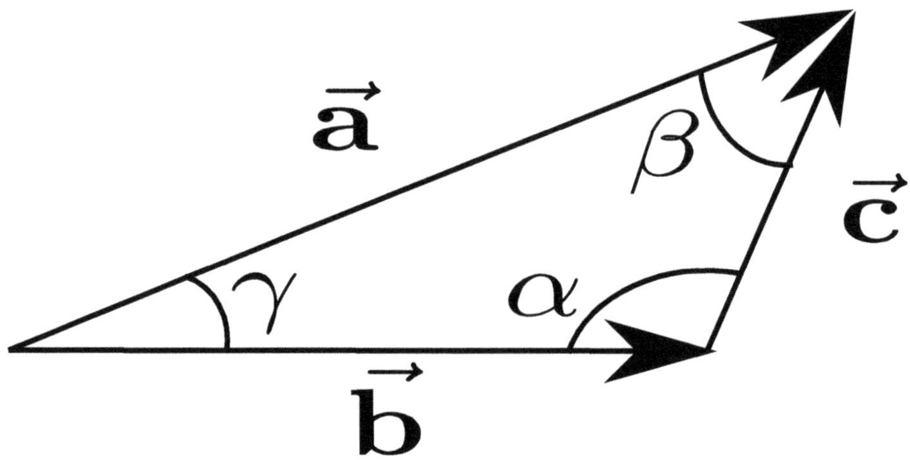


**Figure 2.18.** A triangle constructed by the vectors, $\vec{\mathbf{a}} = \vec{\mathbf{b}} + \vec{\mathbf{c}}$.

using the definition of outer product,

$$\begin{gathered} |\vec{\mathbf{b}}||\vec{\mathbf{a}}|\sin\gamma(\vec{\mathbf{e}}_1 \wedge \vec{\mathbf{e}}_2) = |\vec{\mathbf{a}}||\vec{\mathbf{c}}|\sin\beta(\vec{\mathbf{e}}_1 \wedge \vec{\mathbf{e}}_2) \\ |\vec{\mathbf{b}}|\sin\gamma = |\vec{\mathbf{c}}|\sin\beta, \end{gathered} \tag{2.89}$$

thus,

$$\frac{|\vec{\mathbf{b}}|}{\sin\beta} = \frac{|\vec{\mathbf{c}}|}{\sin\gamma}. \tag{2.90}$$

We use the same thing for the other angle and get the law of sines,

$$\frac{|\vec{\mathbf{a}}|}{\sin\alpha} = \frac{|\vec{\mathbf{b}}|}{\sin\beta} = \frac{|\vec{\mathbf{c}}|}{\sin\gamma}. \tag{2.91}$$

### 2.6.8 Application of Clifford algebras in the law of cosines

Using the same triangle of figure 2.18 with $\vec{\mathbf{a}} = \vec{\mathbf{b}} + \vec{\mathbf{c}}$, we can prove the law of cosine.

We have,

$$\vec{\mathbf{c}} = \vec{\mathbf{a}} - \vec{\mathbf{b}}, \tag{2.92}$$

squaring it,

$$\begin{aligned} \vec{\mathbf{c}}^2 &= (\vec{\mathbf{a}} - \vec{\mathbf{b}})^2 \\ &= (\vec{\mathbf{a}} - \vec{\mathbf{b}})(\vec{\mathbf{a}} - \vec{\mathbf{b}}) \\ &= \vec{\mathbf{a}}\vec{\mathbf{a}} + \vec{\mathbf{b}}\vec{\mathbf{b}} - \vec{\mathbf{a}}\vec{\mathbf{b}} - \vec{\mathbf{b}}\vec{\mathbf{a}} \\ &= |\vec{\mathbf{a}}|^2 + |\vec{\mathbf{b}}|^2 - (\vec{\mathbf{a}}\vec{\mathbf{b}} + \vec{\mathbf{b}}\vec{\mathbf{a}}). \end{aligned} \tag{2.93}$$

Now, remember that we know,

$$\begin{aligned}\frac{1}{2}(\vec{\mathbf{a}}\vec{\mathbf{b}} + \vec{\mathbf{b}}\vec{\mathbf{a}}) &= \vec{\mathbf{a}} \cdot \vec{\mathbf{b}} \\ &= |\vec{\mathbf{a}}||\vec{\mathbf{b}}|\cos\gamma,\end{aligned} \tag{2.94}$$

replacing, we get the law of cosines,

$$|\vec{\mathbf{c}}|^2 = |\vec{\mathbf{a}}|^2 + |\vec{\mathbf{b}}|^2 - 2|\vec{\mathbf{a}}||\vec{\mathbf{b}}|\cos\gamma. \tag{2.95}$$

In conclusion, geometric algebra implements a generalized theory that contains many mathematical objects, such as vectors, bivectors, complex numbers, quaternions, matrix algebra, vectors, tensor and spinor algebras, and the algebra of differential forms. In this treatise, we just focused on the geometric algebra of vectors, bivectors and complex numbers because they are the mathematical objects that we will use to design stigmatic optical systems.

## 2.7 Conclusions

In this chapter, we made a brief review of essential topics in mathematics that are implemented in the design of stigmatic lenses. We focused ourselves more on the topics that are not common for engineers and physicists.

As a summary, we give the central idea that everything in mathematics can be expressed as a set. From sets you can define more interesting sets, and you can also map a set to another set. Sets can also represent physical models, for example, the conic sections that define the conic mirrors. Sets, where the optical systems *live*, are topological spaces, metric spaces and vector spaces. Finally, we studied geometric algebra because it will be remarkably useful in the following chapters. All physical models that include vectors can be represented by geometric algebra.

## Further reading


Baker C W 1991 *Introduction to Topology* (Malabar, FL: Krieger)

Born M and Wolf E 2013 Principles of Optics: Electromagnetic Theory of Propagation *Interference and Diffraction of Light* (Amsterdam: Elsevier)

Boyer C B 2012 *History of Analytic Geometry* (North Chelmsford, MA: Courier Corporation)

Braunecker B, Hentschel R and Tiziani H J 2008 *Advanced Optics Using Aspherical Elements* vol 173 (Bellingham, WA: SPIE)

Chaves J 2016 *Introduction to Nonimaging Optics* 2nd edn (Boca Raton, FL: CRC Press)

Coxeter H S M 1998 *Non-Euclidean Geometry* (Cambridge: Cambridge University Press)

Durbin J R 2008 *Modern Algebra: An Introduction* (New York: Wiley)

Fuerschbach K, Rolland J P and Thompson K P 2014 Theory of aberration fields for general optical systems with freeform surfaces *Opt. Express* **22** 26585–606

Hestenes D 2012 *New Foundations for Classical Mechanics* 15 (Berlin: Springer Science)

Hestenes D and Sobczyk G 2012 *Clifford Algebra to Geometric Calculus: a Unified Language for Mathematics and Physics* vol 5 (Berlin: Springer Science & Business Media)

Kingslake R and Johnson R B 2009 *Lens Design Fundamentals* (New York: Academic)

Kreyszig E 1978 *Introductory Functional Analysis with Applications* vol 1 (New York: Wiley)

Lefaivre J 1951 A new approach in the analytical study of the spherical aberrations of any order *J. Opt. Soc. Am.* **41** 647

Lehmann C H 1984 *Geometría analítica. Analytic Geometry* (New York: Wiley)

Lehmann C H 1942 *Analytic Geometry* (New York: Wiley)

Lin P D and Tsai C-Y 2012 Determination of unit normal vectors of aspherical surfaces given unit directional vectors of incoming and outgoing rays: reply *J. Opt. Soc. Am.* A **29** 1358

Luneburg R K and Herzberger M 1964 *Mathematical Theory of Optics* (Berkeley, CA: University of California Press)

Malacara-Hernández D and Malacara-Hernández Z 2016 *Handbook of Optical Design* (Boca Raton, FL: CRC Press)

Marsden J E and Tromba A 2003 *Vector Calculus* (London: Macmillan)

Matthews P C 2012 *Vector Calculus* (Berlin: Springer Science)

McCoy N H and Janusz G J 1968 *Introduction to Modern Algebra* (Boston, MA: Allyn & Bacon)

Romano A and Cavaliere R 2016 *Geometric Optics: Theory and Design of Astronomical Optical Systems Using Mathematica®* (Basel: Birkhäuser)

Sun H 2016 *Lens Design: A Practical Guide* (Boca Raton, FL: CRC Press)

Swokowski E W 1979 *Calculus with Analytic Geometry* (London: Taylor and Francis)

Wallace A H 2006 *Differential Topology: First Steps* (North Chelmsford, MA: Courier Corporation)

# Chapter 3

# An introduction to geometrical optics

In this chapter, an introduction to geometric optics is presented, explicitly towards the concepts implemented to design stigmatic lenses. It is shown that it is geometric optics and that we study. Lenses and mirrors are considered. The laws of reflection and refraction are introduced. Subsequently, the concept of stigmatism and chromatic and monochromatic aberrations is presented more formally.

## 3.1 Geometrical optics

As we saw in the first chapter, geometrical optics is one of the oldest sciences and it is highly influenced by Euclidean geometry. Here in this chapter, we are going to give a brief review of the physics of geometrical optics, and we will see the most fundamental concepts of geometrical optics that are implemented during this treatise.

The fundamental concept of geometrical optics is the ray. An optical ray is a line perpendicular to the wavefront of the light, and it follows the direction of propagation of light. The ray is an abstraction useful when the optical elements with which the light interacts are much bigger than the wavelength of light. Therefore, geometrical optics, in general, ignores diffraction[1].

The lenses, mirrors and optical systems, proposed in this treatise in general, are assumed to be much bigger than the wavelength of the light with which they interact. Thus, in this treatise, we ignore diffraction, and we follow the premises of geometrical optics.

Geometrical optics assumes that optical elements can modify the trajectory of light, the directions of the light rays, only in two ways, which are reflection and refraction. Reflection of light is the phenomenon of sending back the light rays which strike on the polished surface; usually, these surfaces are mirrors. Refraction is the redirection of a light ray when it enters on a medium where its speed is

[1] Diffraction refers to the events that occur when a wave confronts an obstacle.

doi:10.1088/978-0-7503-5774-6ch3  

different. The refraction index of a medium is a dimensionless number that describes how fast light travels through the medium. The refraction index is defined by

$$\boxed{n \equiv \frac{c}{v},} \tag{3.1}$$

where, $c$ is the speed of light in vacuum, and $v$ is the phase velocity of light in the medium. A medium is called homogeneous if it has the same constant refraction index along with its domain. When the light travels inside homogeneous media, the rays are straight lines.

A refractive surface covers the bodies that have a different index of refraction than the surrounding environment in which they are located. These bodies can be lenses or prisms. Lenses have the objective to diverge or converge light rays. Prisms have the goal to manipulate the light by refraction but without diverging or converging light rays.

Simplifying, geometrical optics deals with mirrors (reflective surfaces), lenses and prisms (refractive bodies). But this does not mean that geometrical optics is easy science. Most of the mathematical results are nonlinear, and many equations that model the light rays are very long and complicated to manipulate. These circumstances led to optical geometers to assume approximations. Approximations come from the definitions of $\sin\theta$ and $\cos\theta$,

$$\sin\theta \equiv \sum_{m=0}^{\infty} \frac{(-1)^m}{(2m+1)!}\theta^{2m+1} = \theta - \frac{\theta^3}{3!} + \frac{\theta^5}{5!} - \frac{\theta^7}{7!} + \cdots \tag{3.2}$$

and,

$$\cos\theta \equiv \sum_{m=0}^{\infty} \frac{(-1)^m}{(2m)!}\theta^{2m} = 1 - \frac{\theta^2}{2!} + \frac{\theta^4}{4!} - \frac{\theta^6}{6!} + \cdots \tag{3.3}$$

The most common approximation is the paraxial approximation. The paraxial approximation applies when the angles are small enough that,

$$\sin\theta \approx \theta, \tag{3.4}$$

and,

$$\cos\theta \approx 1. \tag{3.5}$$

In general, the approximations are called by the number of elements implemented in the sum. For example, the third-order approximation of $\sin\theta$ is when $m = 3$,

$$\sin\theta \approx \theta - \frac{\theta^3}{3!}. \tag{3.6}$$

**In this treatise, we are not interested in any of these approximations, and theory that made the assumptions of this approximations will be dramatically ignored throughout the chapters. Also, we are not going to implement any optimization process.**

In summary, any numerical approach is not in our interest. We are just fascinated with the solutions without this kind of approximation.

Therefore, we are going to use the two main principles in geometrical optics without approximations. These principles are Fermat's principle and Snell's law.

## 3.2 The principle of least action

To understand Fermat's principle, let's first focus on mechanics, that is, the branch of physics that is responsible for studying objects at rest or in motion. With it, once the initial position, the initial velocity and the forces acting on these objects are known, we have full awareness of how it goes.

This formulation is due to the English physicist, mathematician, philosopher, theologian and alchemist Sir Isaac Newton, who could demonstrate the laws governing moving bodies. All this was done with vector calculus.

Although the analysis by vector calculus is comprehensive and allows predictions about the evolution of physical systems, in many cases, it is hard to find the equations of motion through vector analysis. Worse, it may happen that even when it is possible to write various equations, they form a nonlinear system.

To avoid these problems, some physicists devised a new way of studying Nature using the law of conservation of energy, which, as the name implies, tells us that energy is not created or destroyed, but transformed.

The French lawyer and mathematician, Pierre de Fermat, in the year 1662, proposed first ideas of the principle of minimum action. Fermat believed that rays of light travel from one point to another through the path that leads to less time. The principle of least action is the premise nature does everything such as it spends less energy possible.

The principle of least action applied to optics is called Fermat's principle. Fermat's principle can be stated as:

*The optical length of the path followed by light between two fixed different points is the global minima. The optical length is the physical length multiplied by the refractive index of the medium*

Consider the concept *global minimum*. The global minimum is the smallest overall value of a set. So, think that we have a set, such that its elements are all the possible optical paths from one point to another. These paths have their respective optical length. What Fermat's principle says is that the only physically valid path of our set is the one that has the smallest value of an optical path length.

We will have the concept of Fermat's principle throughout the book; this concept is essential for the design of stigmatic systems.

Fermat's principle can be seen as an integral over time. The time light needs to go from a point $x_1$ to a point $x_2$. So, let $t_1$ and $t_2$ be the time when light is at point $x_1$ and $x_2$, respectively,

$$\int_{t_1}^{t_2} dt = \frac{1}{c}\int_{t_1}^{t_2} \frac{c}{v}\frac{ds}{dt}dt = \frac{1}{c}\int_{x_1}^{x_2} nds, \tag{3.7}$$

$c$ is the speed of light in vacuum, $ds$ an infinitesimal displacement along the ray, $v = ds/dt$ the speed of light in a medium and $n = c/v$ the refractive index of the medium. Then, the optical path (OPL) of the light from a point $x_1$ to a point $x_2$ is,

$$\boxed{\mathrm{OPL} = \int_{x_1}^{x_2} n ds.} \tag{3.8}$$

Notice that the optical path is a geometrical quantity, since,

$$\mathrm{OPL} = c \int_{t_1}^{t_2} dt. \tag{3.9}$$

Think that you have a set of all possible travels time between $x_1$ and $x_2$ and also you have the set of all OPLs between $x_1$ and $x_2$. The global minima of the travel time between two points $x_1$ and $x_2$ for the light is equivalent to the global minima of the optical path length between $x_1$ and $x_2$.

## 3.3 Reflection

We mentioned that the reflection of light is the phenomenon of sending back the light rays when they strike on a mirror. Well, we do not indicate that there is a relation between the incident angle $\theta_1$ and the reflected angle $\theta_2$. This relation is an equality and it is called the law of reflection,

$$\boxed{\theta_1 = \theta_2,} \tag{3.10}$$

where $\theta_1$ incident angle with respect to the normal vector of the reflective surface and $\theta_2$ is the reflected angle also concerning the normal. The normal vector $\vec{\mathbf{n}}$ is a vector perpendicular to the surface at the point where the ray hits the surface. For a mirror with curved surfaces, they have different normal vectors on each point.

In figure 3.1, we present a reflection on a flat mirror. The incident angle $\theta_1$ is equal to the reflected angle $\theta_2$.

## 3.4 Refraction

Something that we haven't mentioned yet, but is significant, is why we choose to study geometric algebra. We choose geometric algebra because it is a unified language for scalar vectors and bivectors among other mathematical identities. We can see light rays as vectors if and only if they are moving across homogeneous media[2]. This assumption is very useful because we already have the mathematical machinery to deal with vectors.

We can model refraction correctly with geometric algebra. Refraction is the change of direction of light when it enters a medium with a different refraction index. Let us say that the light travels from a medium with a refraction index $n_1$ to a medium with $n_2$. How the redirection of light is affected by refraction is a question that can be solved using Snell's law, which is given by

[2] If light travels inside a homogeneous medium, its path is a straight line just like a vector.

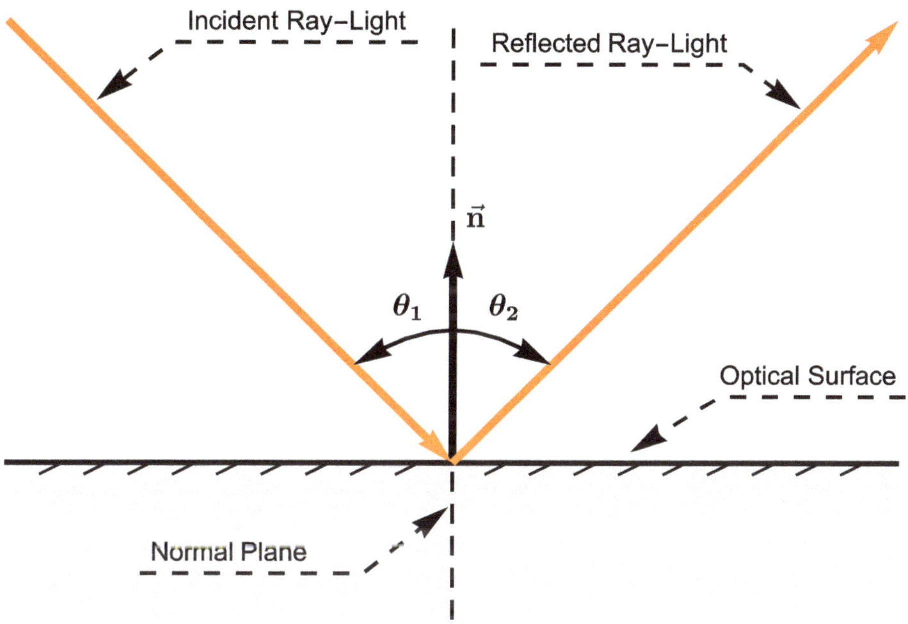


**Figure 3.1.** Law of reflection, the incident angle $\theta_1$ is equal to the reflected angle $\theta_2$.

$$\boxed{n_1 \sin \theta_1 = n_2 \sin \theta_2,} \tag{3.11}$$

where $\theta_1$ is incident angle with respect to the normal vector of the refractive surface and $\theta_2$ is the refracted angle with respect to the negative normal of the refractive surface.

In figure 3.2, we present a diagram of a refraction taking place on a flat interface between two media with different refractive indexes.

Snell's law seems inoffensive, but it is tough to manipulate. It is so hard to manage that paraxial approximations are commonly used. But we decided in the introduction of this chapter that we are going to disregard paraxial approximations at all cost.

## 3.5 Two-dimensional Snell's law in geometric algebra

We need to express Snell's law in a more manageable way, in a vector form. It is only in two dimensions because, in many chapters, we work with rotationally symmetric surfaces.

Let $\{\vec{\mathbf{e}}_1, \vec{\mathbf{e}}_2\}$ be an orthonormal basis for the Euclidean plane $\mathbb{R}^2$. Let $\vec{\mathbf{s}}_1 \in \mathbb{R}^2$ be the incident unit vector, let $\vec{\mathbf{s}}_2 \in \mathbb{R}^2$ be the refracted unit vector and let $\vec{\mathbf{n}} \in \mathbb{R}^2$ be the normal unit vector of the surface. $\theta_1$ is incident angle with respect to $\vec{\mathbf{n}}$ and $\theta_2$ is the refracted angle with respect to $-\vec{\mathbf{n}}$.

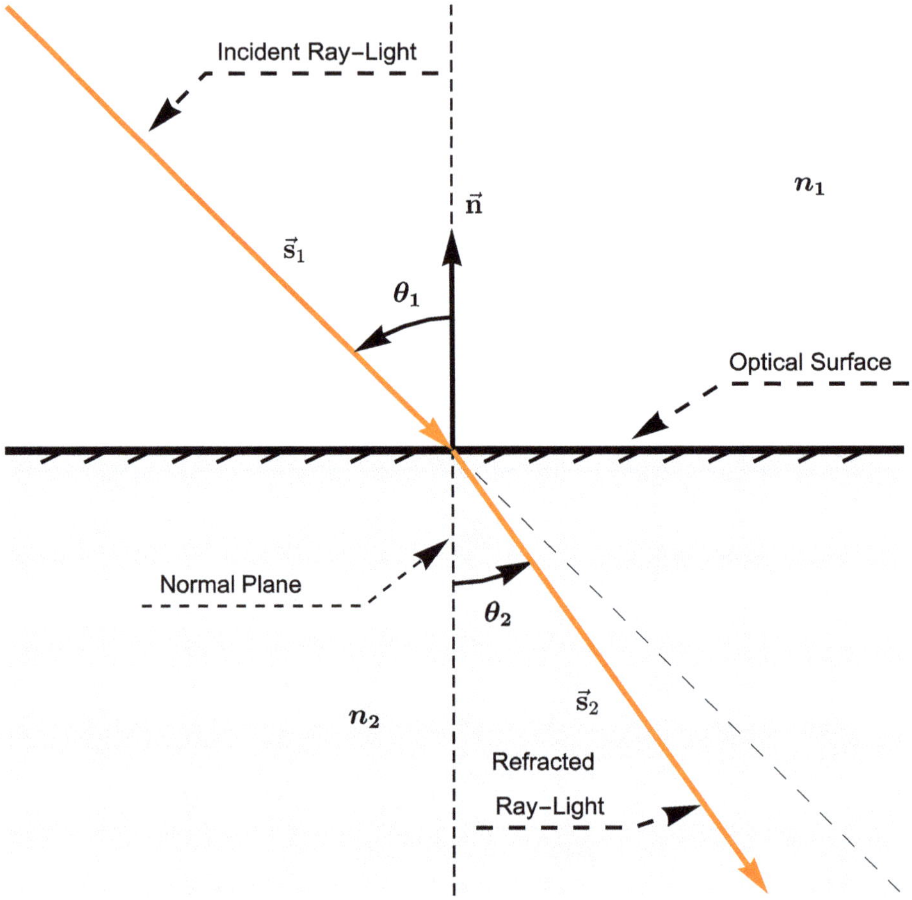


**Figure 3.2.** Refraction taking place on a flat interface between two media with different refractive indexes $n_1$ and $n_2$. The indecent vector is $\vec{s}_1$, the refracted vector is $\vec{s}_2$ and the normal vector of the surface is $\vec{n}$.

Then, the refracted vector is given by

$$\vec{s}_2 = \frac{n_1}{n_2}\vec{s}_1 + \left(\frac{n_1}{n_2}|\vec{n}||\vec{s}_1|\cos\theta_1 - |\vec{n}||\vec{s}_2|\cos\theta_2\right)\vec{n}. \tag{3.12}$$

Looking at figure 3.2, we can see the vertical component of the refracted ray $\vec{s}_2$ is given by

$$\vec{n}\cdot\vec{s}_2 = |\vec{n}||\vec{s}_2|\cos\theta_2. \tag{3.13}$$

Notice that since they are unit vectors $|\vec{n}| = |\vec{s}_2| \equiv 1$,

$$\vec{n}\cdot\vec{s}_2 = \cos\theta_2, \tag{3.14}$$

we square it,

$$(\vec{n}\cdot\vec{s}_2)^2 = \cos\theta_2^2 = 1 - \sin\theta_2^2 = 1 - \frac{n_1^2}{n_2^2}\sin\theta_1^2. \tag{3.15}$$

Now recalling the definition of outer product we have,

$$\vec{\mathbf{n}} \wedge \vec{\mathbf{s}}_1 = |\vec{\mathbf{n}}||\vec{\mathbf{s}}_1|\sin\theta_1(\vec{\mathbf{e}}_1 \wedge \vec{\mathbf{e}}_2), \tag{3.16}$$

also we have, $|\vec{\mathbf{n}}| = |\vec{\mathbf{s}}_1| = 1$.

$$\vec{\mathbf{n}} \wedge \vec{\mathbf{s}}_1 = \sin\theta_1(\vec{\mathbf{e}}_1 \wedge \vec{\mathbf{e}}_2), \tag{3.17}$$

squaring,

$$(\vec{\mathbf{n}} \wedge \vec{\mathbf{s}}_1)^2 = \sin\theta_1^2(\vec{\mathbf{e}}_1 \wedge \vec{\mathbf{e}}_2)^2. \tag{3.18}$$

We found in section 2.6.5 of the last chapter that $(\vec{\mathbf{e}}_1 \wedge \vec{\mathbf{e}}_2)^2 = -1$, thus,

$$(\vec{\mathbf{n}} \wedge \vec{\mathbf{s}}_1)^2 = -\sin\theta_1^2. \tag{3.19}$$

Replacing above's expression in $(-\vec{\mathbf{n}} \cdot \vec{\mathbf{s}}_2)^2 = 1 - (n_1^2/n_2^2)\sin\theta_1^2$, we get,

$$\vec{\mathbf{n}} \cdot \vec{\mathbf{s}}_2 = \sqrt{1 + \frac{n_1^2}{n_2^2}(\vec{\mathbf{n}} \wedge \vec{\mathbf{s}}_1)^2}. \tag{3.20}$$

Now let's see the other two terms of equation (3.12), they are equal to,

$$\frac{n_1}{n_2}[\vec{\mathbf{s}}_1 + |\vec{\mathbf{n}}||\vec{\mathbf{s}}_1|\cos\theta_1\vec{\mathbf{n}}] = \frac{n_1}{n_2}[\vec{\mathbf{s}}_1 - (\vec{\mathbf{n}} \cdot \vec{\mathbf{s}}_1)\vec{\mathbf{n}}]. \tag{3.21}$$

Therefore, the refracted ray is given by

$$\boxed{\vec{\mathbf{s}}_2 = \frac{n_1}{n_2}[\vec{\mathbf{s}}_1 - (\vec{\mathbf{n}} \cdot \vec{\mathbf{s}}_1)\vec{\mathbf{n}}] - \vec{\mathbf{n}}\sqrt{1 + \frac{n_1^2}{n_2^2}(\vec{\mathbf{n}} \wedge \vec{\mathbf{s}}_1)^2} \quad \text{for } \vec{\mathbf{s}}_2, \vec{\mathbf{s}}_1, \vec{\mathbf{n}} \in \mathbb{R}^2.} \tag{3.22}$$

We have expressed Snell's law in terms of inner products and outer products, this with the only fact, that $\vec{\mathbf{e}}_1 \wedge \vec{\mathbf{e}}_2 = i$.

## 3.6 Three-dimensional Snell's law in geometric algebra

Snell's law is almost the same in three dimensions as in two with geometric algebra, the only big difference is that we do not need the bivector $\vec{\mathbf{e}}_1 \wedge \vec{\mathbf{e}}_2$, because we have another orthogonal unit vector $\vec{\mathbf{e}}_3$. The difference arrays in square, $(\vec{\mathbf{e}}_1 \wedge \vec{\mathbf{e}}_2)^2 = -1$ and $\mathbf{e}_3^2 = 1$. So, the procedure is almost the same, it just changes sign.

Let $\{\vec{\mathbf{e}}_1, \vec{\mathbf{e}}_2, \mathbf{e}_3\}$ be an orthonormal basis for the Euclidean space $\mathbb{R}^3$. Let $\vec{\mathbf{s}}_1 \in \mathbb{R}^3$ be the incident unit vector, let $\vec{\mathbf{s}}_2 \in \mathbb{R}^3$ be the refracted unit vector and let $\vec{\mathbf{n}} \in \mathbb{R}^3$ be the normal unit vector of the surface. $\theta_1$ and $\theta_2$ are the angles of the incident and refracted ray, respectively.

The refracted vector in terms of $\theta_1$ and $\theta_2$ is given by

$$\vec{\mathbf{s}}_2 = \frac{n_1}{n_2}\vec{\mathbf{s}}_1 + \left(\frac{n_1}{n_2}|\vec{\mathbf{n}}||\vec{\mathbf{s}}_1|\cos\theta_1 - |\vec{\mathbf{n}}||\vec{\mathbf{s}}_2|\cos\theta_2\right)\vec{\mathbf{n}}. \tag{3.23}$$

Let's focus first on $|\vec{\mathbf{n}}||\vec{\mathbf{s}}_2|\cos\theta_2$, which is equal to,

$$\vec{\mathbf{n}} \cdot \vec{\mathbf{s}}_2 = |\vec{\mathbf{n}}||\vec{\mathbf{s}}_2|\cos\theta_2. \tag{3.24}$$

We choose the unit vectors such that their magnitude is one, thus $|\vec{\mathbf{n}}| = |\vec{\mathbf{s}}_2| \equiv 1$,

$$\vec{\mathbf{n}} \cdot \vec{\mathbf{s}}_2 = \cos\theta_2. \tag{3.25}$$

Let's square it and manipulate it, using trigonometric identities and Snell's law,

$$(-\vec{\mathbf{n}} \cdot \vec{\mathbf{s}}_2)^2 = \cos\theta_2^2 = 1 - \sin\theta_2^2 = 1 - \frac{n_1^2}{n_2^2}\sin\theta_1^2. \tag{3.26}$$

Please remember the definition of outer product,

$$\vec{\mathbf{n}} \wedge \vec{\mathbf{s}}_1 = |\vec{\mathbf{n}}||\vec{\mathbf{s}}_1|\sin\theta_1\vec{\mathbf{e}}_3, \tag{3.27}$$

where, $|\vec{\mathbf{n}}| = |\vec{\mathbf{s}}_1| = 1$, so,

$$\vec{\mathbf{n}} \wedge \vec{\mathbf{s}}_1 = \sin\theta_1\vec{\mathbf{e}}_3, \tag{3.28}$$

squaring,

$$(\vec{\mathbf{n}} \wedge \vec{\mathbf{s}}_1)^2 = \sin\theta_1^2(\vec{\mathbf{e}}_3)^2, \tag{3.29}$$

where $(\mathbf{e}_3)^2 = 1$, this is the sign that is different for the two dimension version, thus,

$$(\vec{\mathbf{n}} \wedge \vec{\mathbf{s}}_1)^2 = \sin\theta_1^2. \tag{3.30}$$

Replacing the above expression in $(-\vec{\mathbf{n}} \cdot \vec{\mathbf{s}}_2)^2 = 1 - (n_1^2/n_2^2)\sin\theta_1^2$, we get,

$$\vec{\mathbf{n}} \cdot \vec{\mathbf{s}}_2 = \sqrt{1 - \frac{n_1^2}{n_2^2}(\vec{\mathbf{n}} \wedge \vec{\mathbf{s}}_1)^2}. \tag{3.31}$$

The other terms are almost the same,

$$\frac{n_1}{n_2}[\vec{\mathbf{s}}_1 + |\vec{\mathbf{n}}||\vec{\mathbf{s}}_1|\cos\theta_1\vec{\mathbf{n}}] = \frac{n_1}{n_2}[\vec{\mathbf{s}}_1 - (\vec{\mathbf{n}} \cdot \vec{\mathbf{s}}_1)\vec{\mathbf{n}}]. \tag{3.32}$$

Therefore, the refracted ray is $\vec{\mathbf{s}}_2 \in \mathbb{R}^3$ given by

$$\boxed{\vec{\mathbf{s}}_2 = \frac{n_1}{n_2}[\vec{\mathbf{s}}_1 - (\vec{\mathbf{n}} \cdot \vec{\mathbf{s}}_1)\vec{\mathbf{n}}] - \vec{\mathbf{n}}\sqrt{1 - \frac{n_1^2}{n_2^2}(\vec{\mathbf{n}} \wedge \vec{\mathbf{s}}_1)^2} \quad \text{for } \vec{\mathbf{s}}_2, \vec{\mathbf{s}}_1, \vec{\mathbf{n}} \in \mathbb{R}^3.} \tag{3.33}$$

With geometric algebra, we can also write Snell's law in four or even more dimensions. Still, it is not necessary since we are only dealing with rotationally symmetric surfaces in $\mathbb{R}^2$ and freeform surfaces in $\mathbb{R}^3$.

## 3.7 Stigmatism

There is an intrinsic relation between Snell's law and Fermat's principle. Snell's law chooses the redirection of light, such that the light travels in the smallest OPL. Fermat's principle tells us that the light travels in the smallest OPL. In both approaches there is something crucial for lens and mirror design, their shape. In any redirection of light, the normal vector of the surface is related $\vec{\mathbf{n}}$. The main goal in mirrors and lenses is to converge light or make it disappear through reflection or refraction, depending on the case. The goal of concentrating light is to unite the light in a single point, like a focus. The goal of dispersing light is to spread the light that comes from a single point source.

If we draw segments that meet the conditions defined for each of the conic sections, we will see that these drawings meet each other in the respective foci.

If we see these drawings as optical paths, we can argue that conic mirrors, mirrors with the shape of conics, can concentrate light in the foci. We can see these drawings as optical paths because it results that they are the paths in which light travels in less time.

A parabolic mirror reflects light from an object in minus infinity in its focus. See figure 3.3.

Elliptic mirrors focus the light of a real point object in a real point image, both lie in the geometrical foci of the ellipse. In figure 3.4 is presented an elliptic mirror where the object is one of the foci and the image in the other.

The elliptical mirror is only stigmatic for images and objects in its focus, if we have the object out of focus, the image will be a point, and we will not have a stigmatic relationship.

Hyperbolic mirrors focus the light of a real point object in a virtual point image, both lie in the geometrical foci of the hyperbola. See figure 3.5.

A spherical mirror forms a perfect point image in the center of the mirror if and only if the point object is located at the center of the mirror, see figure 3.6.

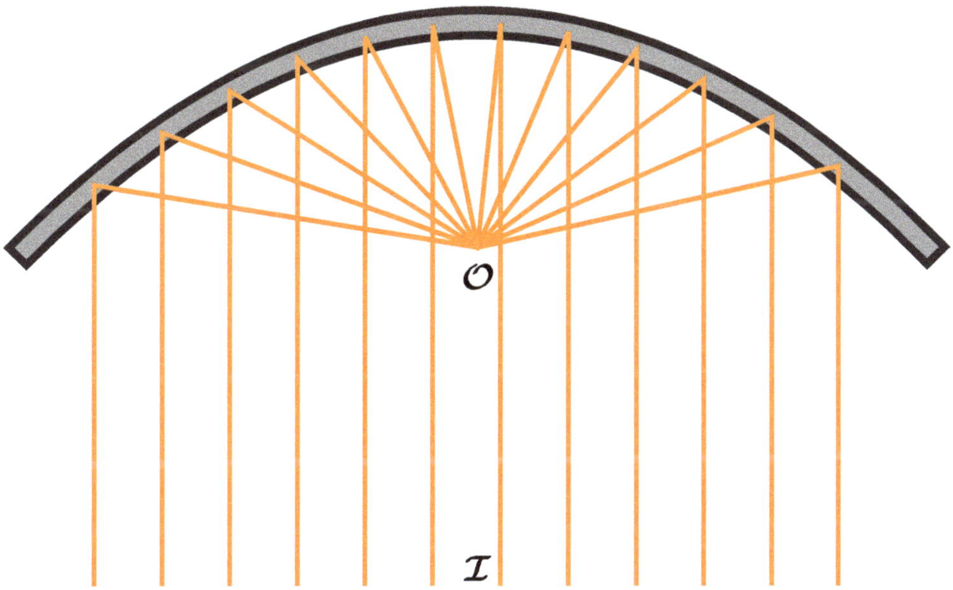


**Figure 3.3.** The parabolic mirror is stigmatic for a point in infinity. The rays coming from that point are reflected in the focus of the parabola.

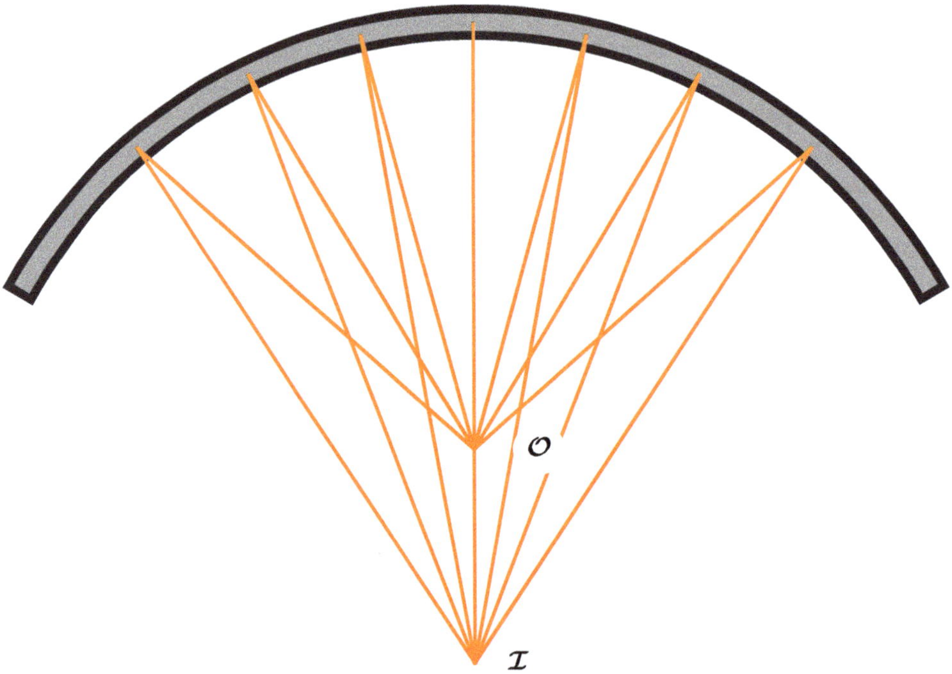


**Figure 3.4.** The ellipse is a plane curve surrounding two focal points, such that for all points on the curve, the sum of the two distances to the focal points is a constant. As such, it generalizes a circle, which is the particular type of ellipse in which the two focal points are the same. Therefore, the two focal points of the ellipse can be seen as the object and the image of the elliptical mirror.

The analogy with the conic sections and the conic mirror is so astonishing because it is a real revelation of the dictatorship of the geometry in geometrical optics. These results were so impressive that for centuries people were obsessed with the concept of focus. This is how the concept of focus is exported to optics, with the conic sections and their amazing intrinsic geometrical properties. Unfortunately, the word *focus* has been maltreated by theories like paraxial approximations theory or third-order paraxial approximations theory. The focus in paraxial approximations differs from the focus of the third-order approximation. Both approaches give functional results for particular circumstances, but they are not as elegant as the property the conic mirrors.

The property is so unique that it deserves a name, and it is called stigmatism. **Stigmatism** refers to the image-formation property of an optical system which focus a point object into a point image. Such points are called a stigmatic pair of the optical system. The stigmatic pair of an optical system is the pair foci of the optical system.

The conic mirrors are not the only optical elements that are stigmatism. The Cartesian oval is a refractive surface such that it is stigmatic. This means it has a point object as a source of light, the light that emerges from it is focused at a single point image, see figure 3.7.

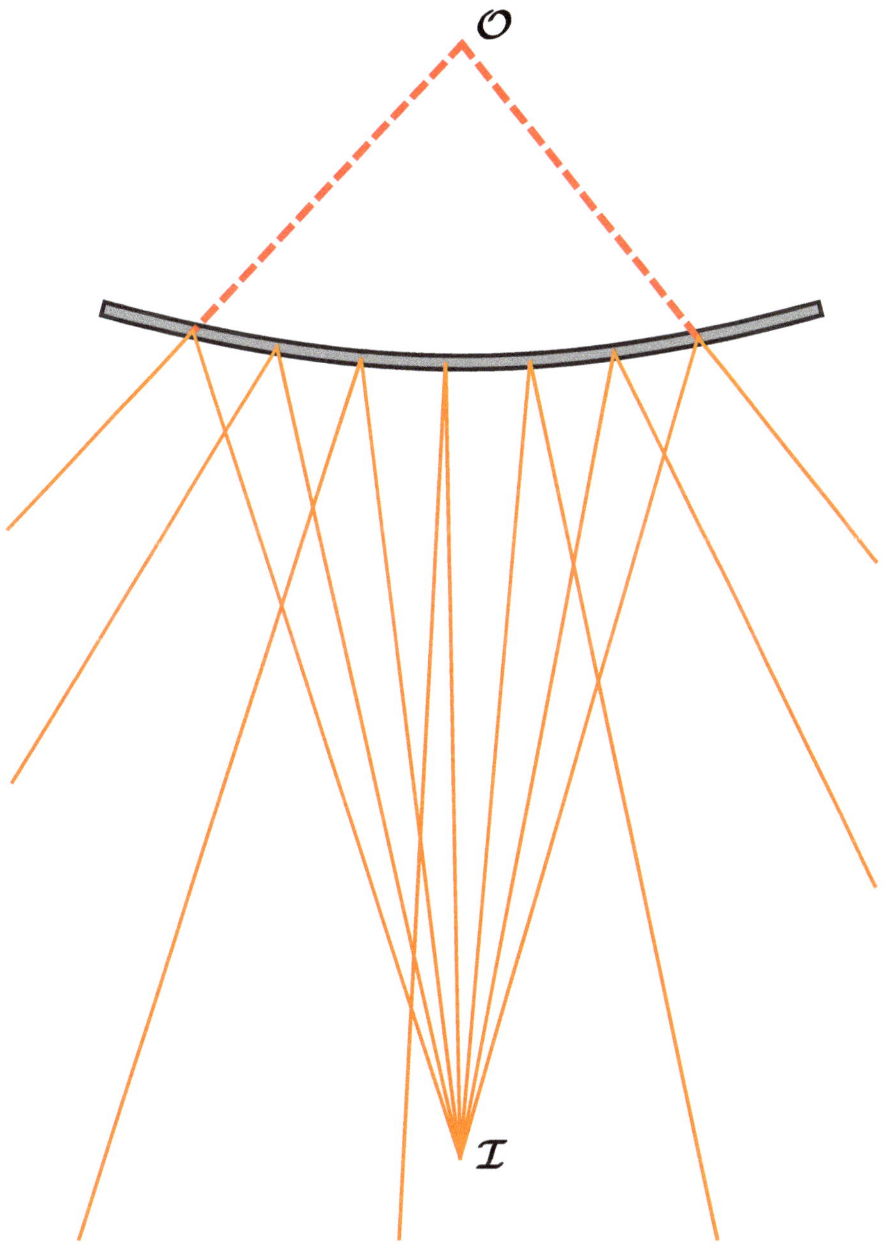


**Figure 3.5.** The hyperbolic mirror is only stigmatic if the image and the object are placed in their respective foci.

The Cartesian oval is such that the OPL of the axial ray is equal to the OPL of any other ray,

$$-t_1n_1 + t_2n_2 = n_1\sqrt{r^2 + (z - t_1)^2} + n_2\sqrt{r^2 + (z - t_2)^2}\,. \tag{3.34}$$

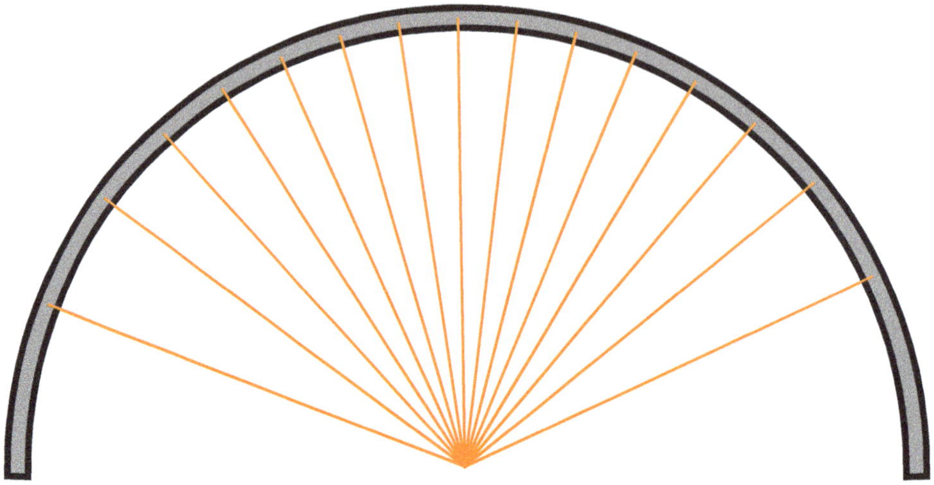


**Figure 3.6.** The spherical mirror is only stigmatic if the object and image are in the center of the spherical mirror.

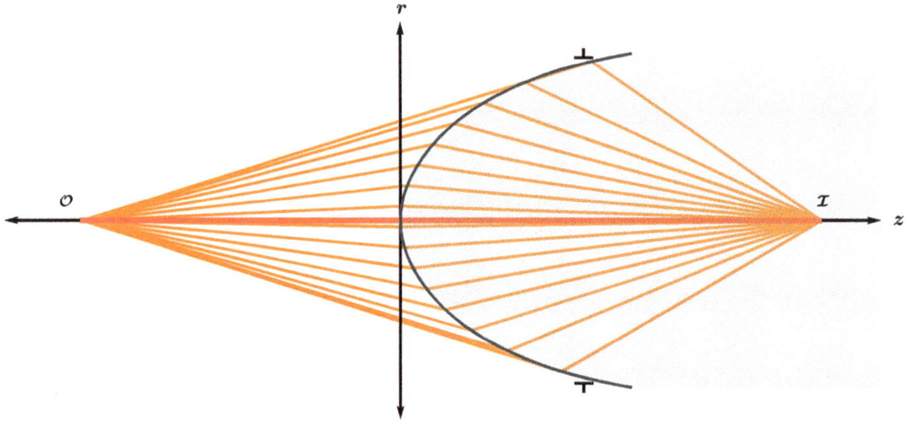


**Figure 3.7.** Cartesian oval is a stigmatic refractive surface.

The left side is the OPL axial ray and the right side is the OPL of any other ray. $-t_1n_1$ has a negative sign since the Cartesian oval is located at the origin. If we solve equation (3.34), we get the Cartesian oval as a function $z(r)$. We can do it, but the expression of $z(r)$ is huge, and it will take several pages to write it. The focal distances are $-t_1$ and $t_2$. See figure 3.7.

There are stigmatic lenses as well, and they are the core of this treatise; we are going to look them in detail in the following chapters.

## 3.8 Optical aberrations

In general, optical systems are not stigmatic. This means there is no corresponding point image for a point object. The image is no longer a point but an area. The size and characteristics of the area depend on the nature of the aberration that the optical systems suffer. Aberrations distort the image and decrease the image quality of an optical system.

Aberrations can be divided into two general categories, polychromatic aberrations and monochromatic aberrations. Polychromatic aberrations are also called chromatic aberrations. The variation of a lens's refractive index concerning the wavelength causes the chromatic aberrations. In other words, for different wavelengths, the lenses have various performances.

The shape of the optical system causes the monochromatic aberrations. This is because in any interaction of light ray with a reflective or refractive surface, the normal vector is related. Snell's law takes account of the normal vector, and Fermat's principle takes account of Snell's law. If both premises are united in a single equation, we can design stigmatic optical elements.

Now we are going to review the primary monochromatic aberrations, which are the spherical aberration, coma, astigmatism, field curvature, and image distortion.

### 3.8.1 Spherical aberration

If the rays that transmit the point object on the optical axis do not concentrate on a point image on the optical axis, then this phenomenon is called spherical aberration. Therefore, spherical aberration is the event in an optical system when a point object located in the optical axis does not have a stigmatic correspondence with a point image.

It is called spherical aberration because of the spherical lens sufferering from it. In figure 3.8 is presented the spherical aberration suffered by a spherical refractive surface. In figure 3.9 there is an example of spherical aberration generated by a spherical lens with refraction index $n$. There are lenses called aspherical because their shape is distinct from a sphere. In most cases, their main goal is to reduce spherical aberration. We will see those stigmatic lenses are aspherical in the following chapters. In figure 3.10 there is an aspherical lens free of spherical aberration.

### 3.8.2 Coma

The coma aberration in an optical system refers to aberration suffered by the image of an off-axis point object. The coma makes the image appear distorted, with a tail, like a coma or a comet. In other words, an optical system with coma has no stigmatic relationship between a point object outside the optical axis and a point image, since this image is not a point but a region.

Although we mentioned that the Cartesian oval is the only stigmatic refractive surface, it has other aberrations. As in figure 3.11, a Cartesian oval is shown for an infinite image and a finite off-axis object. In this case, the Cartesian oval shows a coma.

The coma is very noticeable in this example; please note the crossing of rays that is presented in figure 3.11. Precisely, this crossing of rays is what makes the image look like the aberration of a coma.

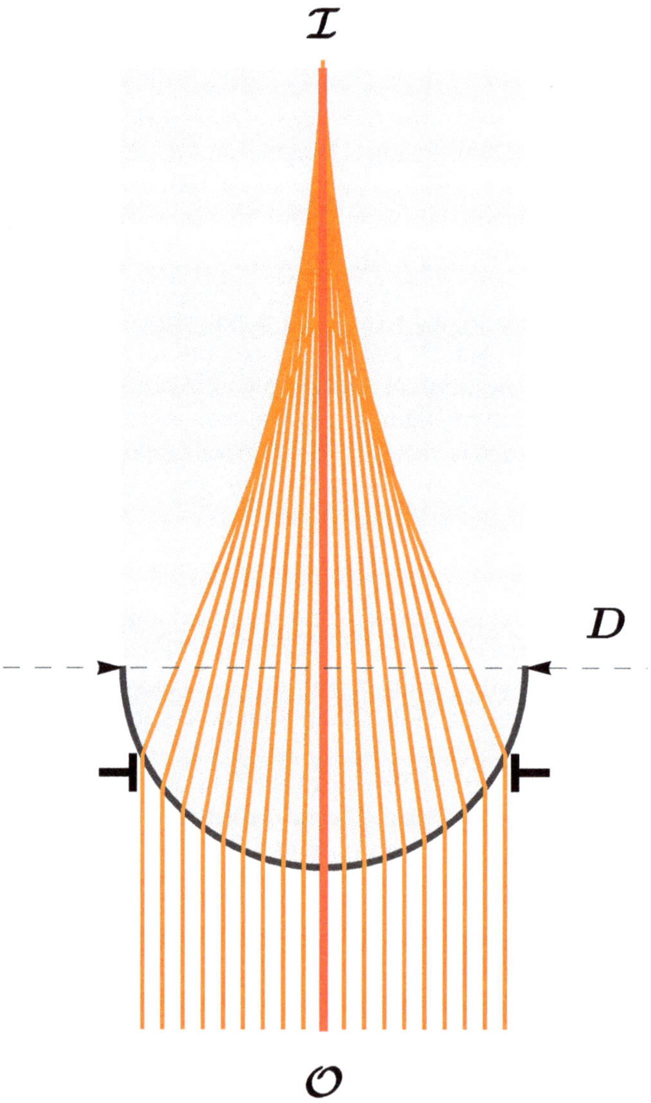


**Figure 3.8.** Spherical aberration suffered by a spherical refractive surface.

Remember that the Cartesian oval is called Cartesian in honor to Descartes but the first person to study it was Ibn Sahl.

### 3.8.3 Astigmatism

The astigmatism of a point object takes place when two perpendicular planes have different image points.

Astigmatism means not stigmatic, therefore the system has two point images, as can be seen in figure 3.12.

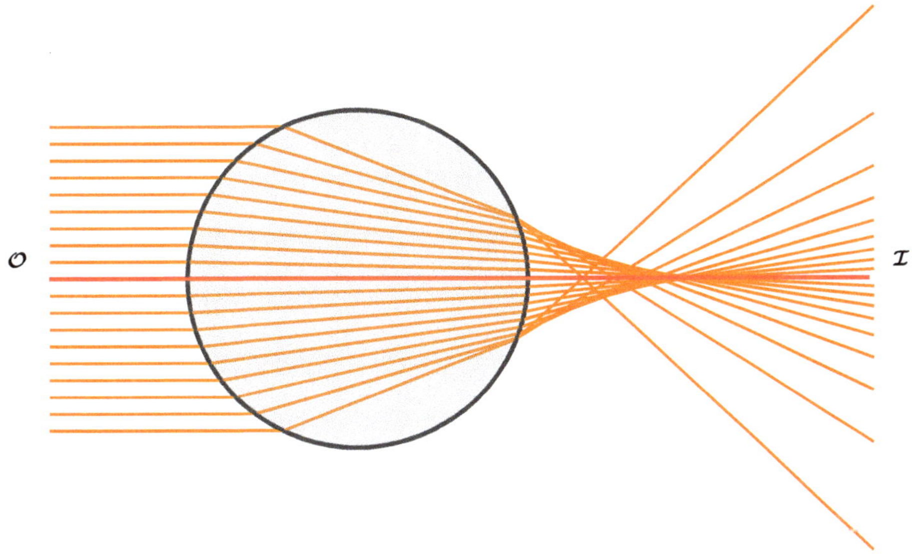


**Figure 3.9.** Spherical aberration suffered by a spherical lens.

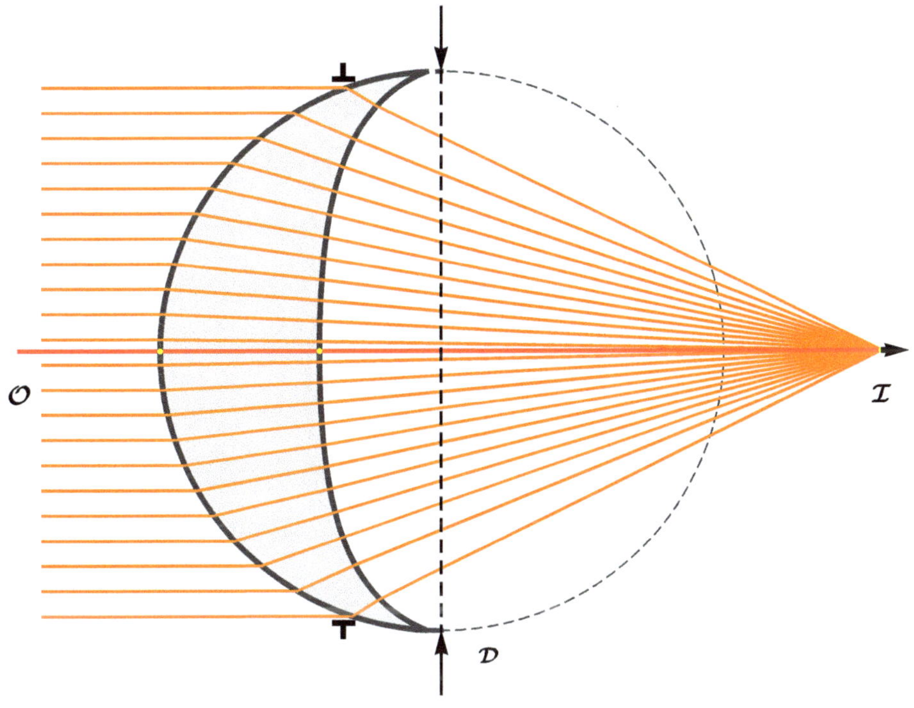


**Figure 3.10.** Aspherical lens free of spherical aberration.

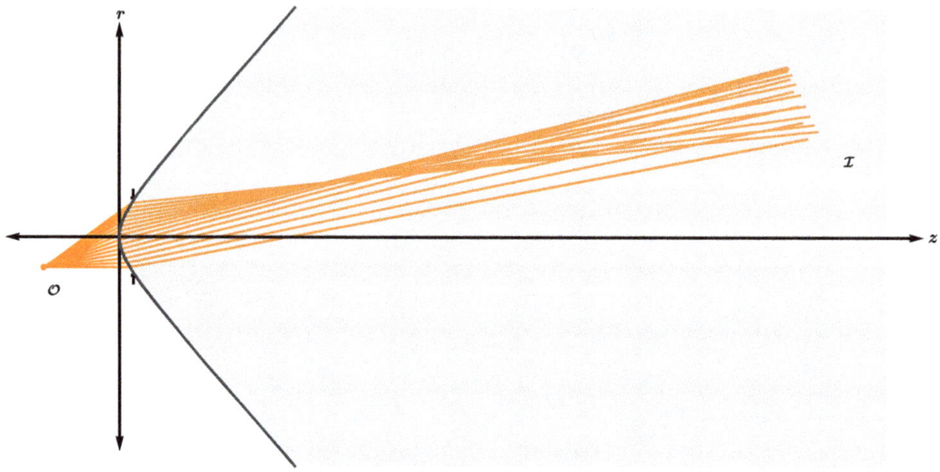


**Figure 3.11.** Cartesian oval suffering from several coma.

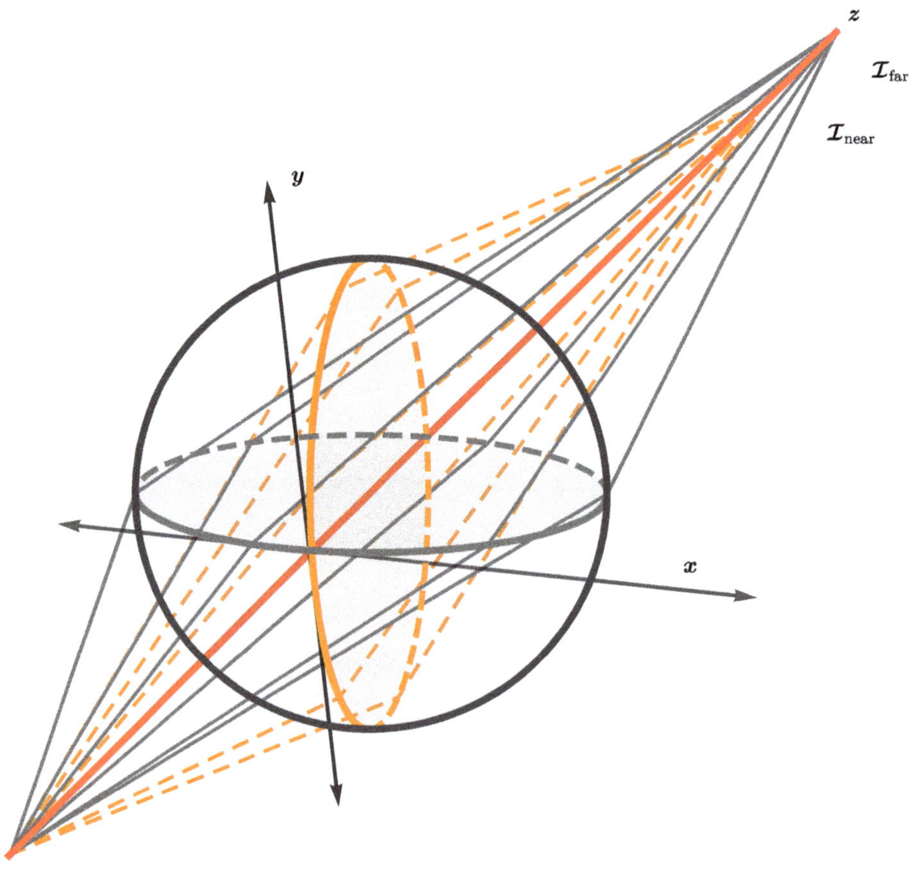


**Figure 3.12.** An astigmatism presented in a lens.

### 3.8.4 Field curvature

Petzval field curvature, named for Joseph Petzval, describes the optical aberration in which a flat object normal to the optical axis is focused as a curved image. In figure 3.13 we present the Petzval field curvature in a Cartesian oval.

As mentioned earlier, the Cartesian oval is the only stigmatic surface for a single point on-axis object and a single on-axis image. If we take a Cartesian oval and apply ray tracing to objects outside the axis, we will have images that are not stigmatic outside the axis. The images will not be points, but regions. The images also converge along a curvature, the Petzval curvature. The idea is to find an optical system where this curve must be a prefect plane, the plane of the image.

### 3.8.5 Image distortion

The distortion in a forming image system is measured with a rectilinear projection. The rectilinear projection is passed through the system, a projection in which straight lines in a scene remain straight in the image if there is no distortion in the system. The two most common types of image distortion can be seen in figures 3.14 and 3.15, pincushion and barrel distortions, respectively.

## 3.9 Conclusions

In this chapter, we briefly reviewed all the concepts needed to start designing stigmatic lenses. In the following chapters, we will focus our efforts on generating stigmatic lenses for both on-axis and off-axis objects.

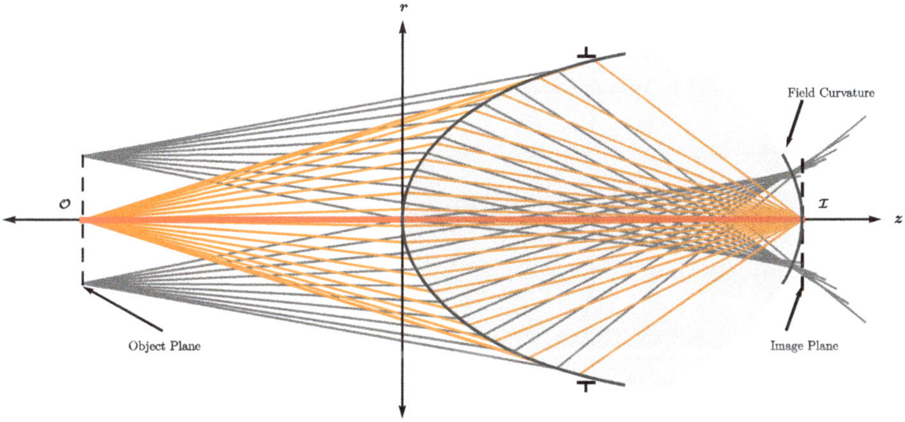


**Figure 3.13.** Petzval field curvature in a Cartesian oval.

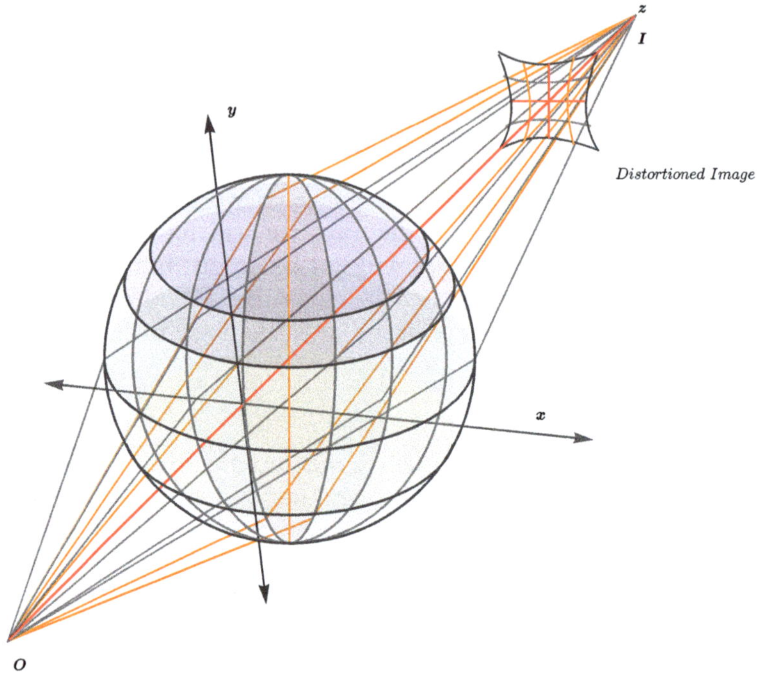


**Figure 3.14.** Pincushion image distortion.

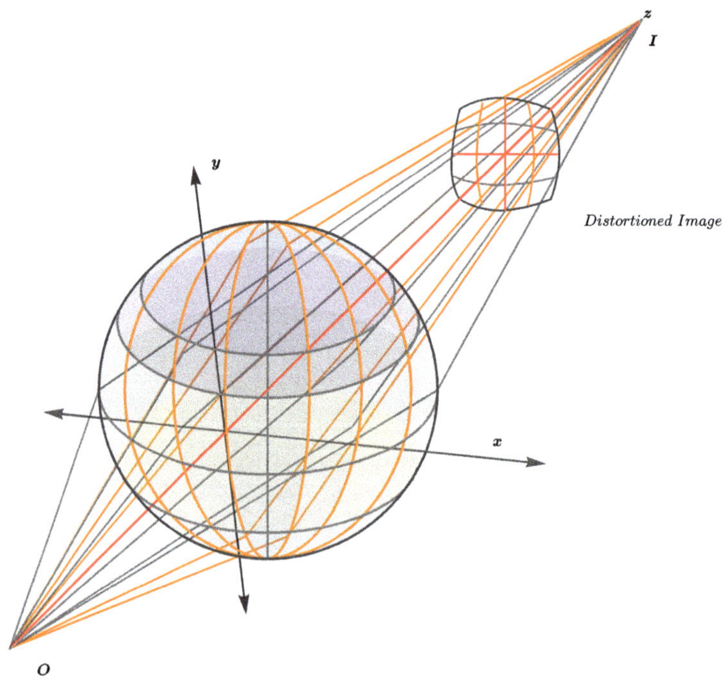


**Figure 3.15.** Barrel image distortion.

## Further reading

Bentley J L and Olson C 2012 *Field Guide to Lens Design* (Bellingham, WA: SPIE)
Born M and Wolf E 2013 *Principles of Optics: Electromagnetic Theory of Propagation,Interference and Diffraction of Light* (Amsterdam: Elsevier)
Chaves J 2016 *Introduction to Nonimaging Optics* 2nd edn (Boca Raton, FL: CRC Press)
Conrady A E 2013 *Applied Optics and Optical Design, Part One* (North Chelmsford, MA: Courier Corporation)
Descartes R 1637a *De La Nature Des Lignes Courbes*
Descartes R 1637b *La Géométrie*
Geary J M 2002 *Introduction to Lens Design: With Practical ZEMAX Examples* (Richmond, VA: Willmann-Bell)
Huygens C 1690 *Traité de la lumière*
Kidger M J 2002 *Fundamental Optical Design* vol 92 (Bellingham, WA: SPIE)
Kingslake R and Johnson R B 2009 *Lens Design Fundamentals* (New York: Academic)
Laikin M 2018 *Lens Design* (Boca Raton, FL: CRC Press)
Luneburg R K and Herzberger M 1964 *Mathematical Theory of Optics* (Berkeley, CA: University of California Press)
Malacara D 2001 *Handbook of Optical Engineering* (Boca Raton, FL: CRC Press)
Malacara D 2003 Color vision and colorimetry: theory and applications *Color Res. Appl.* **28** 77–8
Malacara D 2007 *Optical Shop Testing* vol 59 (New York: Wiley)
Malacara D 2015 *Óptica Básica* (San Diego, CA: Fondo de Cultura Económica)
Malacara D and Malacara Z 1994 *Handbook of Lens Design* (New York: Dekker)
Malacara-Hernández D and Malacara-Hernández Z 2016 *Handbook of Optical Design* (Boca Raton, FL: CRC Press)
Newton I 1704 *Opticks, or, a Treatise of the Reflections, Refractions, Inflections & Colours of Light* (New York: Dover)
Smith W J and Smith W J 1966 Optical Engineering *Modern Optical Engineering* vol 3 (New York: McGraw-Hill)
Smith W J *et al* 2005 *Modern Lens Design* vol 2 (New York: McGraw-Hill)
Stavroudis O N 2006 *The Mathematics of Geometrical and Physical Optics: The k-Function and Its Ramifications* (New York: Wiley)
Sun H 2016 *Lens Design: A Practical Guide* (Boca Raton, FL: CRC Press)

# Part II

## Stigmatic singlets

**IOP** Publishing

Analytical Lens Design (Second Edition)

**Rafael G González-Acuña, Héctor A Chaparro-Romo and Julio C Gutiérrez-Vega**

# Chapter 4

# On-axis stigmatic aspheric lens

In this chapter, we present a rigorous analytical solution for the on-axis stigmatic singlet lens design problem. The general formula entry given here is the first surface of the singlet lens. This surface must be continuous and in such a way that the rays inside the glass do not cross each other. The output is the correcting second surface of the singlet; the second surface is such that the singlet is free of spherical aberration.

## 4.1 Introduction

In the first chapter of this book, we presented a brief introduction to the history of optics, mainly geometrical optics. We saw that geometrical optics started along with geometry. We found that the conic sections are so unique because they are stigmatic. In the last chapter, we saw that there are also stigmatic surfaces called Cartesian ovals, and we mentioned that stigmatic lenses exist. For a long time, people thought they existed, but there was no closed formula to describe them.

Diocles was the first person to adequately study the parabolic mirror. Diocles found the stigmatic property of the parabolic mirror and presented this in his magnum opus, *Burning Mirrors*. Right after demonstrating the mentioned property, Diocles compared the parabolic mirror and the spherical mirror; where he concluded that the parabolic mirror focuses the light much better than the spherical mirror. Then, Diocles mentions that a lens with the same unique property, stigmatic glass/lens, can be made. Diocles, two thousand years ago, proposed a stigmatic lens, just like the parabolic mirror.

We also mentioned that René Descartes was interested in it. He mentioned it in his magnum opus, *De la nature des lignes courbes*, in 1637,

> *I might go farther and explain how, **if one surface of a lens is provided and is not entirely plane nor formed of conic sections or circles, the other surface can be so determined as to carry all the rays from a given point to a***

 

> ***different point, also provided.*** *This is no more complicated than the problems I have just described: indeed it is much easier since the way is now clear; I prefer, however, to leave this for others to struggle out, to the end that they may appreciate the more highly the discovery of those things here explained, through having themselves to face some challenges.*

In 1690, Christian Huygens mentioned that Sir Isaac Newton and Gottfried Wilhelm Leibniz were behind it. The following passage is from the preface of, Huygens's *Traité de la lumière*,

> *I write, and not for the intention of decreasing from the merit of those who, without having seen anything that I have written, may be found to have attended of like matters: as has, in truth, happened to* ***two prominent Geometricians, Messieurs Newton and Leibniz, with regard to the enigma of the shape of glasses for collecting rays when one of the surfaces is provided.***

Actually, Christian Huygens was very interested in the problem,

> *And lastly, I shall approach the several shapes of transparent and reflecting forms by which rays are collected at a point or are directed aside in various ways. From this, it will be seen with what tools, following our new theory, we find not only the Ellipses, Hyperbolas, and other curves which Mr Descartes has ingeniously developed for this purpose;* ***but also those which the surface of a glass lens ought to hold when its other surface is provided as spherical or plane, or of any other shape that may be.***

Huygens, again in *Traité de la lumière*, made another reference to the problem and the opinion of Descartes about it[1].

> *Let us now turn to our way and let us observe how it leads without challenge to the finding of the curves which one side of the glass requires when the other side is of a given figure; a figure, not only plane or spherical, or made by one of the conic sections* ***(which is the limitation with which Descartes introduced this problem, giving the answer to those who should come after him)*** *but generally any figure whatever: that is to say, one created by the revolution of any given curved line to which one must simply know how to draw straight lines as tangents.*

Finally, Huygens, in chapter 6 of *Traité de la lumière* tried to solve it,

> *Let the given figure be that made by the revolution of some curve such as AK regarding the axis AV, and that this side of the glass receives rays*

[1] Huygens refers to the book of Descartes, *De la nature des lignes courbes.*

*coming from the point L. Moreover, let the thickness AB of the center of the glass be given, and the point F at which one aspires the rays to be all perfectly focused, whatever be the initial refraction happening at the surface AK.*

*I announce that for this the single condition is that the outline BDK which composes the other surface shall be such that the path of the light from the point L to the surface AK, and from thence to the surface BDK, and from thence to the point F, shall be traversed wherever in equal times, and in each case in a time equal to that which the light uses to advance along the straight line LF of which the part AB is within the glass.*

*Let LG be a ray descending on the arc AK. Its refraction GV will be provided by means of the tangent which will be carried at the point G. Now, in GV the point D need be determined such that FD together with f of DG and the straight line GL, may be equal to FB together with f of BA and the straight line AL; which, as is plain, make up a given length. Or rather, by deducing from each the length of LG, which is also provided, it will merely be needful to adjust FD up to the straight line VG in such a way that FD together with -f of DG is equal to a provided straight line, which is a quite obvious plane problem: and the point D will be one of those through which the curve BDK ought to pass. And similarly, having drawn another ray LM, and found its refraction MO, the point N will be found in this line, and so on as many occasions as one wants.*

He was very close to doing it, he understood the optical paths must be the same, but he did not report any equation to describe the second surface of the lens.

Huygens with his magnificent mind but his limited environment was able to understand the problem. 'Limited environment' because at the time he lived there was no topology or functional mathematical analysis. If Diocles, Huygens, Newton, Descartes or Leibniz had had the mathematical tools we have now, it would be another story, but there was no such possibility.

It is always surprising to read great minds, read their ideas from their original sources. Fragments from *Traité de la lumière* are shown in figures 4.1–4.3.

Leonard Euler, considered the best mathematician of the eighteenth century and one of the most prolific of all time, was interested in the problem of stigmatic lens design.

In his work *Dioptricae*, in volume one, problem 5, page 28, Euler wondered how the distance between the Ff images should be if the distances EA, AB and BF are known. See figure 4.4.

Where Ff is the distance between images F and f, in other means if Ff is not zero the lens has spherical aberration.

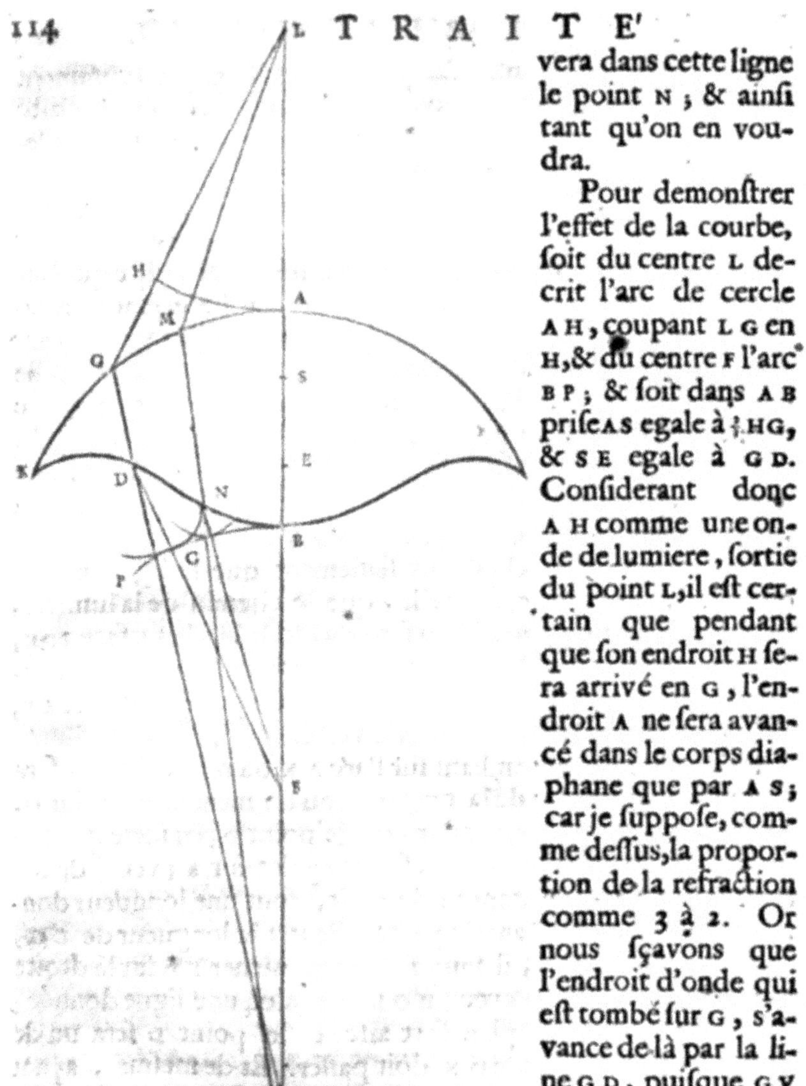

114 TRAITE'

vera dans cette ligne le point N ; & ainſi tant qu'on en voudra.

Pour demonſtrer l'effet de la courbe, ſoit du centre L decrit l'arc de cercle AH, coupant LG en H, & du centre F l'arc BP ; & ſoit dans AB priſe AS egale à ⅔ HG, & SE egale à GD. Conſiderant donc AH comme une onde de lumiere, ſortie du point L, il eſt certain que pendant que ſon endroit H ſera arrivé en G, l'endroit A ne ſera avancé dans le corps diaphane que par AS ; car je ſuppoſe, comme deſſus, la proportion de la refraction comme 3 à 2. Or nous ſçavons que l'endroit d'onde qui eſt tombé ſur G, s'avance de là par la ligne GD, puiſque GV eſt la

**Figure 4.1.** An on-axis stigmatic lens for finite object and finite image proposed by Huygens in *Traité de la lumière.* Source https://archive.org/details/bub_gb_kVxsaYdZaaoC/page/n125.

EA is the distance from the object to the first surface. AB is the central thickness of the lens and BF is the distance between the second surface of the lens and the image F.

Subsequently, on page 31 of the aforementioned book, it is questioned how the shape of said lens should be so that the distance Ff is zero. In other words, how the

116 T R A I T E'

faille trouver l'autre ſurface K D B, qui recevant des rayons paralleles à B A les diviſe en ſorte, qu'aprés eſtre derechef rompus à la ſurface donnée A K, ils s'aſſemblent tous au point L.

Soit du point L menée, à quelque point de la ligne donnée A K, la droite L G; qui eſtant conſiderée comme un rayon de lumiere, on trouvera ſa refraction G D, qui d'un coſté ou d'autre rencontrera, eſtant prolongée, la droite B L, comme icy en V. Soit enſuite erigée ſur A B la perpendiculaire B C, qui repreſentera une onde de lumiere venant du point F infiniment diſtant, par ce que nous avons ſuppoſé des rayons paralleles. Il faut donc que toutes les parties de cette onde B C arrivent en meſme temps au point L; ou bien que toutes les parties d'une onde, émanée du point L, arrivent en meſme temps à la droite B C. Et pour cela il faut trouver, dans la ligne V G D, le point D, en ſorte qu'ayant mené D C parallele à A B, la ſomme de C D & 3/2 de D G & G L ſoit egale à 3/2 A B avec A L: ou bien, en oſtant d'un coſté & d'autre G L qui eſt donnée, il faut que C D avec 3/2 de D G ſoit egale à une ligne donnée: qui eſt un probleme encore plus aiſé que celuy de la conſtruction precedente. Le point D, ainſi trouvé, ſera un de ceux par où la courbe doit paſſer; & la demonſtration

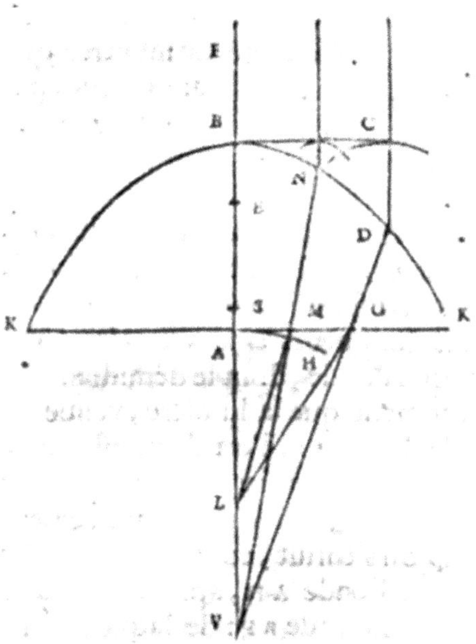


**Figure 4.2.** An on-axis stigmatic collector lens, proposed by Huygens in *Traité de la lumière*. Source https://archive.org/details/bub_gb_kVxsaYdZaaoC/page/n125.

lens shape should be so that it is free of spherical aberration. Euler does not reach a closed or general analytical solution to this problem.

The prominent astronomer John Frederick William Herschel was also interested in the design of the optical system free of spherical aberration. In his book *On the Theory of Light* of 1827, he tried to solve the problem of spherical aberration.

He wrote the following letters describing equation (312) of the mentioned treatise at page 312.

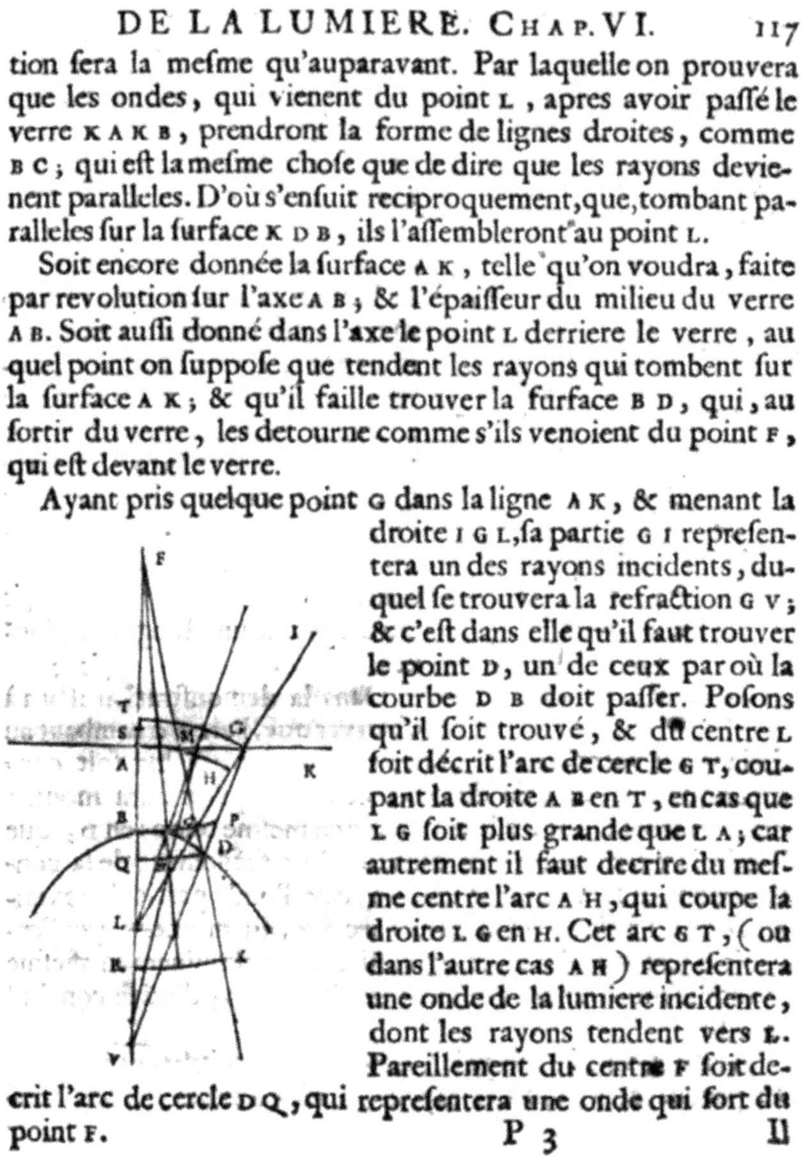

DE LA LUMIERE. CHAP. VI. 117

tion ſera la meſme qu'auparavant. Par laquelle on prouvera que les ondes, qui vienent du point L, apres avoir paſſé le verre KAKB, prendront la forme de lignes droites, comme BC; qui eſt la meſme choſe que de dire que les rayons devienent paralleles. D'où s'enſuit reciproquement, que, tombant paralleles ſur la ſurface KDB, ils l'aſſembleront au point L.

Soit encore donnée la ſurface AK, telle qu'on voudra, faite par revolution ſur l'axe AB; & l'épaiſſeur du milieu du verre AB. Soit auſſi donné dans l'axe le point L derriere le verre, au quel point on ſuppoſe que tendent les rayons qui tombent ſur la ſurface AK; & qu'il faille trouver la ſurface BD, qui, au ſortir du verre, les detourne comme s'ils venoient du point F, qui eſt devant le verre.

Ayant pris quelque point G dans la ligne AK, & menant la droite IGL, ſa partie GI repreſentera un des rayons incidents, duquel ſe trouvera la refraction GV; & c'eſt dans elle qu'il faut trouver le point D, un de ceux par où la courbe DB doit paſſer. Poſons qu'il ſoit trouvé, & du centre L ſoit décrit l'arc de cercle GT, coupant la droite AB en T, en cas que LG ſoit plus grande que LA; car autrement il faut decrire du meſme centre l'arc AH, qui coupe la droite LG en H. Cet arc GT, (ou dans l'autre cas AH) repreſentera une onde de la lumiere incidente, dont les rayons tendent vers L. Pareillement du centre F ſoit decrit l'arc de cercle DQ, qui repreſentera une onde qui ſort du point F.

P 3 Il

**Figure 4.3.** An on-axis stigmatic lens with virtual image and virtual object, proposed by Huygens in *Traité de la lumière*. Source https://archive.org/details/bub_gb_kVxsaYdZaaoC/page/n125.

*the problem of the destruction of the spherical aberration as (it is termed) becomes indeterminate.*

The prominent mathematician, the prince of mathematics, Johann Carl Friedrich Gauss in 1840 in his treatise of optics called *Dioptrische Untersuchungen*, wrote the following words on the state of the art of lens design at that time.

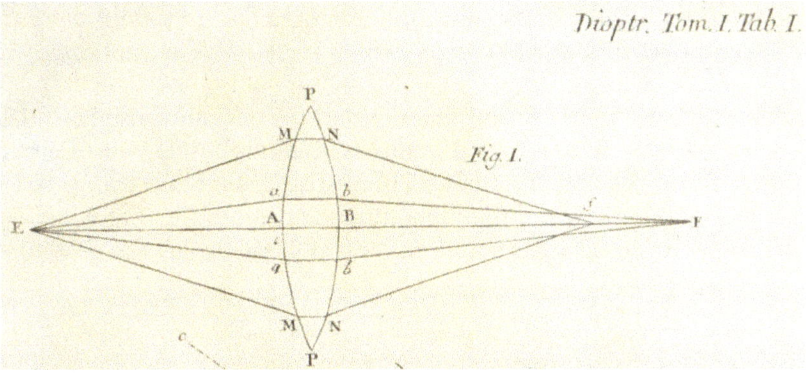


**Figure 4.4.** This image is taken from *Dioptricae* volume 1, figure 1. It corresponds to problem 5 at page 28. Euler wondered how the distance between the Ff images should be if the distances EA, AB and BF are known. Where Ff is the distance between images F and f.EA is the distance from the object to the first surface. AB is the central thickness of the lens and BF is the distance between the second surface of the lens and the image F. Courtesy of The Linda Hall Library of Science, Engineering & Technology.

*The consideration of the path taken by the rays of light through the lenses, which are very little inclined towards their common axis, and the phenomena that depend on them,* ***offers very elegant results, which could seem exhausted by the work of Cotes, Euler, Lagrange and Möbius,*** *but leave more to be desired. An important deficiency in the propositions made by these mathematicians is that the thickness of the lenses is neglected, as a result of which they are imparted their inaccuracy and inadequacy, which decreases their value. Without denying that, for some other dioptric tests, especially for those,* ***taking into account the so-called deviation due to the spherical shape of the lens surfaces,*** *the initial neglect of the lens thickness becomes very useful, in fact necessary, to simpler and more flexible recipes. To win the overturns and the first approximations, one would like to see that such sacrifice is eliminated from all the sharpness where it can be done without all or without a significant loss due to the simplicity of the results.*

In bold letters refers to the spherical aberration suffered by a spherical lens. Then, Gauss criticizes the concept of focal distance.

*We already find a lack of precision, which bothers the mathematical sense, partly from the first definitions of diopters. The terms of axis and focal point of a lens are really clear; but it is not so with the focal length that most writers explain as the distance of the focal point of the lens from its center, either tacitly assuming from the beginning or expressly saying that the thickness of the lens is considered infinitely small, of so that for real lenses the focal length maintains an uncertainty about the order of lens thickness.*

Gauss continues with this treatise, but only for paraxial rays.*Only those light rays that are very close to this axis that form a small angle are considered.*

It has been more than two thousand years to find the general formula of the on-axis stigmatic lens. At least five universal geniuses, Descartes, Newton, Leibniz, Euler and Huygens failed to achieve the solution. The history of this problem is so precious; it has innumerable approximate numerical solutions. But what we want is the general, analytic closed-form solution. Now it's time to get the general equation on-axis stigmatic lens.

## 4.2 Finite object finite image

We have mentioned the problem with the words of Descartes and Huygens, now let's talk with our own words.

Given the position of a point object, the location of a point image, the refractive index of the lens, the first surface of the lens and the central thickness of the lens. How must the second surface be such that the singlet is free of spherical aberration?

So, the essential goal here is to know the shape of the second surface ($z_b$, $r_b$), given a first surface ($z_a$, $r_a$), to get an on-axis stigmatic lens.

More precisely, it is to express ($z_b$, $r_b$) in terms of ($z_a$, $r_a$). Where $r_a$ is the only independent variable, $z_b$, $r_b$ and $z_a$ are functions of $r_a$. Notice that the subscript $a$ refers to the first surface, while the subscript $b$ refers to the second surface.

We placed the first surface of the lens in the origin of the coordinate system, thus $z_a(0) = 0$. We assume that the stigmatic lens is surrounded by air, which has a refractive index of one. The refraction index $n$ of the lens is constant and the singlet is radially symmetric. At the center, the singlet lens has a thickness of $t$. The distance from the object to the first surface is $t_a$. The distance from the second surface to the image is $t_b$, as can be seen in figure 4.5.

### 4.2.1 Fermat principle

As Huygens pointed out in it *Traité de la lumière*, the OPL of all rays that come from a point object on the axis and end up on a point image on the axis must be the same. Therefore, we need to establish a reference ray and compare its OPL with the OPL of any other ray. Keep in mind that the natural reference ray for this problem is the axial ray because we already know its path. It comes from $-t_a$, goes inside the lens and travels a distance $t$ and finally, from the second surface, it moves a distance $t_b$ to meet the point image. Please see figure 4.5. Thus is the optical path is,

$$-t_a + nt + t_b = \text{const.} \tag{4.1}$$

We need to compare this reference OPL with the OPL of any other ray, but since we want the general formula of stigmatic lenses, we need to take into account if the points objects/images are real or virtual.

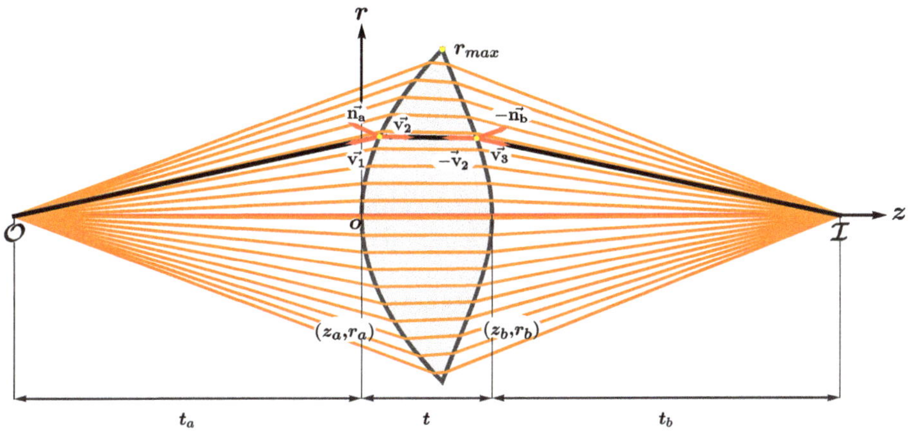


**Figure 4.5.** Diagram of a on-axis stigmatic singlet. The first surface is given by $(z_a, r_a)$, and the second surface is given by $(z_b, r_b)$. The distance between the first surface and the object is $t_a$, the thickness at the center of the lens is $t$, and the distance between the second surface and the image is $t_b$. $\vec{\mathbf{v}}_1$ is unit vector of the incident ray, $\vec{\mathbf{v}}_2$ is the unit vector of the refracted ray inside the lens, $\vec{\mathbf{v}}_3$ is the output $\vec{\mathbf{n}}_a$ is the normal vector of the first surface and $\vec{\mathbf{n}}_b$ is the normal vector of the second surface.

The simplest case to see is when we have a real object and a real image, $t_a < 0$ and $t_b > 0$,

$$
\begin{aligned}
-t_a + nt + t_b = &\sqrt{r_a^2 + (z_a - t_a)^2} + n\sqrt{(r_b - r_a)^2 + (z_b - z_a)^2} \\
&+ \sqrt{r_b^2 + (z_b - t - t_b)^2},
\end{aligned}
\tag{4.2}
$$

for a virtual object and real image we have: $t_a > 0$ and $t_b > 0$,

$$
\begin{aligned}
-t_a + nt + t_b = &-\sqrt{r_a^2 + (z_a - t_a)^2} + n\sqrt{(r_b - r_a)^2 + (z_b - z_a)^2} \\
&+ \sqrt{r_b^2 + (z_b - t - t_b)^2}.
\end{aligned}
\tag{4.3}
$$

Real object and virtual image: $t_a < 0$ and $t_b < 0$,

$$
\begin{aligned}
-t_a + nt + t_b = &\sqrt{r_a^2 + (z_a - t_a)^2} + n\sqrt{(r_b - r_a)^2 + (z_b - z_a)^2} \\
&- \sqrt{r_b^2 + (z_b - t - t_b)^2}.
\end{aligned}
\tag{4.4}
$$

Finally, the case with virtual object and virtual image $t_a > 0$ and $t_b < 0$, we have,

$$
\begin{aligned}
-t_a + nt + t_b = &-\sqrt{r_a^2 + (z_a - t_a)^2} + n\sqrt{(r_b - r_a)^2 + (z_b - z_a)^2} \\
&- \sqrt{r_b^2 + (z_b - t - t_b)^2}.
\end{aligned}
\tag{4.5}
$$

We need to associate the four previous equations, to obtain a general comparison between the OPL of the reference ray and the OPL of any other ray; the equation is as follows,

$$\boxed{\begin{aligned}-t_a+nt+t_b=&-\operatorname{sgn}(t_a)\sqrt{r_a^2+(z_a-t_a)^2}\\&+n\sqrt{(r_b-r_a)^2+(z_b-z_a)^2}\\&+\operatorname{sgn}(t_b)\sqrt{r_b^2+(z_b-t-t_b)^2}.\end{aligned}} \tag{4.6}$$

Note that the sgn(·) functions take the value of the sign of their argument, but if the argument is zero, they are not defined. Please take the time to evaluate the four cases in this last expression and verify its generality.

We have only one equation for two unknowns $(z_b, r_b)$, we need another one and Snell's law gives it.

### 4.2.2 Snell's law

We can use Snell's law at the first or second surface, the result is the same; but if we choose the second surface, we will deal with the derivatives of $(z_b, r_b)$ because the normal vector is in terms of $r_b'$ and $z_b'$. Therefore, it is better to use the Snell on the first surface because $z_a$ is given, so we know $z_a'$. Using Snell's law in two dimensions in geometric algebra we have,

$$\boxed{\vec{\mathbf{v}}_2=\frac{1}{-\operatorname{sgn}(t_a)n}[\vec{\mathbf{v}}_1-(\vec{\mathbf{n}}_a\cdot\vec{\mathbf{v}}_1)\vec{\mathbf{n}}_a]-\vec{\mathbf{n}}_a\sqrt{1+\frac{1}{n^2}(\vec{\mathbf{n}}_a\wedge\vec{\mathbf{v}}_1)^2}\quad\text{for}\quad\vec{\mathbf{v}}_2,\vec{\mathbf{v}}_1,\vec{\mathbf{n}}_a\in\mathbb{R}^2,} \tag{4.7}$$

where $\vec{\mathbf{v}}_1$ is the unit vector of the incident ray, $\vec{\mathbf{v}}_2$ is unit vector of the refracted ray and finally $\vec{\mathbf{n}}_a$ is the normal vector of the first surface. Notice the term $-\operatorname{sgn}(t_a)$, it comes from the fact that the object can be real or virtual. Please see figure 4.5.

The related unit vectors in the refraction suffered by light on the first surface are,

$$\begin{aligned}\vec{\mathbf{v}}_1&=\frac{r_a\vec{\mathbf{e}}_1+(z_a-t_a)\vec{\mathbf{e}}_2}{\sqrt{r_a^2+(z_a-t_a)^2}},\qquad \vec{\mathbf{v}}_2=\frac{(r_b-r_a)\vec{\mathbf{e}}_1+(z_b-z_a)\vec{\mathbf{e}}_2}{\sqrt{(r_b-r_a)^2+(z_b-z_a)^2}},\\ \vec{\mathbf{n}}_a&=\frac{z_a'\vec{\mathbf{e}}_1-\vec{\mathbf{e}}_2}{\sqrt{1+z_a'^2}},\end{aligned} \tag{4.8}$$

where $\vec{\mathbf{e}}_1$ is for the $r$ direction and $\vec{\mathbf{e}}_2$ is for the $z$ direction. We need to replace them in equation (4.7), let's do it step by step. We start with the term, $(\vec{\mathbf{n}}_a\wedge\vec{\mathbf{v}}_1)^2$,

$$(\vec{\mathbf{n}}_a\wedge\vec{\mathbf{v}}_1)=\frac{[r_a+(z_a-t_a)z_a']}{\sqrt{r_a^2+(z_a-t_a)^2}\sqrt{1+z_a'^2}}(\vec{\mathbf{e}}_1\wedge\vec{\mathbf{e}}_2), \tag{4.9}$$

then, we square it, remember that $(\vec{\mathbf{e}}_1\wedge\vec{\mathbf{e}}_2)^2=-1$,

$$(\vec{\mathbf{n}}_a\wedge\vec{\mathbf{v}}_1)^2=-\frac{[r_a+(z_a-t_a)z_a']^2}{[r_a^2+(z_a-t_a)^2](1+z_a'^2)}. \tag{4.10}$$

Therefore, the term with the square can be written as,

$$-\vec{\mathbf{n}}_a\sqrt{1+\frac{1}{n^2}(\vec{\mathbf{n}}_a\wedge\vec{\mathbf{v}}_1)^2}=-\frac{(z_a'\vec{\mathbf{e}}_1-\vec{\mathbf{e}}_2)}{\sqrt{1+z_a'^2}}\sqrt{1-\frac{[r_a+(z_a-t_a)z_a']^2}{n^2[r_a^2+(z_a-t_a)^2](1+z_a'^2)}}. \quad (4.11)$$

Now, let's focus on the term, $\vec{\mathbf{v}}_1-(\vec{\mathbf{n}}_a\cdot\vec{\mathbf{v}}_1)\vec{\mathbf{n}}_a$,

$$\begin{aligned}\vec{\mathbf{v}}_1-(\vec{\mathbf{n}}_a\cdot\vec{\mathbf{v}}_1)\vec{\mathbf{n}}_a=&\frac{r_a\vec{\mathbf{e}}_1+(z_a-t_a)\vec{\mathbf{e}}_2}{\sqrt{r_a^2+(z_a-t_a)^2}}\\&-\left[\frac{r_az_a'-(z_a-t_a)}{\sqrt{r_a^2+(z_a-t_a)^2}\sqrt{1+z_a'^2}}\right]\frac{(z_a'\vec{\mathbf{e}}_1-\vec{\mathbf{e}}_2)}{\sqrt{1+z_a'^2}}.\end{aligned} \quad (4.12)$$

Simplifying, obtaining the least common divisor and multiplying by $1/(-\text{sgn}(t_a)n)$

$$\begin{aligned}\frac{1}{-\text{sgn}(t_a)n}[\vec{\mathbf{v}}_1-(\vec{\mathbf{n}}_a\cdot\vec{\mathbf{v}}_1)\vec{\mathbf{n}}_a]=&\frac{(z_a-t_a)z_a'+r_a}{-\text{sgn}(t_a)n\sqrt{r_a^2+(t_a-z_a)^2}(1+z_a'^2)}\vec{\mathbf{e}}_1\\&+\frac{(r_a+(z_a-t_a)z_a')z_a'}{-\text{sgn}(t_a)n\sqrt{r_a^2+(t_a-z_a)^2}(1+z_a'^2)}\vec{\mathbf{e}_2}.\end{aligned} \quad (4.13)$$

If we sum equations (4.11) and (4.13), we have $\vec{\mathbf{v}}_2$ as,

$$\begin{aligned}\vec{\mathbf{v}}_2=&\frac{(z_a-t_a)z_a'+r_a}{-\text{sgn}(t_a)n\sqrt{r_a^2+(t_a-z_a)^2}(1+z_a'^2)}\vec{\mathbf{e}}_1\\&+\frac{[r_a+(z_a-t_a)z_a']z_a'}{-\text{sgn}(t_a)n\sqrt{r_a^2+(t_a-z_a)^2}(1+z_a'^2)}\vec{\mathbf{e}}_2\\&-\frac{(z_a'\vec{\mathbf{e}}_1-\vec{\mathbf{e}}_2)}{\sqrt{1+z_a'^2}}\sqrt{1-\frac{[r_a+(z_a-t_a)z_a']^2}{n^2[r_a^2+(z_a-t_a)^2](1+z_a'^2)}}.\end{aligned} \quad (4.14)$$

Let's separate the coordinates, $\vec{\mathbf{e}}_1$ and $\vec{\mathbf{e}}_2$,

$$\begin{aligned}\frac{r_b-r_a}{\sqrt{(z_b-z_a)^2+(r_b-r_a)^2}}=&\frac{(z_a-t_a)z_a'+r_a}{-\text{sgn}(t_a)n\sqrt{r_a^2+(t_a-z_a)^2}(1+z_a'^2)}\\&-z_a'\frac{\sqrt{1-\frac{[r_a+(z_a-t_a)z_a']^2}{n^2[r_a^2+(t_a-z_a)^2](1+z_a'^2)}}}{\sqrt{1+z_a'^2}},\end{aligned} \quad (4.15)$$

and,

$$\frac{z_b - z_a}{\sqrt{(z_b - z_a)^2 + (r_b - r_a)^2}} = \frac{[r_a + (z_a - t_a)z_a']z_a'}{-\operatorname{sgn}(t_a)n\sqrt{r_a^2 + (t_a - z_a)^2}(1 + z_a'^2)} + \frac{\sqrt{1 - \frac{[r_a + (z_a - t_a)z_a']^2}{n^2[r_a^2 + (t_a - z_a)^2](1 + z_a'^2)}}}{\sqrt{1 + z_a'^2}}. \tag{4.16}$$

In the left side of equations (4.15) and (4.16) are the unknowns $z_b$ and $r_b$. Please note that the right side of equations (4.15) and (4.16) depends only on parameters that we know, $z_a'$, $z_a$, $t_a$, $t$, and $n$. In the right side of equations (4.15) and (4.16) are the cosine directors of the refracted ray. Let $\wp_z$ be the cosine director of the $z$ direction and let $\wp_r$ be the cosine director of the $r$ direction, then $\wp_r^2 + \wp_z^2 = 1$. Thus,

$$\wp_r = \frac{(z_a - t_a)z_a' + r_a}{-\operatorname{sgn}(t_a)n\sqrt{r_a^2 + (t_a - z_a)^2}(1 + z_a'^2)} - z_a'\frac{\sqrt{1 - \frac{[r_a + (z_a - t_a)z_a']^2}{n^2[r_a^2 + (t_a - z_a)^2](1 + z_a'^2)}}}{\sqrt{1 + z_a'^2}}, \tag{4.17}$$

and,

$$\wp_z = \frac{[r_a + (z_a - t_a)z_a']z_a'}{-\operatorname{sgn}(t_a)n\sqrt{r_a^2 + (t_a - z_a)^2}(1 + z_a'^2)} + \frac{\sqrt{1 - \frac{[r_a + (z_a - t_a)z_a']^2}{n^2[r_a^2 + (t_a - z_a)^2](1 + z_a'^2)}}}{\sqrt{1 + z_a'^2}}. \tag{4.18}$$

For convenience, we assign a name to the distance traveled by each ray within the lens,

$$\vartheta \equiv \sqrt{(z_b - z_a)^2 + (r_b - r_a)^2}. \tag{4.19}$$

Now we can replace $\vartheta$ in (4.15) and (4.16),

$$\frac{z_b - z_a}{\vartheta} = \wp_z, \tag{4.20}$$

and,

$$\frac{r_b - r_a}{\vartheta} = \wp_r. \tag{4.21}$$

Solving for $z_b$ and $r_b$,

$$z_b = z_a + \vartheta\wp_z, \tag{4.22}$$

and,

$$r_b = r_a + \vartheta \wp_r. \tag{4.23}$$

The above equations have the structure of the solution; we just need to know what $\vartheta$ is like. For that, in the next section, we will use the above equations and equation (4.6) to solve it.

Note that we have three equations and two unknowns. This is completely correct since equations (4.22) and (4.23) are not independent.

### 4.2.3 Solution

Let's start the hunt for $\vartheta$, we begin by recalling equation (4.6),

$$\begin{aligned} -t_a + nt + t_b = &- \operatorname{sgn}(t_a)\sqrt{r_a^2 + (z_a - t_a)^2} \\ &+ n\sqrt{(r_b - r_a)^2 + (z_b - z_a)^2} \\ &+ \operatorname{sgn}(t_b)\sqrt{r_b^2 + (z_b - t - t_b)^2}. \end{aligned} \tag{4.6}$$

In the above equation we have $\vartheta$ and the unknowns $z_b$ and $r_b$. We replace them using equations (4.19) (4.22), and (4.23), thus,

$$\begin{aligned} -t_a + nt + t_b = &- \operatorname{sgn}(t_a)\sqrt{r_a^2 + (z_a - t_a)^2} + n\vartheta \\ &+ \operatorname{sgn}(t_b)\sqrt{(r_a + \vartheta\wp_r)^2 + (z_a + \vartheta\wp_z - t - t_b)^2}. \end{aligned} \tag{4.24}$$

We assign the following variables to simplify the manipulation,

$$f \equiv -t_a + nt + t_b + \operatorname{sgn}(t_a)\sqrt{r_a^2 + (z_a - t_a)^2}, \tag{4.25}$$

and,

$$\tau \equiv z_a - t - t_b. \tag{4.26}$$

Let's replace $\tau$ and $f$ in equation (4.24) and square it,

$$(f - n\vartheta)^2 = (r_a + \vartheta\wp_r)^2 + (\tau + \vartheta\wp_z)^2. \tag{4.27}$$

If we expand the above square binomials, we get,

$$f^2 - 2fn\vartheta + n^2\vartheta^2 = r_a^2 + 2r_a\wp_r\vartheta + \wp_r^2\vartheta^2 + \tau^2 + 2\tau\wp_z\vartheta + \wp_z^2\vartheta^2. \tag{4.28}$$

Collecting the terms powered by $\vartheta$,

$$(1 - n^2)\vartheta^2 + [2(fn + r_a\wp_r + \tau\wp_z)]\vartheta + (r_a^2 + \tau^2 - f^2) = 0, \tag{4.29}$$

multiplying by $4(1 - n^2)$, we have,

$$4(1 - n^2)^2\vartheta^2 + 4(1 - n^2)[2(fn + r_a\wp_r + \tau\wp_z)]\vartheta = -4(1 - n^2)(r_a^2 + \tau^2 - f^2). \tag{4.30}$$

Adding $[2(fn + r_a\wp_r + \tau\wp_z)]^2$ in both sides of the equation,

$$\begin{aligned}4(1-n^2)^2\vartheta^2 + 4(1-n^2)[2(fn + r_a\wp_r + \tau\wp_z)]\vartheta + [2(fn + r_a\wp_r + \tau\wp_z)]^2 \\ = [2(fn + r_a\wp_r + \tau\wp_z)]^2 - 4(1-n^2)(r_a^2 + \tau^2 - f^2).\end{aligned} \quad (4.31)$$

Completing the squared binomial,

$$\begin{aligned}\{2(1-n^2)\vartheta + [2(fn + r_a\wp_r + \tau\wp_z)]\}^2 \\ = [2(fn + r_a\wp_r + \tau\wp_z)]^2 - 4(1-n^2)(r_a^2 + \tau^2 - f^2).\end{aligned} \quad (4.32)$$

Applying the square root in both sides,

$$\begin{aligned}2(1-n^2)\vartheta + [2(fn + r_a\wp_r + \tau\wp_z)] \\ = \sqrt{[2(fn + r_a\wp_r + \tau\wp_z)]^2 - 4(1-n^2)(r_a^2 + \tau^2 - f^2)},\end{aligned} \quad (4.33)$$

solving for $\vartheta$,

$$\vartheta = \frac{-2(fn + r_a\wp_r + \tau\wp_z) \pm \sqrt{4(fn + r_a\wp_r + \tau\wp_z)^2 - 4(1-n^2)(r_a^2 + \tau^2 - f^2)}}{2(1-n^2)}. \quad (4.34)$$

Simplifying, we can remove the number two,

$$\vartheta = \frac{-(fn + r_a\wp_r + \tau\wp_z) \pm \sqrt{(fn + r_a\wp_r + \tau\wp_z)^2 - (1-n^2)(r_a^2 + \tau^2 - f^2)}}{(1-n^2)}. \quad (4.35)$$

Finally, we can use the solution to the problem, equation (4.36),

$$\boxed{\begin{aligned}z_b &= z_a + \vartheta\wp_z,\\ r_b &= r_a + \vartheta\wp_r.\end{aligned}} \quad (4.36)$$

Equation (4.36) is the most important equation in the chapter, and it is also the most important equation in the book. It tells how the second surface ($z_b$, $r_b$) should be in terms of the first surface ($z_a$, $r_a$) such that the singlet is on-axis stigmatic.

It works if and only if the rays inside the lens do not cross each other. We will study this condition in chapter 6.

If we expand equation (4.36) we will get an equation of 11 pages, depending on the size of the paper. It may seem difficult due to its length, but trust us, the equation is very noble and we will see it in action. In the next examples, we test all the cases for real/virtual point objects and real/virtual point images.

### 4.2.4 Illustrative examples

In this section, we present a gallery of lenses free of spherical aberration directly using equation (4.36). The figures of the gallery are 4.6–4.8. We do not use any optimization process in the computation. All the information to reproduce the examples is in the captions of the images. All the length units are in millimeters.

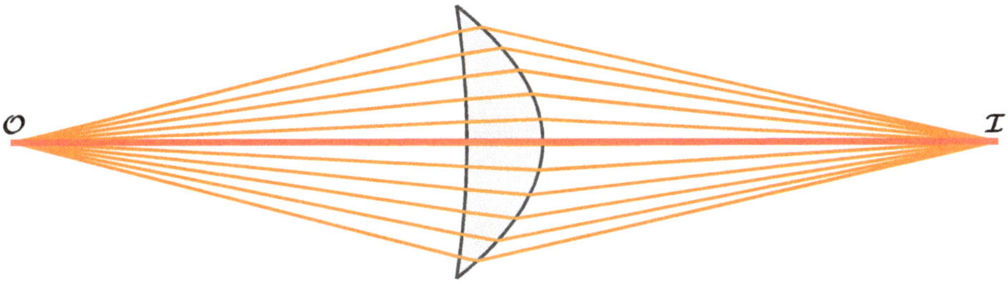


**Figure 4.6.** Design specifications: $n = 1.5$ $t_a = -30$ mm, $t = 5$ mm, $t_b = 30$ mm, $z_a = -r_a^2/120$ and $z_b$ = equation (4.36).

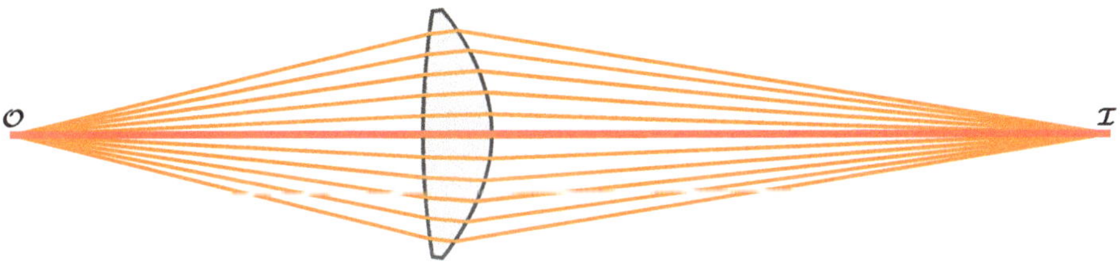


**Figure 4.7.** Design specifications: $n = 1.5$, $t_a = -30$ mm, $t = 5$ mm, $t_b = 45$ mm, $z_a = -r_a^2/120$ and $z_b$ = equation (4.36).

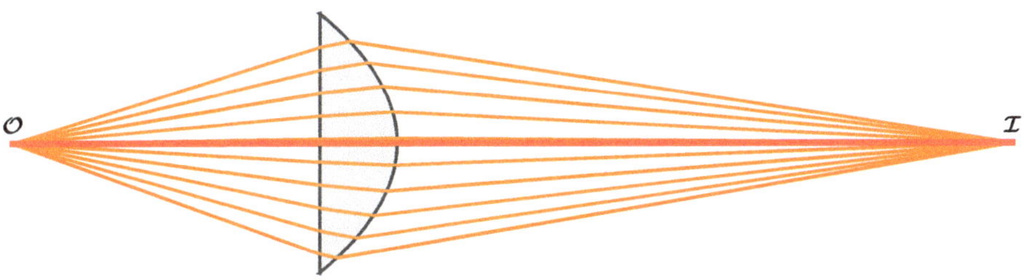


**Figure 4.8.** Design specifications: $n = 1.5$, $t_a = -20$ mm, $t = 5$ mm, $t_b = 40$ mm, $z_a = 0$ and $z_b$ = equation (4.36).

We compute a tribute to Christian Huygens. See the details in the caption of figure 4.9. Also, in figure 4.10 we compute an example of a lens with negative refraction index.

Note that in the following examples, we have lenses with objects and images that can be real or virtual. It all depends on the values of $t_a$ and $t_b$, as shown in section 4.2.1. The figures are 4.11–4.19.

## 4.3 Evolution tables of the shape of on-axis stigmatic lens

We have already presented several examples of spherical aberration-free lenses using equation (4.36). As you can see, the shape of the second surface is modified by the input values, $t_a$, $t$, $t_b$, $n$ and of course $z_a$.

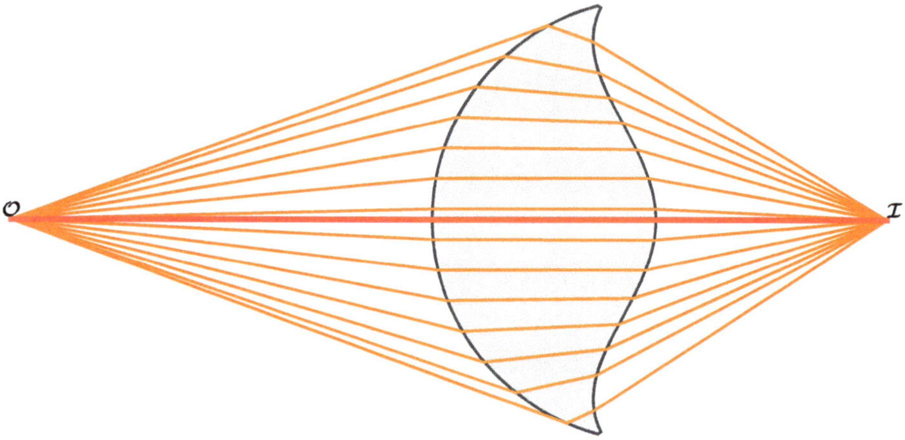


**Figure 4.9.** A tribute to Christian Huygens. An aspherical lens similar to that proposed by Huygens in *Traité de la lumière* over three hundred years ago. The lens is formed with a first spherical refractive surface and a second refractive surface such that the lens is free of spherical aberration. Design specifications: $n = 1.5$, $t_a = -55$ mm, $t = 29$ mm, $t_b = 30$ mm, $z_a = 29 + \sqrt{2962 - r_a^2}$ and $z_b =$ equation (4.36).

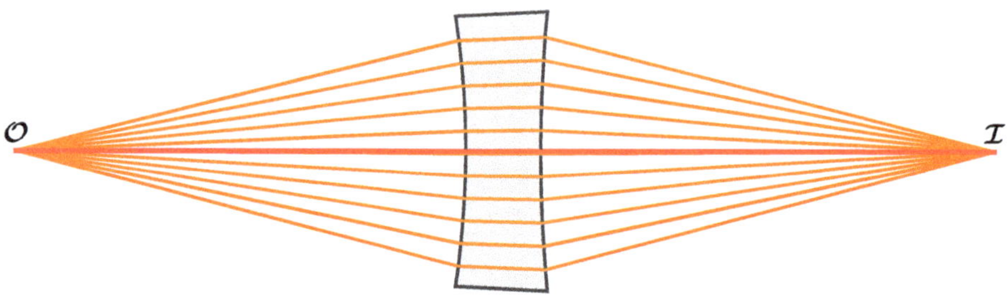


**Figure 4.10.** Equation (4.36) supports negative refractive indexes. Therefore, it can be used in the design of lenses with metamaterials. Design specifications: $n = -1.5$, $t_a = -30$ mm, $t = 5$ mm, $t_b = 30$ mm, $z_a = -r_a^2/120$ and $z_b =$ equation (4.36).

The following tables show the evolution of the spherical aberration-free lens shape when one of the input parameters is modified, and the others are kept constant.

The tables only show the cases where the objects are finite, and the images are also finite. The objects are real, and the images are also real, or the objects are virtual, and the images are virtual.

The tables are in figures 4.20–4.29.

Note in figure 4.20 that the maximum radius of the lens becomes larger when the object is far from the first surface. It is also shown that the curvature of the second surface increases its curvature when the object is closer to the first surface. This is because the second surface has to compensate for the spherical aberration generated by the first refractive surface.

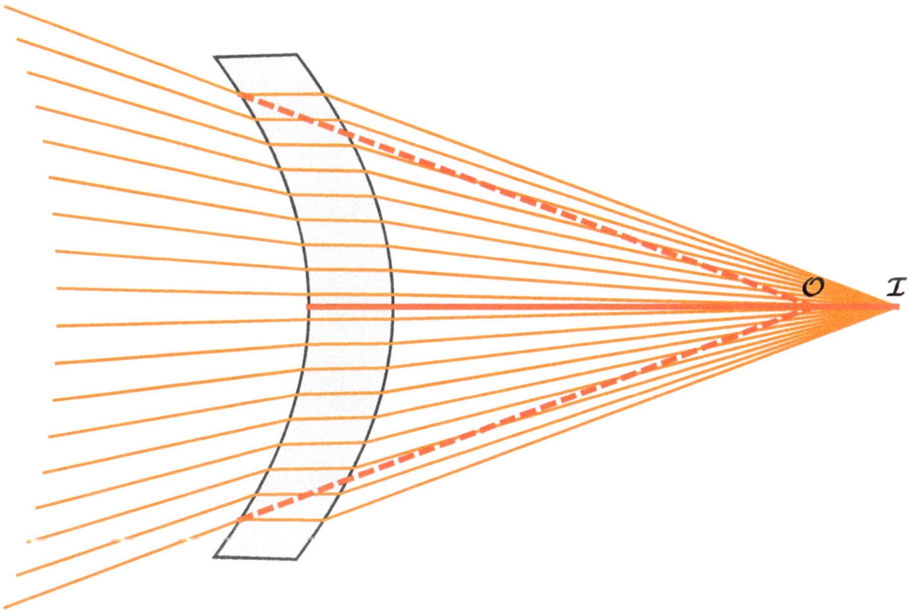


**Figure 4.11.** Design specifications: $n = 1.5$, $t_a = 30$ mm, $t = 5$ mm, $t_b = 30$ mm, $z_a = -r_a^2/40$ and $z_b$ = equation (4.36).

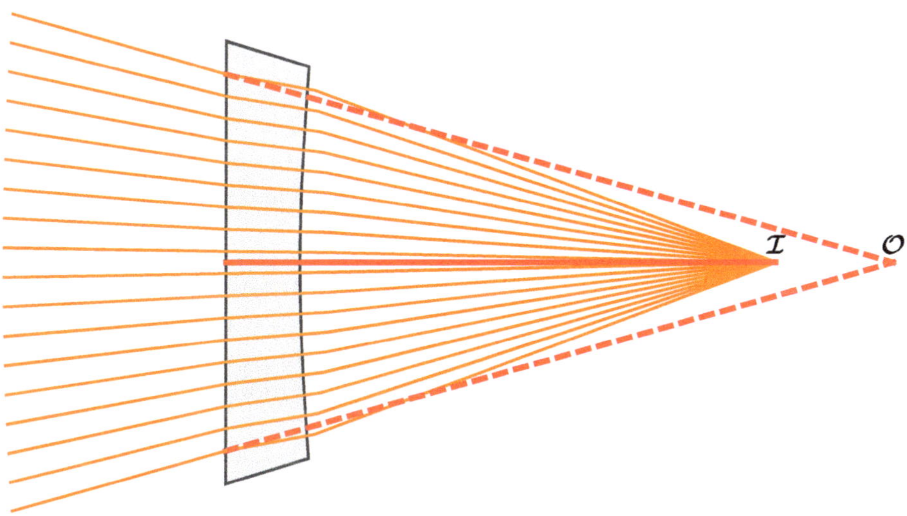


**Figure 4.12.** Design specifications: $n = 1.5$, $t_a = 45$ mm, $t = 7$ mm, $t_b = 30$ mm, $z_a = 0$ and $z_b$ = equation (4.36).

Form the tables we learn the following:

- The pronunciation of the second surface and the diameter of the lens are proportional to the position of image/object.

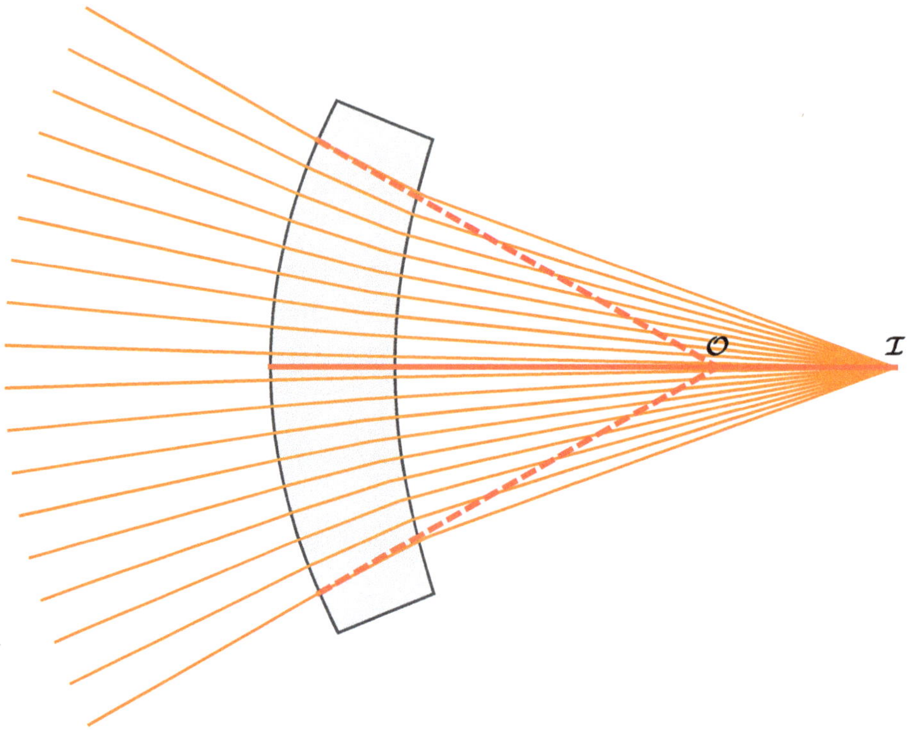


**Figure 4.13.** Design specifications: $n = 1.5$, $t_a = 25$ mm, $t = 5$ mm, $t_b = 30$ mm, $z_a = r_a^2/60$ and $z_b$ = equation (4.36).

- The higher the refractive index, the greater the diameter of the lens.
- The thicker the lens, the higher the width of the lens.

## 4.4 Stigmatic aspheric collector

Our next target is the lens analogous to the parabola as a tribute to Diocles. The parabolic mirror has a real object at minus infinity. So, what we need to do is to apply the limit when $t_a \to -\infty$ in the terms inside equation (4.36). Let's recall the parameters inside equation (4.36),

$$\wp_r \equiv \frac{(z_a - t_a)z_a' + r_a}{-\text{sgn}(t_a)n\sqrt{r_a^2 + (t_a - z_a)^2}(1 + z_a'^2)} - z_a' \frac{\sqrt{1 - \frac{[r_a + (z_a - t_a)z_a']^2}{n^2[r_a^2 + (t_a - z_a)^2](1 + z_a'^2)}}}{\sqrt{1 + z_a'^2}}, \tag{4.17}$$

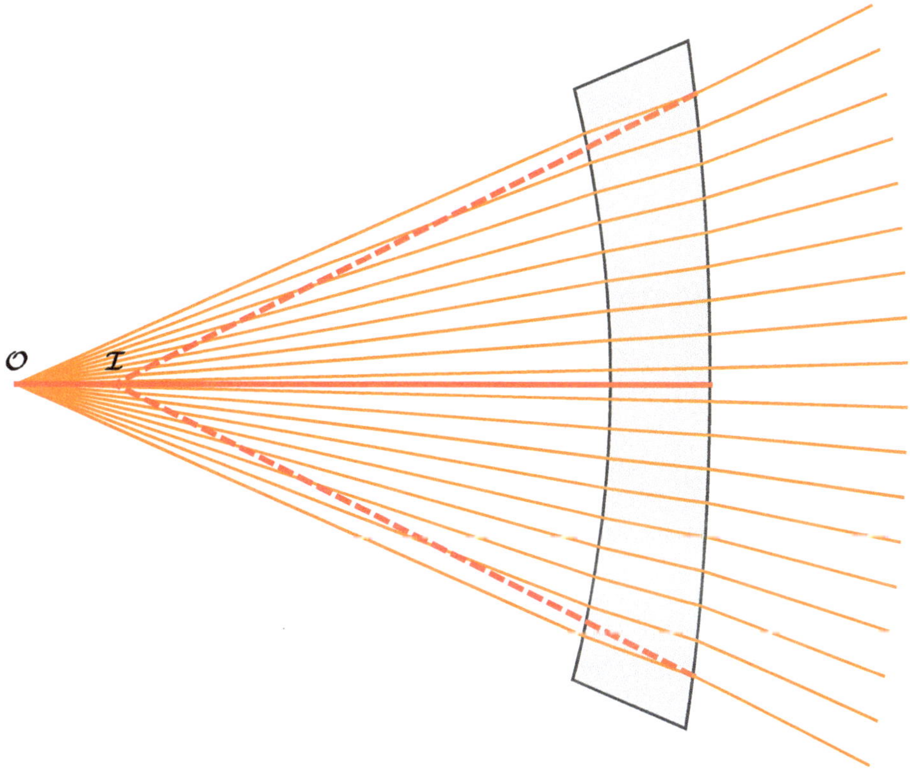


**Figure 4.14.** Design specifications: $n = 1.5$, $t_a = -30$ mm, $t = 5$ mm, $t_b = -25$ mm, $z_a = -r_a^2/120$ and $z_b =$ equation (4.36).

$$\wp_z \equiv \frac{[r_a + (z_a - t_a)z_a']z_a'}{-\mathrm{sgn}(t_a)n\sqrt{r_a^2 + (t_a - z_a)^2}(1 + z_a'^2)} + \frac{\sqrt{1 - \frac{[r_a + (z_a - t_a)z_a']^2}{n^2[r_a^2 + (t_a - z_a)^2](1 + z_a'^2)}}}{\sqrt{1 + z_a'^2}}, \tag{4.18}$$

and,

$$f \equiv -t_a + nt + t_b + \mathrm{sgn}(t_a)\sqrt{r_a^2 + (z_a - t_a)^2}, \tag{4.25}$$

applying the $t_a \to -\infty$ on them

$$\lim_{t_a \to -\infty} \wp_r = -\frac{z_a'\left(n\sqrt{(z_a')^2 + 1}\sqrt{\frac{(n^2 - 1)(z_a')^2 + n^2}{n^2((z_a')^2 + 1)}} - 1\right)}{n[(z_a')^2 + 1]}, \tag{4.37}$$

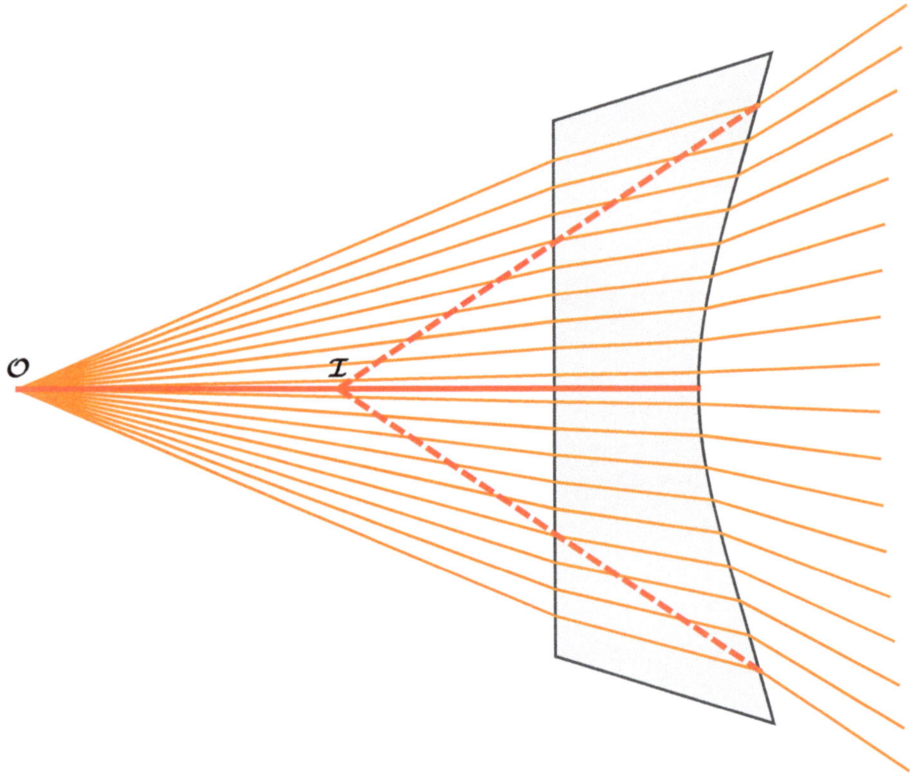


**Figure 4.15.** Design specifications: $n = 1.5$, $t_a = -30$ mm, $t = 8$ mm, $t_b = -20$ mm, $z_a = 0$ and $z_b =$ equation (4.36).

$$\lim_{t_a \to -\infty} \wp_z = \frac{\sqrt{\dfrac{(n^2-1)(z_a')^2 + n^2}{n^2[(z_a')^2+1]}}}{\sqrt{(z_a')^2+1}} + \frac{(z_a')^2}{n(z_a')^2 + n}, \tag{4.38}$$

and,

$$\lim_{t_a \to -\infty} f = nt + t_b - z_a. \tag{4.39}$$

Therefore, from equation (4.35), $\vartheta$ when $t_a \to -\infty$ is

$$\lim_{t_a \to -\infty} (\vartheta) = \frac{-\beta \pm \sqrt{\beta^2 - (1-n^2)\left\{r_a^2 + \tau^2 - \left[\lim_{t_a \to -\infty} (f)\right]^2\right\}}}{(1-n^2)}, \tag{4.40}$$

where,

$$\beta \equiv \left\{\left[\lim_{t_a \to -\infty} (f)\right] n + r_a\left[\lim_{t_a \to -\infty} (\wp_r)\right] + \tau\left[\lim_{t_a \to -\infty} (\wp_z)\right]\right\}. \tag{4.41}$$

If we want to design collector lenses, we just need to use the aboves equations, equations (4.37), (4.38), (4.39), (4.40),and (4.41), and use them in equation (4.42),

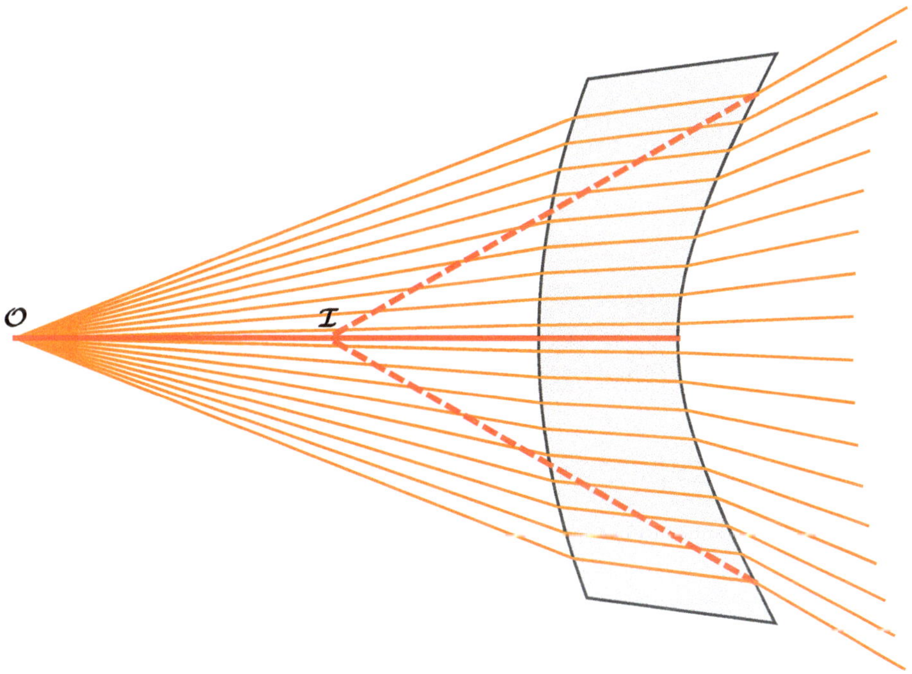


**Figure 4.16.** Design specifications: $n = 1.5$, $t_a = -30$ mm, $t = 8$ mm, $t_b = -20$ mm, $z_a = r_a^2/80$ and $z_b =$ equation (4.36).

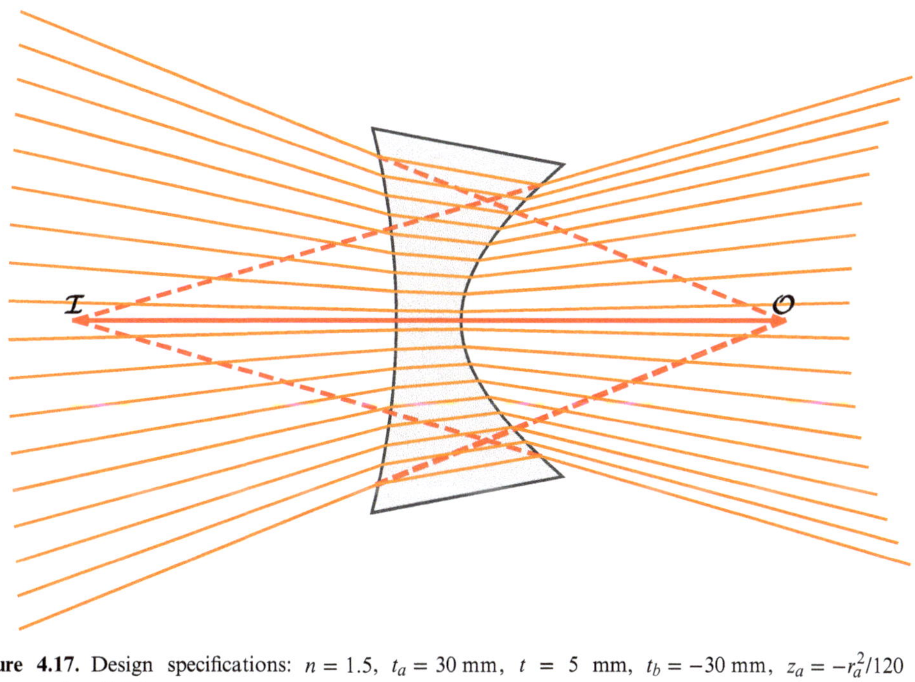


**Figure 4.17.** Design specifications: $n = 1.5$, $t_a = 30$ mm, $t = 5$ mm, $t_b = -30$ mm, $z_a = -r_a^2/120$ and $z_b =$ equation (4.36).

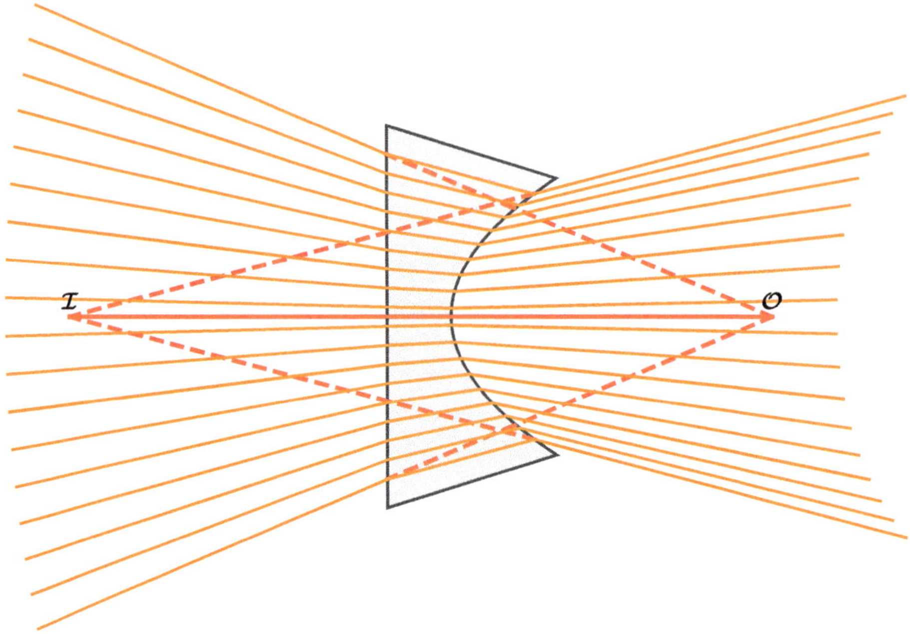


**Figure 4.18.** Design specifications: $n = 1.5$, $t_a = 30$ mm, $t = 5$ mm, $t_b = -30$ mm, $z_a = 0$ and $z_b$ = equation (4.36).

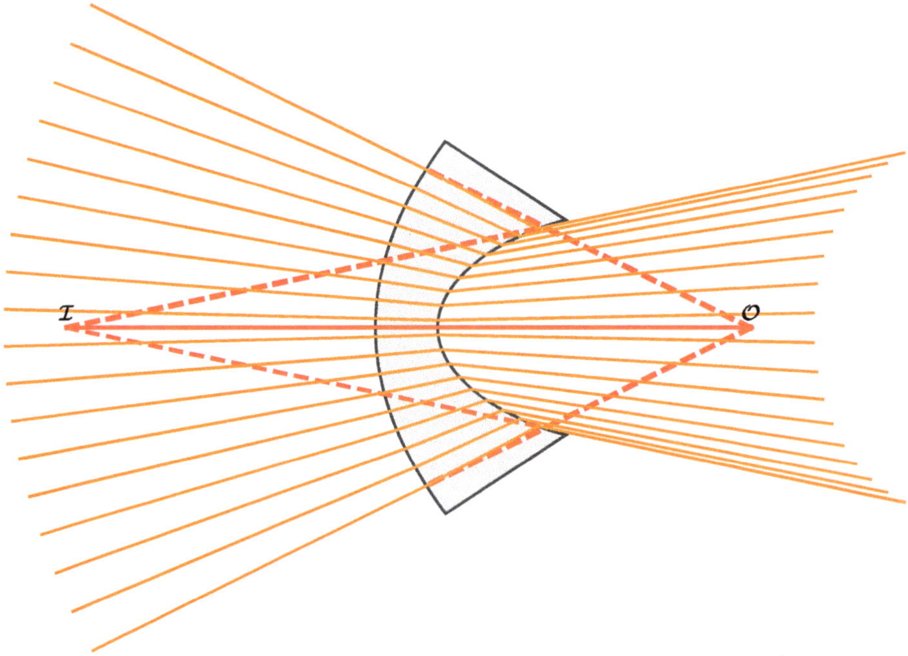


**Figure 4.19.** Design specifications: $n = 1.5$, $t_a = 30$ mm, $t = 5$ mm, $t_b = -30$ mm, $z_a = -r_a^2/40$ and $z_b$ = equation (4.36).

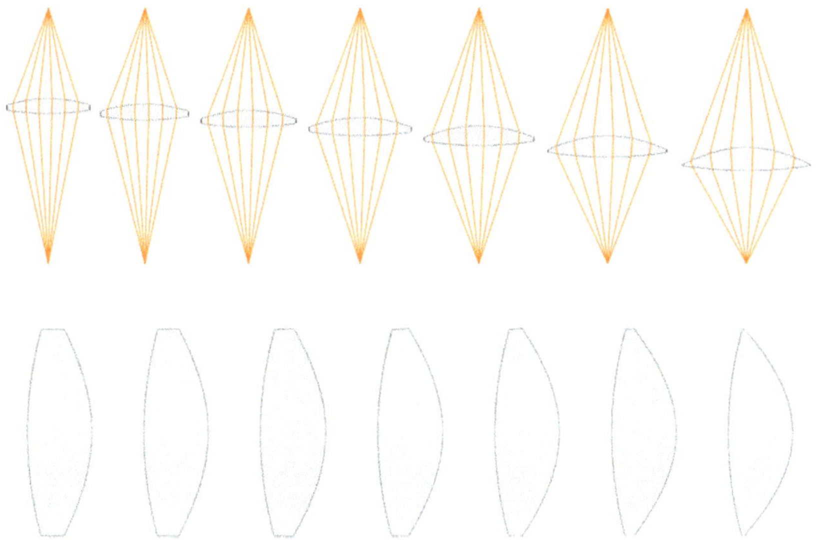

**Figure 4.20.** This table shows the evolution of an aberration-free spherical lens designed with equation (4.36). The input parameters that remain constant are $n = 1.5$, $t = 5$ mm, $t_b = 30$ mm, and $z_a = r_a^2/60$. The variable that changes from left to right is $t_a$, from −50 mm to −20 mm in of steps of 5 mm.

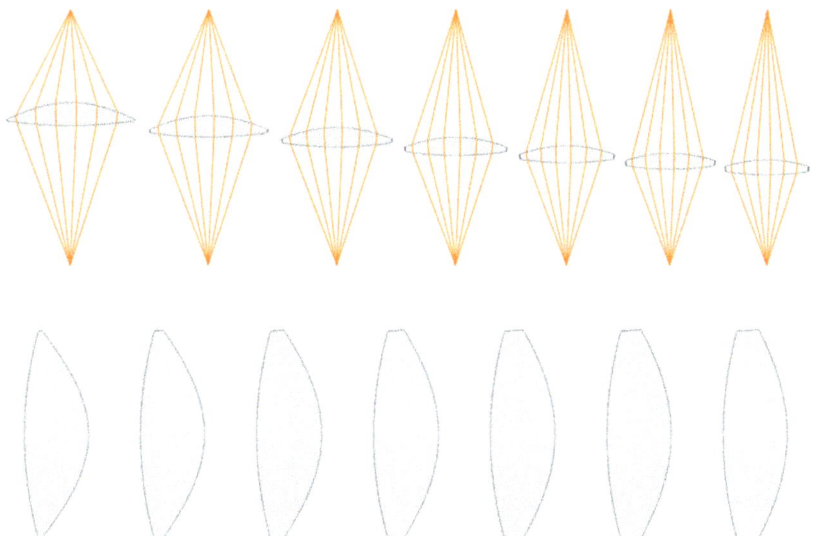

**Figure 4.21.** This table presents the change of an aberration-free spherical lens designed with equation (4.36). The input parameters that remain constant are $n = 1.5$, $t = 5$ mm, $t_a = -30$ mm, and $z_a = r_a^2/60$. The input value that varies from left to right is $t_b$, from 20 mm to 50 mm in of steps of 5 mm.

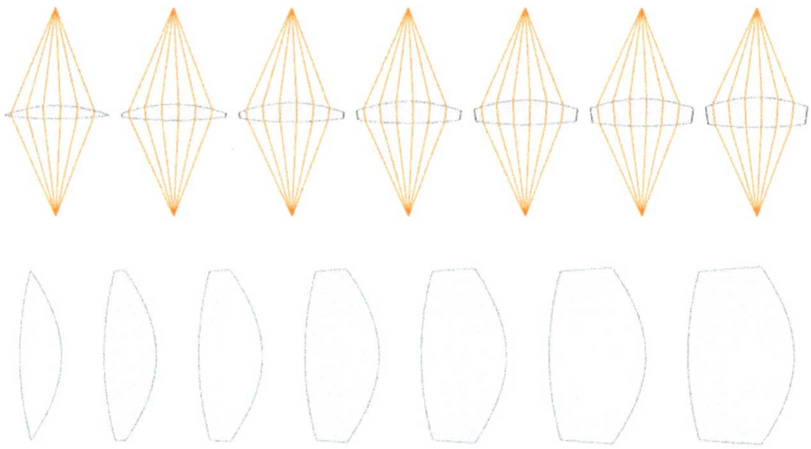

**Figure 4.22.** The table shows the profile of an on-axis stigmatic lens described by equation (4.36).. The input parameters that remain constant are $t_a = -30$ mm, $n = 1.5$, $t_b = 30$ mm, and $z_a = r_a^2/60$. The variable that changes from left to right is $t$, from 4 mm to 10 mm in of steps of 1 mm.

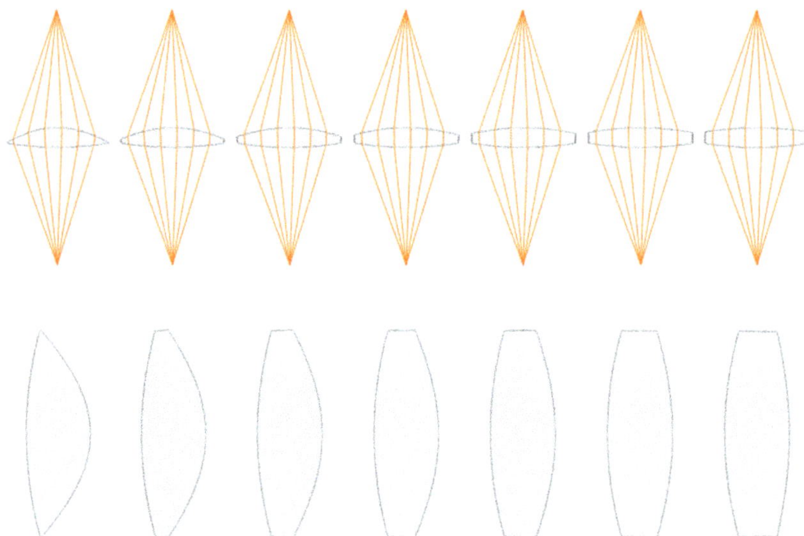

**Figure 4.23.** The table shows the profile of an lens described by equation (4.36). The input parameters that remain constant are $t_a = -30$ mm, $t = 5$, $t_b = 30$ mm, and $z_a = r_a^2/60$. The input value that changes from left to right is $n$, from 1.4 to 2 in of steps of 0.1.

$$\boxed{\begin{aligned} \lim_{t_a \to -\infty} (z_b) &= z_a + \left[\lim_{t_a \to -\infty} (\vartheta)\right]\left[\lim_{t_a \to -\infty} (\wp_z)\right], \\ \lim_{t_a \to -\infty} (r_b) &= r_a + \left[\lim_{t_a \to -\infty} (\vartheta)\right]\left[\lim_{t_a \to -\infty} (\wp_r)\right]. \end{aligned}} \tag{4.42}$$

Equation (4.42) expresses how the second surface of the singlet must be, for an object at minus infinity, such that the singlet is free of spherical aberration. In the next section, we present several examples of stigmatic collector lenses.

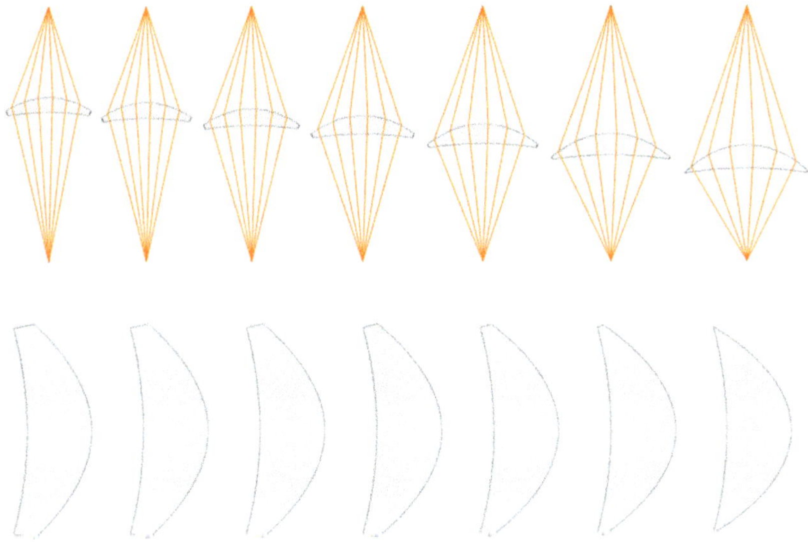

**Figure 4.24.** This table shows lenses designed with equation (4.36). The input constants are $n = 1.5$, $t = 5$ mm, $t_b = 30$ mm, and $z_a = -r_a^2/60$. The variable in question from left to right is $t_a$, from −50 mm to −20 mm in of steps of 5 mm.

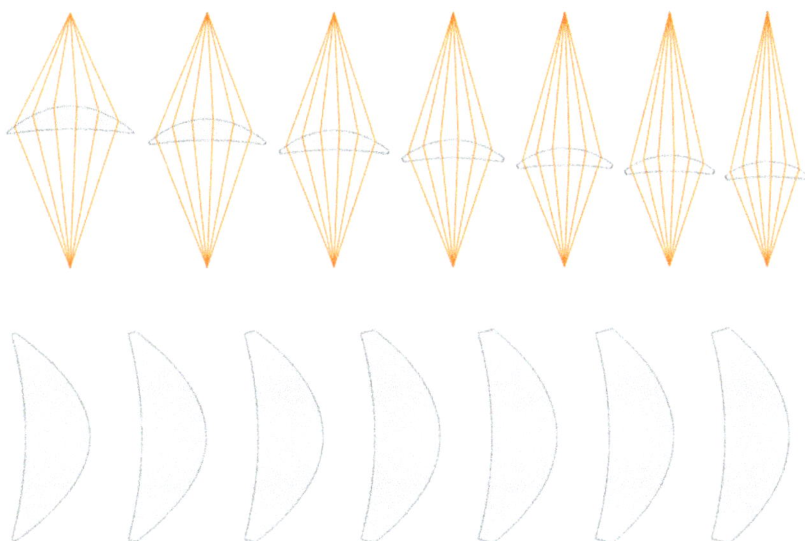

**Figure 4.25.** We present several lenses computed with equation (4.36). The input parameters that remain constant are $n = 1.5$, $t = 5$ mm, $t_a = -30$ mm, and $z_a = -r_a^2/60$. The variable that varies from left to right is $t_b$, from 20 mm to 50 mm in of steps of 5 mm.

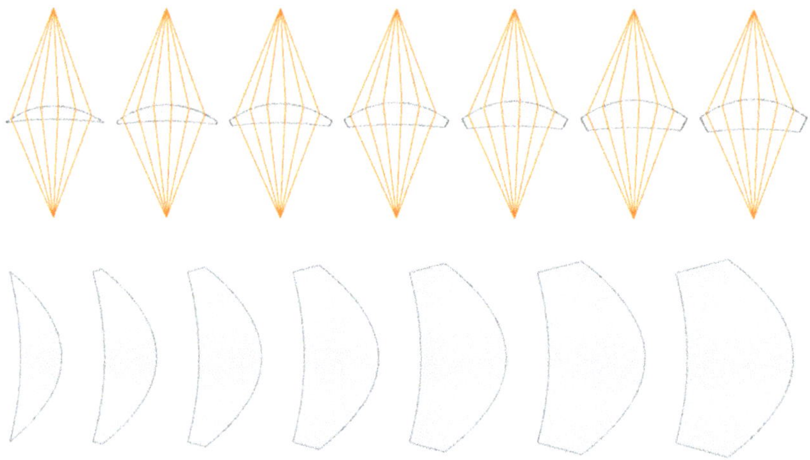

**Figure 4.26.** The table shows lenses computed by equation (4.36). The input constants are $t_a = -30$ mm, $n = 1.5, t_b = 30$ mm, and $z_a = -r_a^2/60$. The parameter that changes from left to right is $t$, from 4 mm to 10 mm in of steps of 1 mm.

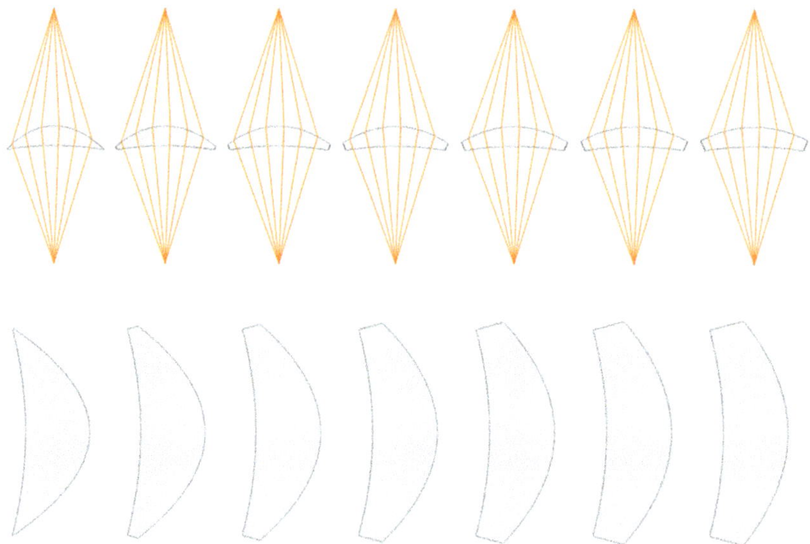

**Figure 4.27.** We present several lenses described by equation (4.36). The input parameters that remain constant are $t_a = -30$ mm, $t = 5, t_b = 30$ mm, and $z_a = -r_a^2/60$. The parameter that changes from left to right is $n$, from 1.4 to 2 in of steps of 0.1.

### 4.4.1 Examples

Here we presenter a gallery of stigmatic collector lenses for real and virtual images. The specification of each design is in its respective figure. The figures of the gallery are 4.30–4.35.

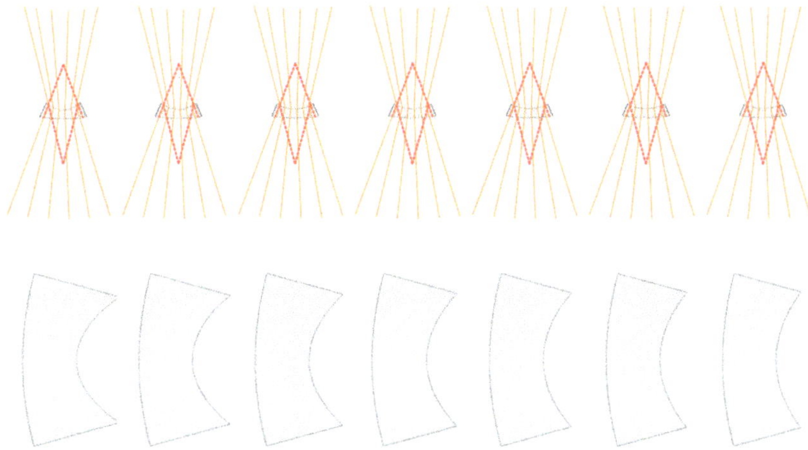

**Figure 4.28.** Several lenses of equation (4.36) are shown. The input constants are $t_a = 30$ mm, $t = 5$, $t_b = -30$ mm, and $z_a = r_a^2/60$. The parameter that changes from left to right is $n$, from 1.4 to 2 in of steps of 0.1.

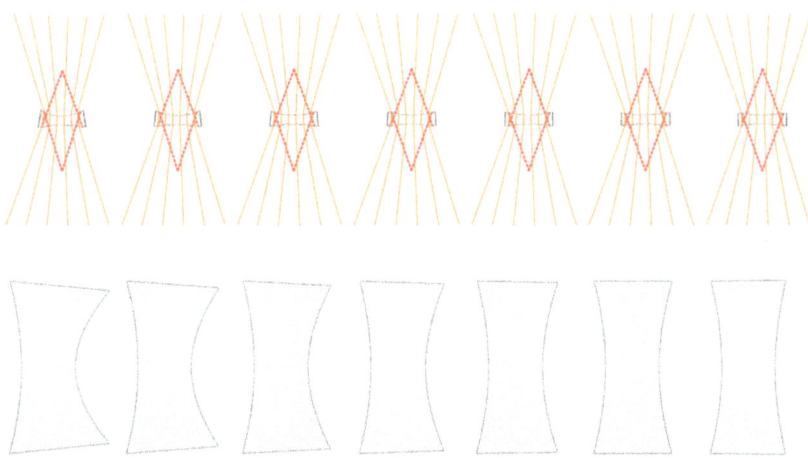

**Figure 4.29.** Several lenses of equation (4.36) are presented. The inputs are $t_a = 30$ mm, $t = 5$, $t_b = -30$ mm, and $z_a = -r_a^2/60$. The parameter that changes from left to right is $n$, from 1.4 to 2 in of steps of 0.1.

## 4.5 Stigmatic aspheric collimator

Now, let's work on the opposite case when the image is at infinity $t_b \to \infty$. Observe that for this case, the point object is finite. It can be real/virtual, but it is finite. Thus, we will use the same cosine directors $\wp_r$ and $\wp_z$ given equations (4.17) and (4.18).

For the Fermat principle of equation we need to compute the $t_b \to \infty$; we start by calling it, from equation (4.6),

$$\begin{aligned} -t_a + nt + t_b = &- \operatorname{sgn}(t_a)\sqrt{r_a^2 + (z_a - t_a)^2} + n\sqrt{(r_b - r_a)^2 + (z_b - z_a)^2} \\ &+ \operatorname{sgn}(t_b)\sqrt{r_b^2 + (z_b - t - t_b)^2}. \end{aligned}$$

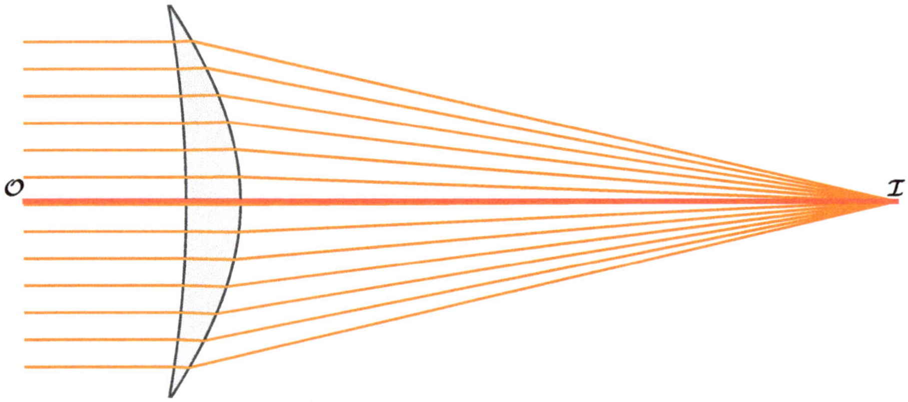


**Figure 4.30.** Design specifications: $n = 1.5$, $t = 5$ mm, $t_b = 60$ mm, $z_a = -r_a^2/200$ and $z_b =$ equation (4.42).

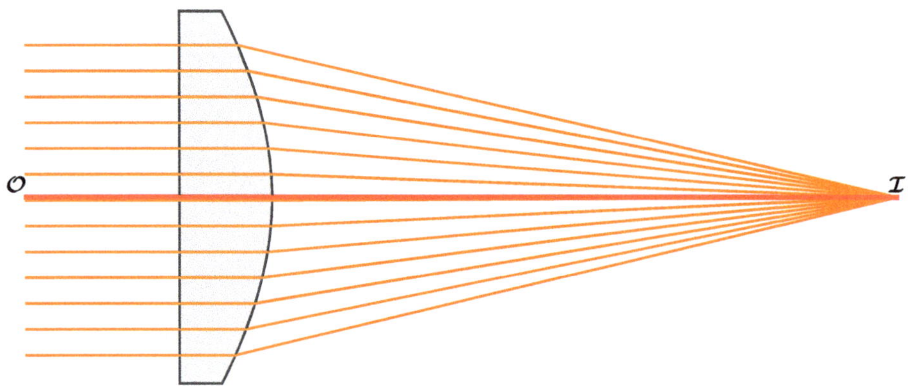


**Figure 4.31.** Design specifications: $n = 1.5$, $t = 9$ mm, $t_b = 60$ mm, $z_a = 0$ and $z_b =$ equation (4.42).

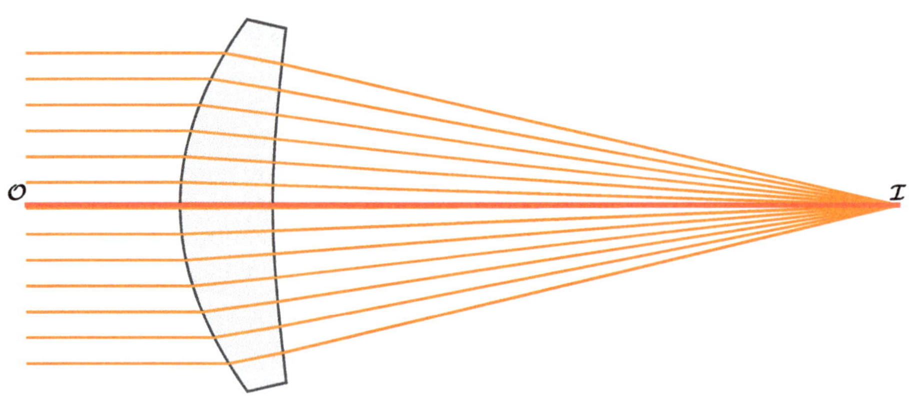


**Figure 4.32.** Design specifications: $n = 1.5$, $t = 9$ mm, $t_b = 60$ mm, $z_a = r_a^2/50$ and $z_b =$ equation (4.42).

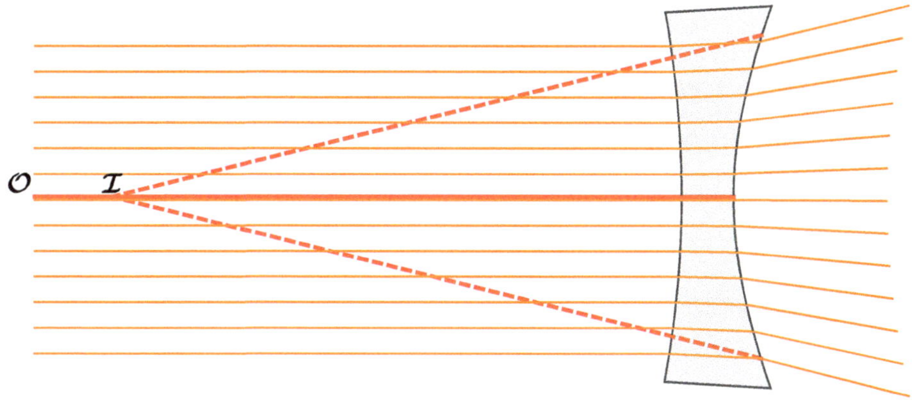


**Figure 4.33.** Design specifications: $n = 1.5$, $t = 5$ mm, $t_b = -60$ mm, $z_a = -r_a^2/200$ and $z_b$ = equation (4.42).

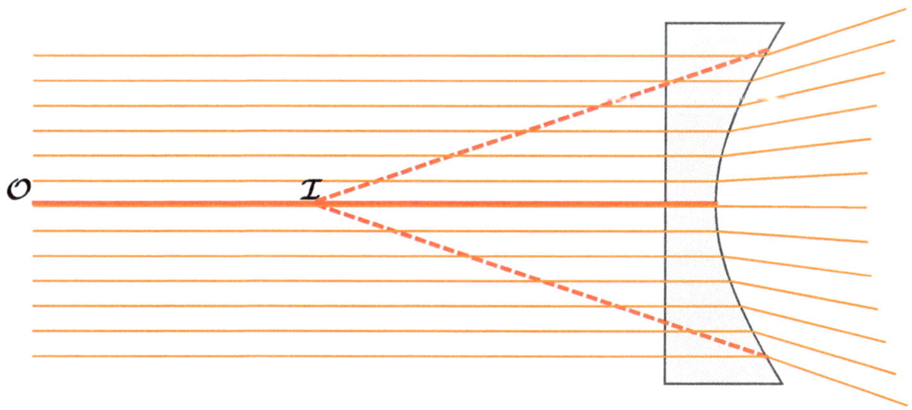


**Figure 4.34.** Design specifications: $n = 1.5$, $t = 5$ mm, $t_b = -40$ mm, $z_a = 0$ and $z_b$ = equation (4.42).

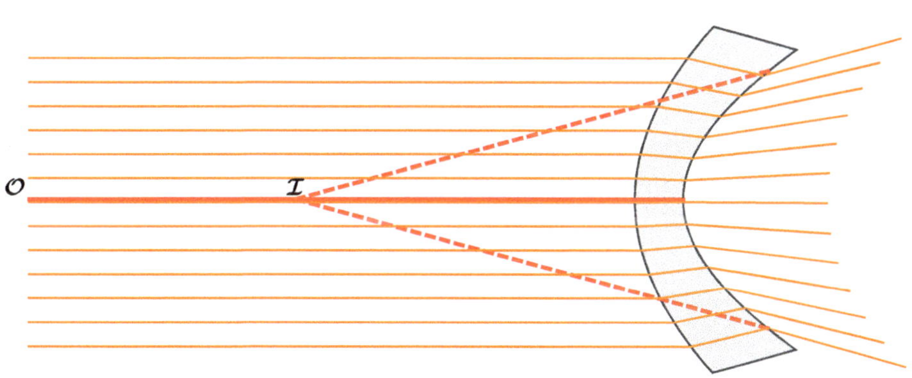


**Figure 4.35.** Design specifications: $n = 1.5$, $t = 5$ mm, $t_b = -40$ mm, $z_a = r_a^2/40$ and $z_b$ = equation (4.42).

Let's put $\vartheta \equiv \sqrt{(r_b - r_a)^2 + (z_b - z_a)^2}$ in equation (4.6),

$$\begin{aligned} &-t_a + nt + t_b - \mathrm{sgn}(t_b)\sqrt{r_b^2 + (z_b - t - t_b)^2} - n\vartheta \\ &= -\mathrm{sgn}(t_a)\sqrt{r_a^2 + (z_a - t_a)^2}. \end{aligned} \tag{4.43}$$

Observe that only on the right side are terms that depend on $t_b$. Thus we only need to apply the limit when $t_b \to \infty$ is on the right side,

$$\begin{aligned} &\lim_{t_b\to\infty}\left[-t_a + nt + t_b - \mathrm{sgn}(t_b)\sqrt{r_b^2 + (z_b - t - t_b)^2} - n\vartheta\right] \\ &= (-1 + n)t - t_a + [\lim_{t_b\to\infty}(z_b)] - n[\lim_{t_b\to\infty}(\vartheta)]. \end{aligned} \tag{4.44}$$

Thus, when $t_b \to \infty$, in equation (4.44) we have,

$$(-1 + n)t - t_a + [\lim_{t_b\to\infty}(z_b)] - n[\lim_{t_b\to\infty}(\vartheta)] = -\mathrm{sgn}(t_a)\sqrt{r_a^2 + (z_a - t_a)^2}, \tag{4.45}$$

replacing $z_b = z_a + \vartheta\wp_z$,

$$\begin{aligned} &(-1 + n)t - t_a + [\lim_{t_b\to\infty}(z_a + \vartheta\wp_z)] - n[\lim_{t_b\to\infty}(\vartheta)] \\ &= -\mathrm{sgn}(t_a)\sqrt{r_a^2 + (z_a - t_a)^2}, \end{aligned} \tag{4.46}$$

$z_a$ and $\wp_z$ do not depend on $t_b$, therefore they are not affected by the limit when $t_b \to \infty$,

$$\begin{aligned} &(-1 + n)t - t_a + z_a + [\lim_{t_b\to\infty}(\vartheta)]\wp_z \\ &= -\mathrm{sgn}(t_a)\sqrt{r_a^2 + (z_a - t_a)^2} + n[\lim_{t_b\to\infty}(\vartheta)]. \end{aligned} \tag{4.47}$$

We are now in a direct position to solve for $\left[\lim_{t_b\to\infty}(\vartheta)\right]$,

$$[\lim_{t_b\to\infty}(\vartheta)] = \frac{(-1 + n)t - t_a + z_a + \mathrm{sgn}(t_a)\sqrt{r_a^2 + (z_a - t_a)^2}}{n - \wp_z}. \tag{4.48}$$

Thus,

$$\boxed{\begin{aligned} \lim_{t_b\to\infty}(z_b) &= z_a + [\lim_{t_b\to\infty}(\vartheta)]\wp_z, \\ \lim_{t_b\to\infty}(r_b) &= r_a + [\lim_{t_b\to\infty}(\vartheta)]\wp_r. \end{aligned}} \tag{4.49}$$

Equation (4.49) is the second surface of the lens collimator no matter if the object is real or virtual. To properly use equation (4.49),we need equation (4.48) for $\left[\lim_{t_b\to\infty}(\vartheta)\right]$ equation (4.17) and equation (4.18), for $\wp_r$ and $\wp_z$ respectively.

### 4.5.1 Illustrative examples

In the following gallery, several stigmatic collimator lenses are presented on the axis. The specifications and ray tracing are presented in the corresponding figure of each example. The figures of the gallery are 4.36–4.41.

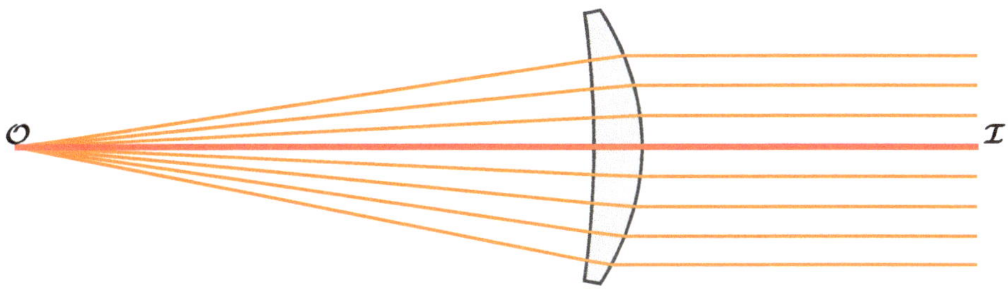


**Figure 4.36.** Design specifications: $n = 1.5$, $t = 5$ mm, $t_a = -60$ mm, $z_a = -r_a^2/200$ and $z_b$ = equation (4.49).

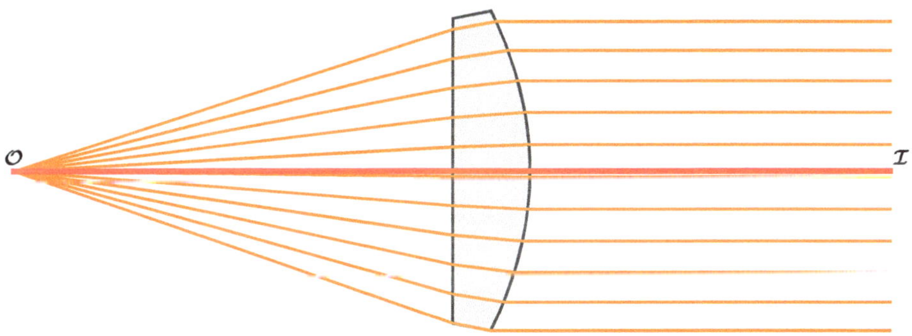


**Figure 4.37.** Design specifications: $n = 1.7$, $t = 7$ mm, $t_a = -60$ mm, $z_a = 0$ and $z_b$ = equation (4.49).

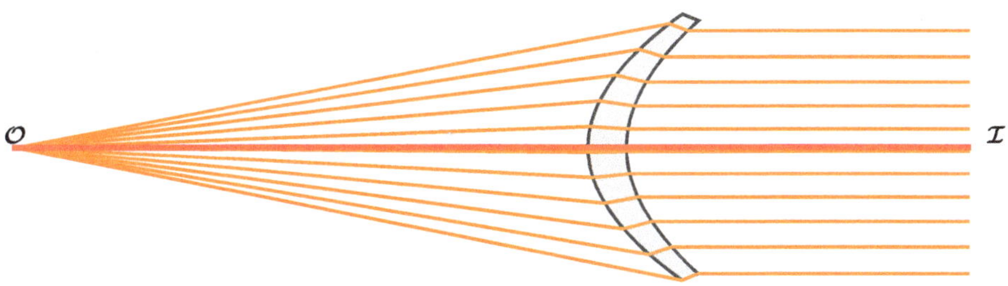


**Figure 4.38.** Design specifications: $n = 1.7$, $t = 4$ mm, $t_a = -60$ mm, $z_a = r_a^2/20$ and $z_b$ = equation (4.49).

## 4.6 The single-lens telescope

The last case to cover is when $t_a \to -\infty$ and $t_b \to +\infty$. The input of this lens is collimated light and the output as well, like a telescope.

These lenses can be applied as telescope or as beam expander. Usually, telescopes and beam expanders are designed with two or more lenses. That's why we called it the single-lens telescope.

To get the single-lens telescope general equation is simple. We need to use equations (4.37), (4.38), and (4.39). Which are the equations of the cosine directors and the parameter $f$ when $t_a \to -\infty$.

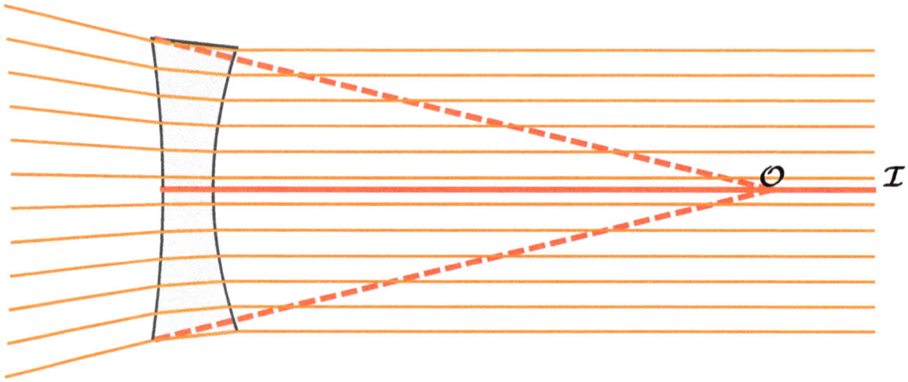


**Figure 4.39.** Design specifications: $n = 1.5$, $t = 6$ mm, $t_a = 60$ mm, $z_a = -r_a^2/200$ and $z_b$ = equation (4.49).

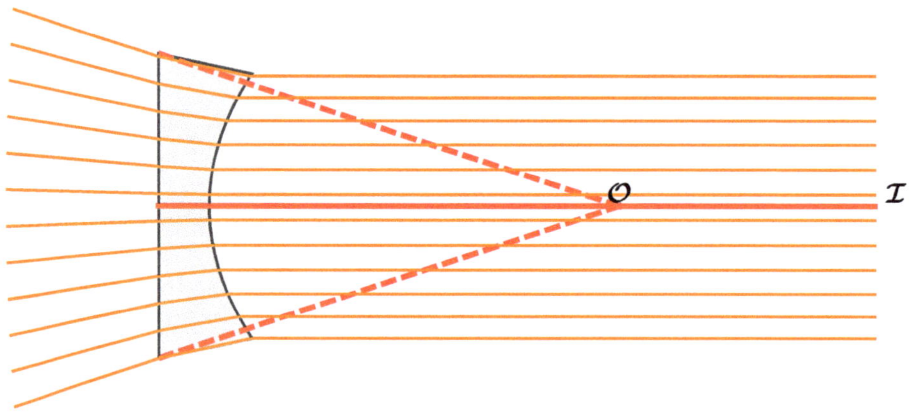


**Figure 4.40.** Design specifications: $n = 1.5$, $t = 5$ mm, $t_a = 45$ mm, $z_a = 0$ and $z_b$ = equation (4.49).

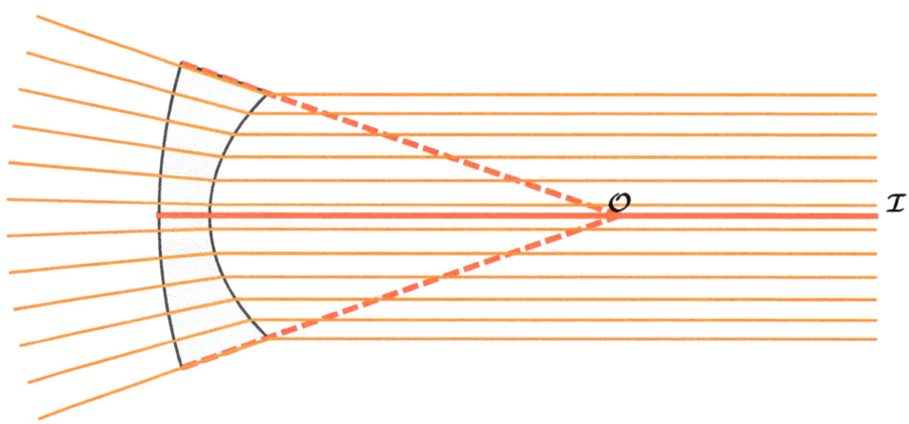


**Figure 4.41.** Design specifications: $n = 1.5$, $t = 5$ mm, $t_a = 45$ mm, $z_a = r_a^2/200$ and $z_b$ = equation (4.49).

Then we need to apply the limit when $t_a \to -\infty$ over $\left[\lim_{t_b\to\infty}(\vartheta)\right]$, equation (4.48),

$$\lim_{t_a\to-\infty}[\lim_{t_b\to\infty}(\vartheta)] = \lim_{t_a\to-\infty}\left[\frac{(-1+n)t - t_a + z_a + \text{sgn}(t_a)\sqrt{r_a^2 + (z_a - t_a)^2}}{n - \wp_z}\right], \tag{4.50}$$

thus, when we apply the limit on $\lim_{t_a\to-\infty}\left[\lim_{t_b\to\infty}(\vartheta)\right]$. we have,

$$\lim_{t_a\to-\infty}[\lim_{t_b\to\infty}(\vartheta)] = \frac{(n-1)t}{n - \left[\lim_{t_a\to-\infty}(\wp_z)\right]}. \tag{4.51}$$

Thus, the second surface of the single-lens telescope is given by,

$$\boxed{\begin{aligned}\lim_{t_a\to-\infty}[\lim_{t_b\to\infty}(z_b)] &= z_a + \left\{\lim_{t_a\to-\infty}[\lim_{t_b\to\infty}(\vartheta)]\right\}\left[\lim_{t_a\to-\infty}(\wp_z)\right],\\ \lim_{t_a\to-\infty}[\lim_{t_b\to\infty}(r_b)] &= r_a + \left\{\lim_{t_a\to-\infty}[\lim_{t_b\to\infty}(\vartheta)]\right\}\left[\lim_{t_a\to-\infty}(\wp_r)\right].\end{aligned}} \tag{4.52}$$

### 4.6.1 Examples

Examples of single-lens telescopes are presented in the following gallery. Each figure shows the input values with the respective ray tracing. The figures of the gallery are 4.42–4.44.

## 4.7 Conclusions

Finally in this chapter, we solve the spherical aberration lens conjecture. We derive it step by step until we obtain the general formula of lenses free of spherical aberrations.

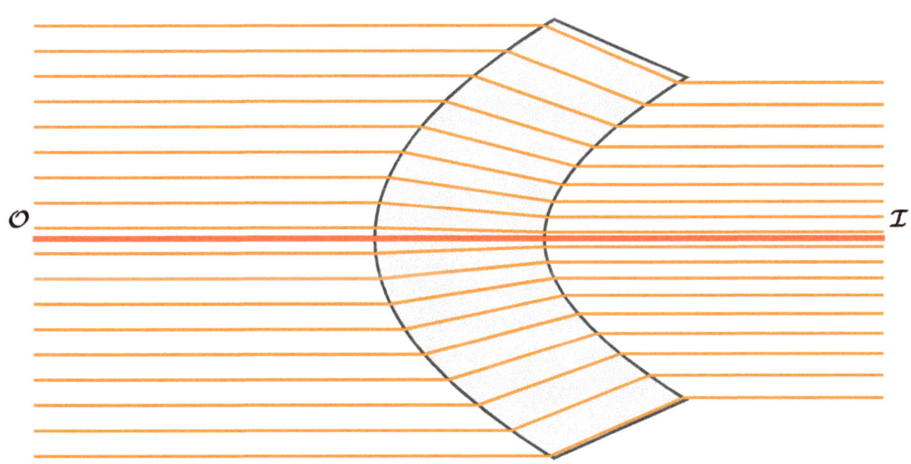


**Figure 4.42.** Design specifications: $n = 1.5$, $t = 10$ mm, $z_a = r_a^2/16$ and $z_b$ = equation (4.52).

**Figure 4.43.** This is a very interesting example, if $z_a = 0$ it does not matter what is the value of $n$, we will get a flat glass if we use equation (4.52).

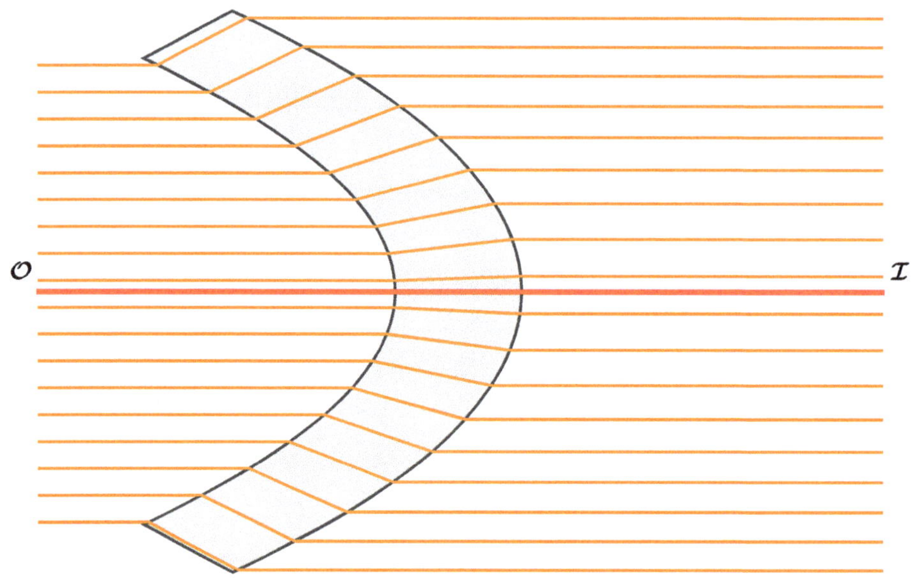


**Figure 4.44.** Design specifications: $n = 1.5$, $t = 7$ mm, $z_a = -r_a^2/12$ and $z_b$ = equation (4.52).

We presented several examples, including all the relevant cases. For real/virtual point objects and point images. Also, for the instance when the point object is at minus infinity and when point image is at infinity. In the following chapters, we are going to study the geometrical and topological properties of equation (4.36).

## Further reading

Acuña R G G and Gutiérrez-Vega J C 2019 General formula of the refractive telescope design free spherical aberration *Novel Optical Systems, Methods, and Applications XXII* vol 11105 (Bellingham, WA: International Society for Optics and Photonics)

Born M and Wolf E 2013 Principles of Optics: Electromagnetic Theory of Propagation *Interference and Diffraction of Light* (Amsterdam: Elsevier)

Braat J and Török P 2019 *Imaging Optics* (Cambridge: Cambridge University Press)

Chaves J 2016 *Introduction to Nonimaging Optics* 2nd edn (Boca Raton, FL: CRC Press)

Conrady A E 2013 *Applied Optics and Optical Design, Part One* (North Chelmsford, MA: Courier Corporation)

Descartes R 1637a *De la Nature des Lignes Courbes*

Descartes R 1637b *La Géométrie*

Duerr F, Benítez P, Miñano J C, Meuret Y and Thienpont H 2012 Analytic free-form lens design in 3D: coupling three ray sets using two lens surfaces *Opt. Express* **20** 10839–46

Estrada J C V, Calle Á H B and Hernández D M 2013 Explicit representations of all refractive optical interfaces without spherical aberration *J. Opt. Soc. Am.* A **30** 1814–24

Euler L 1770 *Dioptricae*

González-Acuña R G and Chaparro-Romo H A 2018 General formula for bi-aspheric singlet lens design free of spherical aberration *Appl. Opt.* **57** 9341–5

González-Acuña R G and Guitierrez-Vega J C 2018 Generalization of the axicon shape: the gaxicon *J. Opt. Soc. Am.* A **35** 1915–8

González-Acuña R G and Gutiérrez-Vega J C 2019a Analytic formulation of a refractive-reflective telescope free of spherical aberration *Opt. Eng.* **58** 085105

González-Acuña R G and Gutiérrez-Vega J C 2019b General formula for aspheric collimator lens design free of spherical aberration *Current Developments in Lens Design and Optical Engineering XX* vol 11104 (Bellingham, WA: International Society for Optics and Photonics)

González-Acuña R G, Avendaño-Alejo M and Gutiérrez-Vega J C 2019a Singlet lens for generating aberration-free patterns on deformed surfaces *J. Opt. Soc. Am.* A **36** 925–9

González-Acuña R G, Chaparro-Romo H A and Gutiérrez-Vega J C 2019b General formula to design freeform singlet free of spherical aberration and astigmatism *Appl. Opt.* **58** 1010–5

Gross H, Singer W, Totzeck M, Blechinger F and Achtner B 2005 *Handbook of Optical Systems* vol 1 (New York: Wiley Online Library)

Gross H, Zügge H, Peschka M and Blechinger F 2007 Handbook of Optical Systems *Aberration Theory and Correction of Optical Systems* vol 3 (New York: Wiley-VCH)

Gross H, Blechinger F and Achtner B 2008 Handbook of Optical Systems *Survey of Optical Instruments* vol 4 (New York: Wiley-VCH)

Herschel J F W 1827 *On the Theory of Light*

Holladay J T, Piers P A, Koranyi G, van der Mooren M and Norrby N E S 2002 A new intraocular lens design to reduce spherical aberration of pseudophakic eyes *J. Refract. Surg.* **18** 683–91

Huygens C 1690 *Traité de la lumière*

Kingslake R and Johnson R B 2009 *Lens Design Fundamentals* (New York: Academic)

Laikin M 2018 *Lens Design* (Boca Raton, FL: CRC Press)

Lozano-Rincón N d C and Valencia-Estrada J C 2017 Paraboloid-aspheric lenses free of spherical aberration *J. Mod. Opt.* **64** 1146–57

Luneburg R K and Herzberger M 1964 *Mathematical Theory of Optics* (Berkeley, CA: University of California Press)

Newton I 1704 *Opticks, or, a Treatise of the Reflections, Refractions, Inflections & Colours of Light*

Stavroudis O N 2006 *The Mathematics of Geometrical and Physical Optics: The k-Function and Its Ramifications* (New York: Wiley)

Sun H 2016 *Lens Design: A Practical Guide* (Boca Raton, FL: CRC Press)

Vaskas E M 1957 Note on the Wasserman-Wolf method for designing aspheric surfaces *J. Opt. Soc. Am.* **47** 669–70

Wassermann G D and Wolf E 1949 On the theory of aplanatic aspheric systems *Proc. Phys. Soc. Sect.* B **62** 2

Wolf E 1948 On the designing of aspheric surfaces *Proc. Phys. Soc.* **61** 494

Wolf E and Preddy W S 1947 On the determination of aspheric profiles *Proc. Phys. Soc.* **59** 704

Wynne C G and Wormell P M J H 1963 Lens design by computer *Appl. Opt.* **2** 1233–8

**IOP** Publishing

Analytical Lens Design (Second Edition)

**Rafael G González-Acuña, Héctor A Chaparro-Romo and Julio C Gutiérrez-Vega**

# Chapter 5

## Geometry of on-axis stigmatic lenses

In this chapter, we present a study of the entrance pupil of the stigmatic singlet lens on the axis. We focus on the nature of the entrance pupil, we examine its behavior when the refractive index is positive or negative, and if the object/image is finite or infinite.

### 5.1 Introduction

Spherical aberration is caused by off-center light rays striking a lens surface. These off-center rays are refracted depending on how close they approach the center. Spherical aberration reduces the quality of the images produced by a lens and, therefore, in optical systems in general.

Spherical aberration is drastically reduced when the angles of incidence on lenses or mirrors are small. This has laid the foundations for the generation of several theories about ray tracing. The problem with these theories is that if the angles are not small enough, they fail. In addition, to obtain the small angle, the entrance pupil of the lens must be proportionally small with respect to the angle of incidence, which affects the final magnification of the lens. When the small-angle approximation is not implemented then usually brute-force optimization techniques are used leading to a variety of results and methodologies which have proved to be useful for particular cases. However, no perspective explores the natural properties of the lens pupil.

The lenses presented in chapter 4 possess the particular property that the entrance pupil of the system is the lens itself. In this chapter, we explore the features of the entrance pupil and study the variables that affect its maximum size. The study presented here differs from the approximated ray tracing paradigms or the brute-force optimization techniques. The paradigm presented is entirely analytic.

doi:10.1088/978-0-7503-5774-6ch5

## 5.2 Lens free of spherical aberration finite–finite case

In the last chapter, we found how the second surface $(z_b, r_b)$ of a lens must be to correct the spherical aberration produced by the first surface $(z_a, r_a)$. See figures 5.1 and 5.2.

We take $n$ as the refraction index of the lens. The central thickness of the lens is $t$. The distance from the object to the first surface is $t_a$. The distance from the second surface to the image is $t_b$. Let's recall the expression of $(z_b, r_b)$,

$$\begin{cases} z_b = z_a + \vartheta \wp_z, \\ r_b = r_a + \vartheta \wp_r, \end{cases} \tag{4.36}$$

where,

$$\vartheta \equiv \frac{-(fn + r_a\wp_r + \tau\wp_z) \pm \sqrt{(fn + r_a\wp_r + \tau\wp_z)^2 - (1 - n^2)(r_a^2 + \tau^2 - f^2)}}{(1 - n^2)}, \tag{4.35}$$

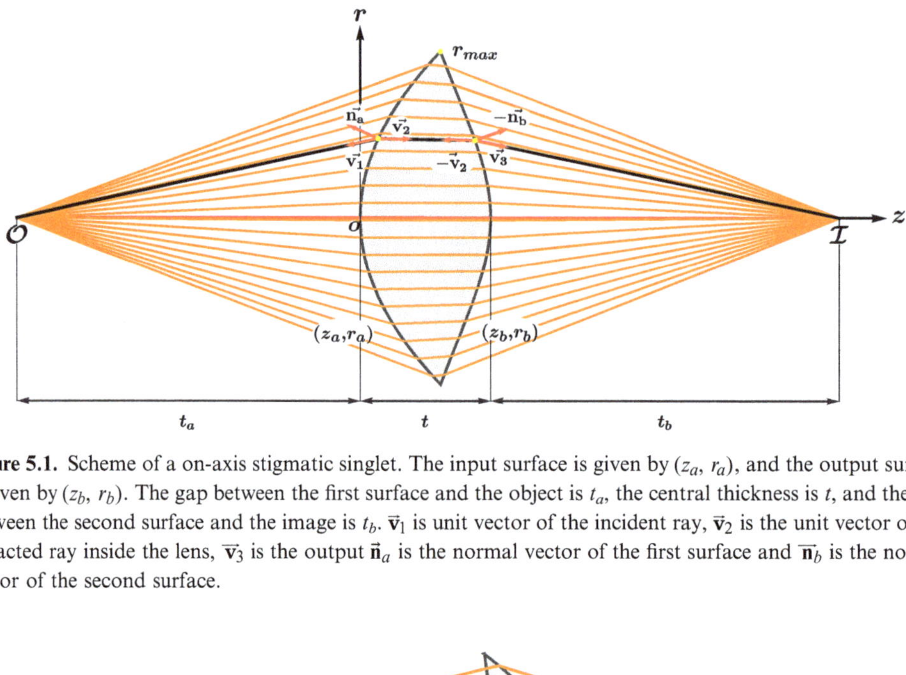


**Figure 5.1.** Scheme of a on-axis stigmatic singlet. The input surface is given by $(z_a, r_a)$, and the output surface is given by $(z_b, r_b)$. The gap between the first surface and the object is $t_a$, the central thickness is $t$, and the gap between the second surface and the image is $t_b$. $\vec{\mathbf{v}}_1$ is unit vector of the incident ray, $\vec{\mathbf{v}}_2$ is the unit vector of the refracted ray inside the lens, $\vec{\mathbf{v}}_3$ is the output $\vec{\mathbf{n}}_a$ is the normal vector of the first surface and $\vec{\mathbf{n}}_b$ is the normal vector of the second surface.

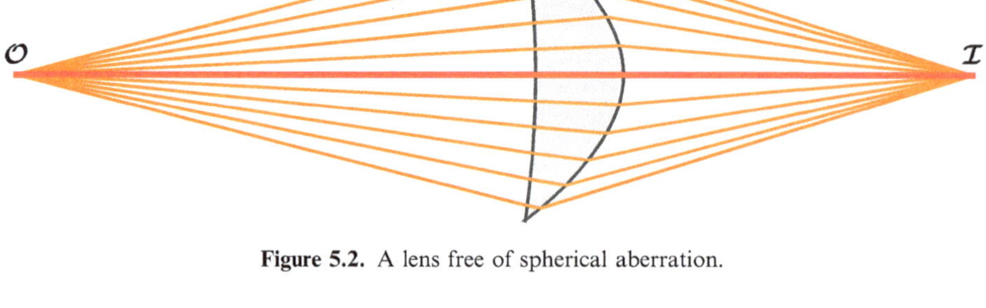


**Figure 5.2.** A lens free of spherical aberration.

$$f \equiv -t_a + nt + t_b + \text{sgn}(t_a)\sqrt{r_a^2 + (z_a - t_a)^2}, \tag{4.25}$$

and,

$$\tau \equiv z_a - t - t_b. \tag{4.26}$$

Finally, $\wp_r$, $\wp_z$ are the cosine director of the ray inside the lens,

$$\wp_r \equiv \frac{(z_a - t_a)(z_a') + r_a}{-\text{sgn}(t_a)n\sqrt{r_a^2 + (t_a - z_a)^2}\,(1 + (z_a')^2)} - (z_a')\frac{\sqrt{1 - \dfrac{[r_a + (z_a - t_a)(z_a')]^2}{n^2[r_a^2 + (t_a - z_a)^2](1 + (z_a')^2)}}}{\sqrt{1 + (z_a')^2}}, \tag{4.17}$$

and,

$$\wp_z \equiv \frac{[r_a + (z_a - t_a)(z_a')](z_a')}{-\text{sgn}(t_a)n\sqrt{r_a^2 + (t_a - z_a)^2}\,(1 + (z_a')^2)} + \frac{\sqrt{1 - \dfrac{[r_a + (z_a - t_a)(z_a')]^2}{n^2[r_a^2 + (t_a - z_a)^2](1 + (z_a')^2)}}}{\sqrt{1 + (z_a')^2}}. \tag{4.18}$$

They fulfill the following condition, $\wp_r^2 + \wp_z^2 = 1$.

### 5.2.1 The condition of maximum aperture for finite–finite case

The displacement of each ray inside the lens is given by $\vartheta$, for example for the central ray: $\vartheta|_{r_a \to 0} = t$. $\vartheta$ is directly related to the entrance pupil. Because the entrance pupil for this kind of lens is the lens itself. The maximum pupil entrance happens when both refractive surfaces touch each other, $z_a|_{r_a \to r_{\max}} = z_b|_{r_a \to r_{\max}}$, where $r_{\max}$ is the radius of the maximum pupil entrance. In this case the displacement is $\vartheta|_{r_a \to r_{\max}} = 0$, since there is no displacement in the border. Therefore,

$$0 = \left.\frac{-(fn + r_a\wp_r + \tau\wp_z) \pm \sqrt{(fn + r_a\wp_r + \tau\wp_z)^2 - (1 - n^2)(r_a^2 + \tau^2 - f^2)}}{(1 - n^2)}\right|_{r_a \to r_{\max}}. \tag{5.1}$$

Multiplying by $(1 - n^2)$,

$$0 = -(fn + r_a\wp_r + \tau\wp_z) \pm \left.\sqrt{(fn + r_a\wp_r + \tau\wp_z)^2 - (1 - n^2)(r_a^2 + \tau^2 - f^2)}\right|_{r_a \to r_{\max}}, \tag{5.2}$$

so,

$$-(fn + r_a\wp_r + \tau\wp_z)|_{r_a \to r_{\max}} = \pm\left.\sqrt{(fn + r_a\wp_r + \tau\wp_z)^2 - (1 - n^2)(r_a^2 + \tau^2 - f^2)}\right|_{r_a \to r_{\max}}, \tag{5.3}$$

thus,

$$(fn + r_a\wp_r + \tau\wp_z)^2|_{r_a\to r_{\max}} = (fn + r_a\wp_r + \tau\wp_z)^2 - (1-n^2)(r_a^2 + \tau^2 - f^2)\Big|_{r_a\to r_{\max}}, \tag{5.4}$$

eliminating terms,

$$0 = (1-n^2)(r_a^2 + \tau^2 - f^2)|_{r_a\to r_{\max}}, \tag{5.5}$$

divining by $(1 - n^2)$, we get the condition of max entrance pupil,

$$\boxed{0 = (r_a^2 + \tau^2 - f^2)|_{r_a\to r_{\max}}.} \tag{5.6}$$

Equation (5.6) is the main result of this chapter. It is the expression that relates the maximum entrance pupil of the lens and the other variables of the system. There is no general solution to it since $z_a$ is given by the user, and it depends on $r_a$. However, note that equation (5.6) does not depend on the direction cosines $\wp_r$, $\wp_z$. In the following subsections, we show several illustrative examples.

*5.2.1.1 Example when the first surface is flat*

The simplest case is when the first surface of a lens is flat $z_a = 0$. This example is interesting because a relatively small expression of $r_{\max}$ can be obtained. Recalling equation (5.6) with $z_a = 0$,

$$0 = (r_a^2 + \tau^2 - f^2)|_{r_a\to r_{\max}}, \tag{5.7}$$

where,

$$f\Big|_{r_a\to r_{\max}} = -t_a + nt + t_b + \operatorname{sgn}(t_a)\sqrt{r_{\max}^2 + t_a^2}, \tag{5.8}$$

and,

$$\tau|_{r_a\to r_{\max}} = -t - t_b. \tag{5.9}$$

Thus,

$$0 = r_{\max}^2 + \tau^2 - \left(-t_a + nt + t_b + \operatorname{sgn}(t_a)\sqrt{r_{\max}^2 + t_a^2}\right)^2, \tag{5.10}$$

solving for $r_{\max}$ we obtain,

$$r_{\max} = \pm\frac{\sqrt{n-1}\sqrt{t}\sqrt{2t_b + nt + t}\sqrt{((n-1)t - 2t_a)(-2t_a + 2t_b + nt + t)}}{2(-t_a + t_b + nt)}, \tag{5.11}$$

where the $\pm$ sign comes from the fact that there is a maximum positive radius and maximum non-positive radius.

Note, that if the refractive index is negative, the maximum radius will be imaginary $r_{\max} \in \mathbb{I}$. If the maximum radius is imaginary, the refractive surfaces never touch each other; this is a significant result, it means that the aperture entrance can also be infinite. Theoretically, it means that there can be infinite amplification for infinite aperture entrance size. For obvious reasons, this will never happen physically. Nevertheless, if the negative refraction index is well developed, gigantic amplification can be expected by increasing the lens diameter. Also it is very important to note that if $n = 1$, then $r_{max} = 0$.

It is exciting to see that with these equations a lens design can be made for microscope objectives and we can assume that the object is very close to the first surface, which represents the limit $t_a \to 0$ that applies to the equation (5.11),

$$r_{max} = \frac{(n-1)\,t\,(2t_b + nt + t)}{2(t_b + nt)}. \tag{5.12}$$

The last equation is when the object is to close to the first surface.

## 5.3 Lens free of spherical aberration infinite–finite case

The lens free of spherical aberration or an object very far away are in terms of the following equations,

$$\begin{cases} \lim\limits_{t_a \to -\infty} (z_b) = z_a + \left[\lim\limits_{t_a \to -\infty} (\vartheta)\right]\left[\lim\limits_{t_a \to -\infty} (\wp_z)\right], \\ \lim\limits_{t_a \to -\infty} (r_b) = r_a + \left[\lim\limits_{t_a \to -\infty} (\vartheta)\right]\left[\lim\limits_{t_a \to -\infty} (\wp_r)\right], \end{cases} \tag{4.42}$$

where,

$$\lim_{t_a \to -\infty} (\vartheta) = \frac{-\beta \pm \sqrt{\beta^2 - (1-n^2)\left\{r_a^2 + \tau^2 - \left[\lim\limits_{t_a \to -\infty} (f)\right]^2\right\}}}{(1-n^2)}, \tag{4.40}$$

where,

$$\beta \equiv \left\{\left[\lim_{t_a \to -\infty} (f)\right] n + r_a\left[\lim_{t_a \to -\infty} (\wp_r)\right] + \tau\left[\lim_{t_a \to -\infty} (\wp_z)\right]\right\}, \tag{4.41}$$

$$\lim_{t_a \to -\infty} f = nt + t_b - z_a, \tag{4.39}$$

the cosine directors are,

$$\lim_{t_a \to -\infty} \wp_r = -\frac{(z_a')\left(n\sqrt{(z_a')^2 + 1}\sqrt{\dfrac{(n^2-1)(z_a')^2 + n^2}{n^2((z_a')^2 + 1)}} - 1\right)}{n((z_a')^2 + 1)}, \tag{4.37}$$

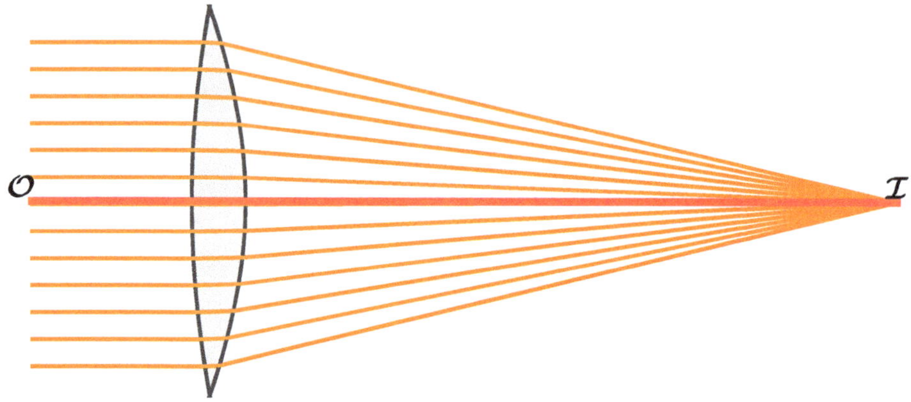


**Figure 5.3.** An on-axis collector lens.

and,

$$\lim_{t_a \to -\infty} \wp_z = \frac{\sqrt{\frac{(n^2-1)(z_a')^2+n^2}{n^2((z_a')^2+1)}}}{\sqrt{(z_a')^2+1}} + \frac{(z_a')^2}{n(z_a')^2+n}. \tag{4.38}$$

In the next subsection, we are going to study its maximum diameter condition. See figure 5.3.

### 5.3.1 The condition of maximum aperture for infinite–finite case

For the on-axis stigmatic collector lens, the maximum aperture condition is almost the same as the maximum aperture condition for the finite–finite case. The only difference is in the parameter $f$, as we can see in the following expression,

$$\boxed{0 = \left( r_a^2 + \tau^2 - \lim_{t_a \to -\infty} f^2 \right) \Big|_{r_a \to r_{\max}},} \tag{5.13}$$

where $\lim_{t_a \to -\infty} f = nt + t_b - z_a$. Therefore, this expression is simple to see how the value of $r_{\max}$ must be in order to get $\vartheta \to 0$.

#### *5.3.1.1 Example when the first surface is flat*

Let's explore the case when $z_a = 0$, thus we have that the conditions are the following,

$$0 = (r_a^2 + \tau^2 - (nt + t_b)^2)|_{r_a \to r_{\max}}. \tag{5.14}$$

Remember that $\lim_{t_a \to -\infty} \tau = -t - t_b$, since $z_a = 0$. Thus,

$$0 = r_{\max}^2 + (t + t_b)^2 - (nt + t_b)^2. \tag{5.15}$$

We can directly solve for $r_{\max}$ and we get,

$$r_{\max} = \pm\sqrt{(nt + t_b)^2 - (t + t_b)^2}, \tag{5.16}$$

where $r_{\max}$ is the maximum radius of a lens with $t_a \to -\infty$ and $z_a = 0$. As we can see, this maximum radius depends on $n$, $t$ and $t_b$. If these parameters increase $r_{\max}$ increases as well.

*5.3.1.2 Example when the first surface is parabolic*

For the collector case, the expression of $r_{\max}$ is considerably simplified. Here we take $t_a \to -\infty$ and $z_a = r_a^2/p$, where $p$ is a real constant, $p \in \mathbb{R}$. Thus, we have,

$$0 = \left( r_a^2 + \tau^2 - \lim_{t_a \to -\infty} f^2 \right)\Bigg|_{r_a \to r_{\max}}, \tag{5.17}$$

with $\lim_{t_a \to -\infty} f = nt + t_b - r_{\max}^2/p$

$$0 = r_{\max}^2 + \tau^2 - (nt + t_b - r_{\max}^2/p)^2. \tag{5.18}$$

Expanding and manipulating, we get a quadratic equation,

$$\frac{2r_{\max}^2(t_b + nt)}{p} - (t_b + nt)^2 - \frac{r_{\max}^4}{p^2} + r_{\max}^2 + \tau^2 = 0, \tag{5.19}$$

whose solution,

$$r_{\max} = \pm\frac{\sqrt{(nt - t)(2t_b + nt + t)}}{\sqrt{\dfrac{2(t_b + nt) - 2(t_b + t) + p}{p}}}. \tag{5.20}$$

Please observe that if $0 > n$, $r_{\max} \in \mathbb{I}$, if $0 < n$, $r_{\max} \in \mathbb{R}$. Again, if both surfaces never touch, this means that we can predict a gigantic amplification.

*5.3.1.3 Example when the first surface is spheric*

Our last example for the on-axis stigmatic collector lens is when $z_a = R - \sqrt{R^2 - r_{\max}^2}$, therefore,

$$0 = \left( r_a^2 + \tau^2 - \lim_{t_a \to -\infty} f^2 \right)\Bigg|_{r_a \to r_{\max}}, \tag{5.21}$$

with $\lim_{t_a \to -\infty} f = nt + t_b - \left(R - \sqrt{R^2 - r_{\max}^2}\right)$ and $\lim_{t_a \to -\infty} \tau = R - \sqrt{R^2 - r_{\max}^2} - t - t_b$, replacing,

$$0 = r_{\max}^2 + \left(R - \sqrt{R^2 - r_{\max}^2} - t - t_b\right)^2 - \left(nt + t_b - R + \sqrt{R^2 - r_{\max}^2}\right)^2, \tag{5.22}$$

we can solve for $r_{max}$ using symbolic algebra software and we get,

$$r_{max} = \mp\sqrt{\frac{2\sqrt{(n-1)^2t^2[-2(n-1)t(t_b+t)+2(n-1)Rt+R^2]}}{+(n-1)t[2t_b-(n-3)t-2R]}}. \tag{5.23}$$

$r_{max}$ is such that $\vartheta$ evaluated in $r_{max}$ is equal to zero, $\vartheta = 0$. The $\mp$ defines the two radii. Remember that the lens is centered at the origin.

## 5.4 Lens free of spherical aberration finite–infinite case

Now, let's focus on the on-axis stigmatic collimator, we recall the equations that described it,

$$\begin{cases} \lim\limits_{t_b\to\infty}(z_b) = z_a + \left[\lim\limits_{t_b\to\infty}(\vartheta)\right]\wp_z, \\ \lim\limits_{t_b\to\infty}(r_b) = r_a + \left[\lim\limits_{t_b\to\infty}(\vartheta)\right]\wp_r, \end{cases} \tag{4.49}$$

where, the distance inside the lens that the rays travel is $\vartheta$,

$$\left[\lim_{t_b\to\infty}(\vartheta)\right] = \frac{(-1+n)t - t_a + z_a + \operatorname{sgn}(t_a)\sqrt{r_a^2+(z_a-t_a)^2}}{n-\wp_z}, \tag{4.48}$$

the cosine directors of the refracted ray inside the lens are,

$$\wp_r \equiv \frac{(z_a-t_a)(z_a') + r_a}{-\operatorname{sgn}(t_a)n\sqrt{r_a^2+(t_a-z_a)^2}(1+(z_a')^2)} - (z_a')\frac{\sqrt{1-\dfrac{[r_a+(z_a-t_a)(z_a')]^2}{n^2[r_a^2+(t_a-z_a)^2](1+(z_a')^2)}}}{\sqrt{1+(z_a')^2}}, \tag{4.17}$$

and,

$$\wp_z \equiv \frac{[r_a+(z_a-t_a)(z_a')](z_a')}{-\operatorname{sgn}(t_a)n\sqrt{r_a^2+(t_a-z_a)^2}(1+(z_a')^2)} + \frac{\sqrt{1-\dfrac{[r_a+(z_a-t_a)(z_a')]^2}{n^2[r_a^2+(t_a-z_a)^2](1+(z_a')^2)}}}{\sqrt{1+(z_a')^2}}. \tag{4.18}$$

The next step is to study its condition of maximum aperture, see figure 5.4.

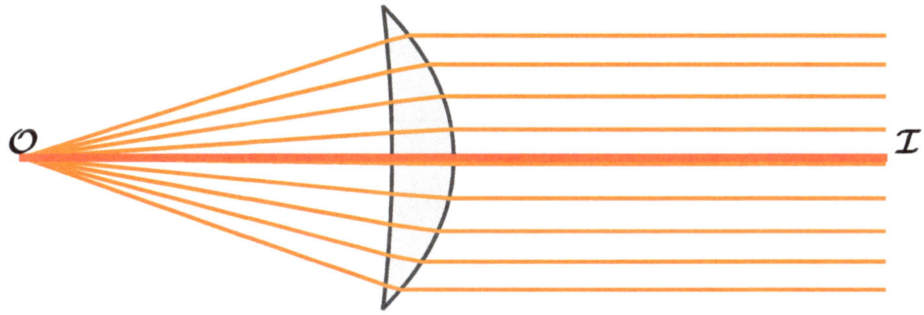


**Figure 5.4.** An on-axis collimator lens.

### 5.4.1 The condition of maximum aperture for finite–infinite case

The maximum aperture condition express that $\vartheta = 0$, in other means, when both surfaces meet each other, therefore,

$$0 = \left.\frac{(-1+n)t - t_a + z_a + \text{sgn}(t_a)\sqrt{r_a^2 + (z_a - t_a)^2}}{n - \wp_z}\right|_{r_a \to r_{\max}}. \tag{5.24}$$

Multiplying both side by $n - \wp_z$,

$$\boxed{0 = \left.\left(-t + nt - t_a + z_a + \text{sgn}(t_a)\sqrt{r_a^2 + (z_a - t_a)^2}\right)\right|_{r_a \to r_{\max}}}. \tag{5.25}$$

Finally, the above expression is the maximum aperture condition for the on-axis collimator singlet. Let's now pay attention in some example of how must be $r_{\max}$ when the first surface is flat or is a parabola.

#### *5.4.1.1 Example when the first surface is flat*

We begin with $z_a = 0$, therefore we have,

$$0 = -t + nt - t_a + \text{sgn}(t_a)\sqrt{r_{\max}^2 + t_a^2}, \tag{5.26}$$

manipulating,

$$(t - nt + t_a)^2 = r_{\max}^2 + t_a^2, \tag{5.27}$$

we can solve for $r_{\max}$,

$$r_{\max} = \sqrt{(t - nt + t_a)^2 - t_a^2}. \tag{5.28}$$

We already have the result, but we can manipulate such that we get,

$$r_{\max} = \sqrt{n-1}\sqrt{t}\sqrt{(n-1)t - 2t_a}. \tag{5.29}$$

In the above expression it is easy to see that if $0 > n$, $r_{\max} \in \mathbb{I}$, if $0 < n$, $r_{\max} \in \mathbb{R}$.

*5.4.1.2 Example when the first surface is parabolic*

Now let's focus on when the first surface is parabolic, this means that the maximum aperture condition is,

$$-t_a + nt + \frac{r_{\max}^2}{p} - t = -\text{sgn}(t_a)\sqrt{\left(\frac{r_{\max}^2}{p} - t_a\right)^2 + r_{\max}^2}, \tag{5.30}$$

squaring,

$$\left(-t_a + nt + \frac{r_{\max}^2}{p} - t\right)^2 = \left(\frac{r_{\max}^2}{p} - t_a\right)^2 + r_{\max}^2, \tag{5.31}$$

simplifying,

$$-2(n-1)tt_a - \frac{r_{\max}^2(-2nt + p + 2t)}{p} + (n-1)^2t^2 = 0. \tag{5.32}$$

We can solve for $r_{\max}$ directly,

$$r_{\max} = \frac{i\sqrt{n-1}\sqrt{p}\sqrt{t}\sqrt{nt - t - 2\text{ta}}}{\sqrt{2nt - p - 2t}}. \tag{5.33}$$

Notice that if $0 > n$, $r_{\max} \in \mathbb{I}$, if $0 < n$, $r_{\max} \in \mathbb{R}$. The same that happens in the other examples.

## 5.5 Lens free of spherical aberration infinite–infinite case

Finally, we recall the single-lens telescope, see figure 5.5. its second surface is given by,

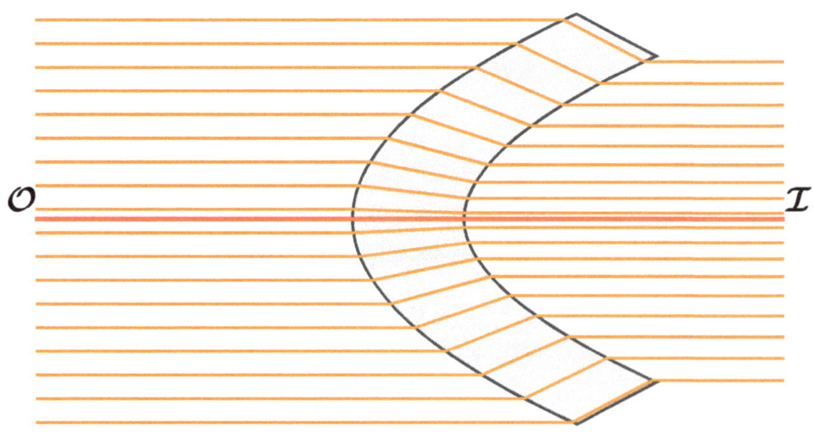


**Figure 5.5.** An on-axis stigmatic singlet for infinite object/image.

$$\begin{cases} \lim_{t_a \to -\infty}\left[\lim_{t_b \to \infty}(z_b)\right] = z_a + \left\{\lim_{t_a \to -\infty}\left[\lim_{t_b \to \infty}(\vartheta)\right]\right\}\left[\lim_{t_a \to -\infty}(\wp_z)\right], \\ \lim_{t_a \to -\infty}\left[\lim_{t_b \to \infty}(r_b)\right] = r_a + \left\{\lim_{t_a \to -\infty}\left[\lim_{t_b \to \infty}(\vartheta)\right]\right\}\left[\lim_{t_a \to -\infty}(\wp_r)\right]. \end{cases} \tag{4.52}$$

The distance each ray travels inside the single-lens telescope is,

$$\lim_{t_a \to -\infty}\left[\lim_{t_b \to \infty}(\vartheta)\right] = \frac{(n-1)t}{n - \left[\lim_{t_a \to -\infty}(\wp_z)\right]}. \tag{4.51}$$

The cosine directors of the refracted ray inside the singlet lens telescope are,

$$\lim_{t_a \to -\infty} \wp_r = -\frac{(z_a')\left(n\sqrt{(z_a')^2 + 1}\sqrt{\dfrac{(n^2-1)(z_a')^2 + n^2}{n^2((z_a')^2 + 1)}} - 1\right)}{n((z_a')^2 + 1)}, \tag{4.37}$$

and

$$\lim_{t_a \to -\infty} \wp_z = \frac{\sqrt{\dfrac{(n^2-1)(z_a')^2 + n^2}{n^2((z_a')^2 + 1)}}}{\sqrt{(z_a')^2 + 1}} + \frac{(z_a')^2}{n(z_a')^2 + n}. \tag{4.38}$$

The single-lens telescope has a fantastic maximum aperture condition for positive and negative refractive indices.

### 5.5.1 The condition of maximum aperture for infinite–infinite case

We mentioned that the condition of maximum aperture for the other lenses must be that the distance they travel inside the lens when both surfaces touch each other should be zero, therefore,

$$0 = \left.\frac{(n-1)t}{n - \left[\lim_{t_a \to -\infty}(\wp_z)\right]}\right|_{r_a \to r_{\max}}. \tag{5.34}$$

If we multiply by $n - \left[\lim_{t_a \to -\infty}(\wp_z)\right]$ in both sides, we have

$$\boxed{0 \neq (n-1)t.} \tag{5.35}$$

This is because $(n-1)t$ is zero if and only if $n = 1$ since $t > 0$, but in general $n \neq 0$. This means that no matter if $n$ is positive or negative, both surfaces never touch each other.

## 5.6 Conclusions

We studied how to obtain the maximum entrance pupil of the lens. This property is given by the nature of the lens, which offers a physically natural aperture in which the lens is free of spherical aberration. Using the expression of the maximum entrance pupil for several illustrative cases, we found that when the refraction index is negative the radius of the lens tends to infinity and when the refraction index is positive the radius tends to be finite.

## Further reading


Acuña R G G and Gutiérrez-Vega J C 2019 General formula of the refractive telescope design free spherical aberration *Novel Optical Systems, Methods, and Applications XXII* vol 11105 (Bellingham, WA: International Society for Optics and Photonics)

Bartle R G and Sherbert D R 2000 *Introduction to Real Analysis* vol 2 (New York: Wiley)

Born M and Wolf E 2013 Principles of Optics: Electromagnetic Theory of Propagation *Interference and Diffraction of Light* (Amsterdam: Elsevier)

Chaves J 2016 *Introduction to Nonimaging Optics* 2nd edn (Boca Raton, FL: CRC Press)

González-Acuña R G and Chaparro-Romo H A 2018 General formula for bi-aspheric singlet lens design free of spherical aberration *Appl. Opt.* **57** 9341–5

González-Acuña R G and Guitiérrez-Vega J C 2018 Generalization of the axicon shape: the gaxicon *J. Opt. Soc. Am.* A **35** 1915–8

González-Acuña R G and Gutiérrez-Vega J C 2019a Analytic formulation of a refractive-reflective telescope free of spherical aberration *Opt. Eng.* **58** 085105

González-Acuña R G and Gutiérrez-Vega J C 2019b General formula for aspheric collimator lens design free of spherical aberration *Current Developments in Lens Design and Optical Engineering XX* vol 11104 (Bellingham, WA: International Society for Optics and Photonics)

González-Acuña R G, Avendaño-Alejo M and Gutiérrez-Vega J C 2019a Singlet lens for generating aberration-free patterns on deformed surfaces *J. Opt. Soc. Am.* A **36** 925–9

González-Acuña R G, Chaparro-Romo H A and Gutiérrez-Vega J C 2019b General formula to design freeform singlet free of spherical aberration and astigmatism *Appl. Opt.* **58** 1010–5

Kingslake R and Johnson R B 2009 *Lens Design Fundamentals* (New York: Academic)

Kolmogorov A 1975 *Introductory Real Analysis* (North Chelmsford, MA: Courier Corporation)

Luneburg R K and Herzberger M 1964 *Mathematical Theory of Optics* (Berkeley, CA: University of California Press)

Malacara D 2001 *Handbook of Optical Engineering* (Boca Raton, FL: CRC Press)

Malacara D and Malacara Z 1994 *Handbook of Lens Design* (New York: Dekker)

Malacara-Hernández D and Malacara-Hernández Z 2016 *Handbook of Optical Design* (Boca Raton, FL: CRC Press)

Royden H L and Fitzpatrick P 1988 *1988Real Analysis* vol 32 (New York: Macmillan)

Stavroudis O N 2006 *The Mathematics of Geometrical and Physical Optics: The k-Function and Its Ramifications* (New York: Wiley)

Sun H 2016 *Lens Design: A Practical Guide* (Boca Raton, FL: CRC Press)

**Rafael G González-Acuña, Héctor A Chaparro-Romo and Julio C Gutiérrez-Vega**

# Chapter 6

# Topology of on-axis stigmatic lenses

In this chapter, we study the topological properties of the on-axis stigmatic singlet lens. We recall the definition of topological space, we explore the homeomorphism between the surfaces of the lens and, we examine the connectivity and compactness of the lens free of spherical aberration.

## 6.1 Introduction

All curves resulting from the different intersections between a cone and a plane are called conic sections. They are classified into four types: ellipse, parabola, hyperbola and circumference. The first known definition of conic section arises in ancient Greece, about 340 BC due to Apollonius of Perge, a Greek geometer famous for his work on conic sections.

Thanks to the intrinsic geometric properties of cones, conical shaped mirrors are free of spherical aberrations for specific cases. Conics have been studied for centuries, geometrically and topologically, since they are unique curves.

In chapter 4, we found lenses such that they are free of spherical aberration to do their intrinsic geometric properties. The formula presented in chapter 4 is the general equation to design on-axis stigmatic lenses. In this chapter, we study the topology of these lenses, see figure 6.1.

First, we recall the analytical solution of the problem of how the second surface $(z_b, r_b)$ is such that the lens is free of spherical aberration when the first surface $(z_a, r_a)$ is given,

$$\begin{cases} z_b = z_a + \vartheta\wp_z, \\ r_b = r_a + \vartheta\wp_r, \end{cases} \tag{4.36}$$

where,

$$\vartheta \equiv \frac{-(fn + r_a\wp_r + \tau\wp_z) \pm \sqrt{(fn + r_a\wp_r + \tau\wp_z)^2 - (1 - n^2)(r_a^2 + \tau^2 - f^2)}}{(1 - n^2)} \tag{4.35}$$

 

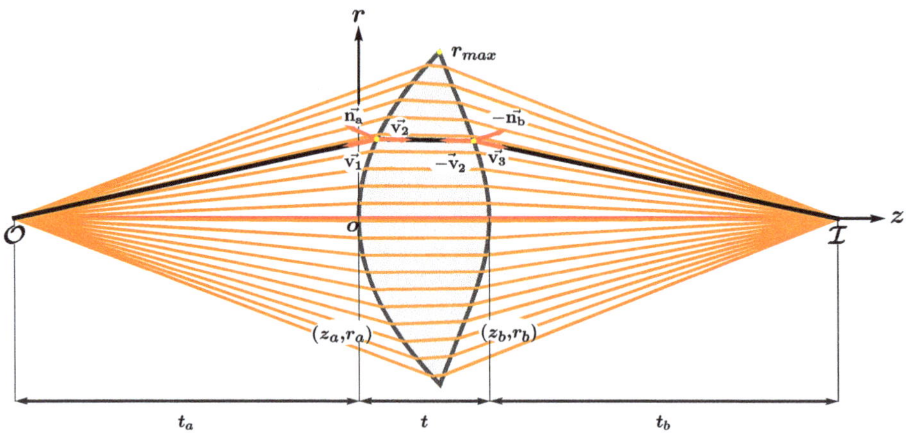


**Figure 6.1.** Diagram of an on-axis stigmatic singlet. The first surface is given by $(z_a, r_a)$, and the second surface is given by $(z_b, r_b)$. The distance between the first surface and the object is $t_a$, the thickness at the center of the lens is $t$, and the distance between the second surface and the image is $t_b$. $\vec{\mathbf{v}}_1$ is unit vector of the incident ray, $\vec{\mathbf{v}}_2$ is the unit vector of the refracted ray inside the lens, $\vec{\mathbf{v}}_3$ is the output $\vec{\mathbf{n}}_a$ is the normal vector of the first surface and $\vec{\mathbf{n}}_b$ is the normal vector of the second surface.

$$f \equiv -t_a + nt + t_b + \operatorname{sgn}(t_a)\sqrt{r_a^2 + (z_a - t_a)^2}\,. \tag{4.25}$$

and,

$$\tau \equiv z_a - t - t_b. \tag{4.26}$$

The cosine directors $\wp_r$, $\wp_z$ are given by,

$$\wp_r \equiv \frac{(z_a - t_a)z_a{}' + r_a}{-\operatorname{sgn}(t_a)n\sqrt{r_a^2 + (t_a - z_a)^2}\,(1 + z_a{}'^2)} - z_a{}'\frac{\sqrt{1 - \dfrac{[r_a + (z_a - t_a)z_a{}']^2}{n^2[r_a^2 + (t_a - z_a)^2](1 + z_a{}'^2)}}}{\sqrt{1 + z_a{}'^2}}, \tag{4.17}$$

and,

$$\wp_z \equiv \frac{[r_a + (z_a - t_a)z_a{}']z_a{}'}{-\operatorname{sgn}(t_a)n\sqrt{r_a^2 + (t_a - z_a)^2}\,(1 + z_a{}'^2)} + \frac{\sqrt{1 - \dfrac{[r_a + (z_a - t_a)z_a{}']^2}{n^2[r_a^2 + (t_a - z_a)^2](1 + z_a{}'^2)}}}{\sqrt{1 + z_a{}'^2}}, \tag{4.18}$$

where $t_a$ is the distance between the object and the first surface, $t$ is the central thickens of the lens. $t_b$ is the distance from the second surface to the image. $n$ is the refraction index. $\wp_r$ and $\wp_z$ are the cosine directors of the rays traveling inside the lens in the $r$ and $z$ directions, respectively. $z_a$ is the sagitta of the first surface, and finally $r_a$ is the only independent variable.

## 6.2 The topology of on-axis stigmatic lens

The on-axis stigmatic lenses can be expressed as two curves that touch in both extremes; in terms of set theory both surfaces can be seen as,

$$R_a = \{r_a \in \mathbb{R} | z_a < z_b\}, \tag{6.1}$$

$$Z_a = \{z_a \in \mathbb{R} | z_a \text{ is continuous}, \ z_a < z_b\}, \tag{6.2}$$

$$R_b = \{r_b \in \mathbb{R} | r_b = r_a + \vartheta\wp_r, \ z_b > z_a\}, \tag{6.3}$$

$$Z_b = \{z_b \in \mathbb{R} | z_b = z_a + \vartheta\wp_z, \ z_b > z_a\}. \tag{6.4}$$

Let's explore the topological properties of these sets. In chapter 2, we studied the basic definition of topology. Remember that topology can be thought of as the study of nearness and continuity. So let's pick a point $p$ from Euclidean space. The neighborhood $U$ of $p$ is such that it should be a set of points near $p$, entirely surrounding $p$. See figure 6.2. The definition of the neighborhood is formulated in this way to be as free as possible. Taking the meaning of the neighborhood in Euclidean space, we can recall the definition of a topological space:

1. $p$ belongs to any neighborhood of $p$.
2. if $U$ is a neighborhood of $p$ and $V \subset U$, then $V$ neighborhood of $p$.
3. if $U$ and $V$ are neighborhoods of $p$, then $p \in U \cap V$.
4. If $U$ is a neighborhood of $p$, then there is a neighborhood of $p$ such as $V \subset U$, $V$ neighborhood of $p$.

**Definition 1.** A topological space is a set $E$, along with assignment of each $p \in E$ of a collection of subsets of $E$, called *neighborhoods* of $p$ that satisfy the four properties mentioned.

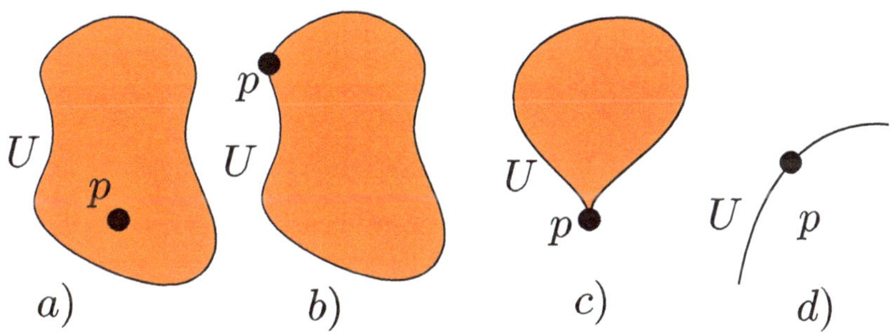


**Figure 6.2.** Example of neighborhoods in the Euclidean plane. The only valid neighborhood is (a), the other examples, (b), (c) and (d) are not well defined neighborhoods.

**Theorem 1.** $R_a$, $R_b$, $Z_a$ and $Z_b$ are topological spaces.

***Proof.*** It is easy to see that for every $p$ of them, all four properties are maintained. Since they are open intervals of real numbers and since real numbers are continuous. Please look the definition of $R_a$, $R_b$, $Z_a$ and $Z_b$, and its condition $z_a < z_b$. This includes all the values of before $z_a$ and $z_b$ before they meet each other. This condition defines the maximum entrance pupil of the system which is given by $z_a = z_b$ or by $\vartheta = 0$ and it defines the sets $R_a$, $R_b$, $Z_a$ and $Z_b$ as open sets.

For derivation proposes, it is essential to define what is an open set and prove its properties. The open and closed sets are defined below.

**Definition 2.** In a topological space $E$, a subset $U \in E$, is called *open*, if for each $p$ in $U$, $U$ is a neighborhood of $p$.

**Definition 3.** In a topological space $E$, a subset $F \in E$, is called *closed*, if $E - F$ is open.

Now let's recall a well-known theorem that will be very useful to study the topological properties of the lenses free of spherical aberration.

**Theorem 2.** For open sets and closet sets the following sentences are true,

1. The union of any collection of open sets in a topological space is open.
2. The intersection of finite collection of open sets is open.
3. The intersection of any collection of closed sets is closed.
4. The union of a finite collection of closed sets is closed.

***Proof* 1.** Let $\cup_i U_i$ be a collection of open sets in $E$. The members of the collection are $U_i$, where $i$ is some index. Let $U$ be the union of the members of the collection, thus $U = \cup_i U_i$. Now let's take $p \in U$. This follows to $p \in U_i$ for some $i$. Therefore, $U_i$ is a neighborhood of $p$, so $p \in U_i \subset U$. Since $U$ is a neighborhood of $p$, then $U$ is a neighborhood of each of its points. Therefore, by definition $U$ is open.

***Proof* 2.** Let $U_1$ and $U_2$ be open sets. Now let $p$ be a point in their interaction, thus $p \in U_1 \cap U_2$. Therefore, $U_1$ and $U_2$ are open and both contain $p \in U_1$ $p \in U_2$ Thus, $U_1$ and $U_2$ are neighborhoods of $p$. $U_i \cap U_2$ is a neighborhood of $p$ and it is a neighborhood of each of its points and thus $U_i \cap U_2$ is open.

***Proofs* 3** and **4** can be obtained using the complements of proofs 1 and 2.

Once we have defined the properties of the neighborhood, we can move further on other topological properties. These kinds of lenses are continuous; we recall the definition of continuity.

**Definition 4.** The map $g$: $E \to F$ is *continuous* at $p$, if given any neighborhood $V$ of $g(p) \in F$, there is a neighborhood $U$ of $p \in E$, such that $g(U) \subset V$. Therefore, $g$ is *continuous* if it is continuous at each $p \in E$.

**Theorem 3.** $z_b$ is continuous.

***Proof.*** By definition $z_a$ is continuous, this means that parting from values of the real line that $r_a$ takes, $z_a$: $R_a \to Z_a$, this means that the neighborhood $U \subset R_a$ is mapped to $z_a(R_a) \subset Z_a$, know $z_b$ is a composite function obtained by $z_a$ and $z_a$ is continuous, the neighborhood $z_a(R_a) \subset Z_a$ is mapped to the neighborhood $z_b(z_a(R_a)) \subset Z_b$.

Not let's recall the definition of homeomorphism.

**Definition 5.** Let $g$ be a one-o-one of $E$ onto $F$. Thus, there is an inverse map $g^{-1}$ of $F$ onto $E$, If $g$ and its inverse are continuous, then $g$ is a *homeomorphism* and $E$ and $F$ are *homeomorphic*.

From the topological point of view, two sets that are homeomorphic are the same; this means the properties of one set are the equal of another set that is homeomorphic to the first one.

**Theorem 4.** $Z_a$ is homeomorphic to $Z_b$ and $R_a$ is homeomorphic to $R_b$.

***Proof.*** They are homeomorphic since there is a one-to-one continuous function whose inverse is also continuous. The function is equation (4.36) and since geometrical optics is invertible, equation (4.36) has an inverse continuous function.

All this time, we have been focusing on curves given by equation (4.36). But lenses are not curves; they are surfaces, so we need to pay attention to topological products.

**Definition 6.** $E \times F$ is the Cartesian product of $E$ and $F$. We studied the Cartesian product in chapter 2.

**Theorem 5.** $R_b \times Z_b$ is a topological space.

***Proof.*** The set $R_b \times Z_b$ is defined as the set of ordered pairs $(p, q)$ where $p \in R_b$ and $q \in Z_b$. This is made into a topological space as follows: if $(p, q) \in R_b \times Z_b$, then a neighborhood of $(p, q)$ is any set containing a set of the form $U \times V$, where $U$ is a neighborhood of $p \in R_b$ and where $V$ is a neighborhood of $q \in Z_b$. It is easy to see that the neighborhoods satisfy the axioms (1) to (4) of definition 1.

**Theorem 6.** $R_a \times Z_a$ is a topological space.

This means that the second surface of the lens is a topological space and the first surface is also a topological space. Then by theorem 4, 5 and theorem 6, $R_b \times Z_b$ is homeomorphic to $R_a \times Z_a$. We are going to see two important topological properties of $R_b \times Z_b$ and $R_a \times Z_a$. The first one is connectedness, that is, the property of being all in one piece.

**Definition 7.** A space $E$ is connected if it cannot be expressed as the union of two non-empty disjoint sets open in $E$.

**Theorem 7.** Let $g$: $Z_a \to Z_b$ be a continuous map given by equation (4.36) of a connected space $Z_a$ onto a space $Z_b$. Then $Z_b$ is connected.

***Proof.*** Since $Z_b$ is in terms of $Z_a$, we can take a $Z_a$ that is connected with a continuous $z_a$. Now let's prove it by contradiction. Therefore, $Z_b = A \cup B$ where $A$

and $B$ are non-empty, disjoint and open sets. Then $Z_a = g^{-1}(A) \cup g^{-1}(B)$ and $g^{-1}(A)$ and $g^{-1}(B)$ are non-empty, disjoint and open sets. This contradicts the connectedness of $Z_a$, so $Z_b$ must be connected.

**Theorem 8.** If $R_b$ and $Z_b$ are connected spaces, then $R_b \times Z_b$ is connected.

***Proof.*** By contradiction, let's assume that $R_b \times Z_b$ is not connected. Therefore, we can say that $R_b \times Z_b = A \cup B$ where $A$ and $B$ are open, disjoint and non-empty sets. Take $(x, y)$ in $A$. The set $R_b \times \{y\}$ is homeomorphic to $R_b$ and so it is connected. Thus $R_b \times \{y\}$ is contained in $A$. Otherwise its intersection with $A$ and $B$ is a decomposition into open, disjoint non-empty sets. Observe that we can do the same for $\{x'\} \times Z_b$ would be in $A$ for each $x'$ in $E$, so $R_b \times Z_b$ would be in $A$, then $B$ is not necessary, $B$ becomes the empty set $B = \varnothing$, thus $R_b \times Z_b = A \cup B = A \cup \varnothing = A$. The contradiction is that $B$ was previously defined as non-empty. Therefore, $R_b \times Z_b$ is connected.

The second topological property of lenses free of spherical aberration is compactness. The idea of compactness is a generalization of the feature of being closed and bounded in the Euclidean space. Please note that here we are going to take $n > 0$ then $z_b$ and $z_a$ touch each other at two points, one in the first quadrant and the second in the second quadrant. If $n$ is negative then $z_b$ and $z_a$ never touch each other, so before exploring the concepts of compactness, it is useful first to examine the Hausdorff separation axiom.

**Definition 8.** A topological space will be called *Hausdorff*, if for any two distinct points $p$ and $q$, there are neighborhoods $U$ of $p$ and $V$ of $q$ that are disjoint, $U \cap V = \varnothing$. Thus the distinct points are separated by disjoint neighborhoods.

Other preliminary definitions are needed.

**Definition 9.** A *covering* of a topological space $E$ is a collection of sets in $E$ whose union is $E$. If the collection that forms the cover is composed only by open sets, then it is called *open cover*. It is called *finite open cover* if the collection of sets that forms the cover is finite.

**Definition 10.** Given a cover of a topological space, a *subcover* is a cover whose sets all belong to the given covering.

**Definition 11.** A *compact space* is Hausdorff space with the property that any open cover contains a finite subcover.

**Theorem 9.** Let $g$: $Z_a \rightarrow Z_b$ be a continuous map given by equation (4.36) of a compact space $Z_a$ onto a Hausdorff space $Z_b$. Then $Z_b$ is compact.

***Proof.*** Let $\cup_i U_i$ be an open covering of $Z_b$. The individual sets being denoted by $U_i$, where $i$ is the running index. Then the sets $g^{-1}(U_i)$ from a covering of $Z_a$ that by the properties of maps of opens sets, it is open. Since $Z_a$ is compact, a finite collection of sets coverts it, let us call them $g^{-1}(U_1)$, $g^{-1}(U_2)$, … , $g^{-1}(U_m)$ from a finite subcover of the given cover $Z_b$. Since we assume from the beginning that $Z_b$ is a Hausdorff space then, $Z_b$ is compact. This case is taking $n$ as positive. If $n$ is negative, $z_a$ will never be equal to $z_b$, then $z_a$ tends to infinity and it is homeomorphic to the real line and the real line is not closed nor bounded, thus it is not compact and since $z_a$ is homeomorphic to $z_b$, then all the topological properties $z_a$ are preservedr in $z_b$. $z_b$ is not compact when $n < 0$.

## 6.3 Example of the topological properties

The properties of the first surface are inherent in the second surface. Since the second is in terms of the first, they are homeomorphic to each other. For example, if the first surface is connected, the second will be connected. If the first is a piecewise function, the second one will be a piecewise function. So, if the first surface is a Fresnel lens which is divided into $m$ pieces of continuous functions, then the second surface will be a piecewise function with $m$ pieces. The singlet will be free of spherical aberration only in the neighborhoods where the first and second surfaces are continuous. See for example figures 6.3 and 6.4.

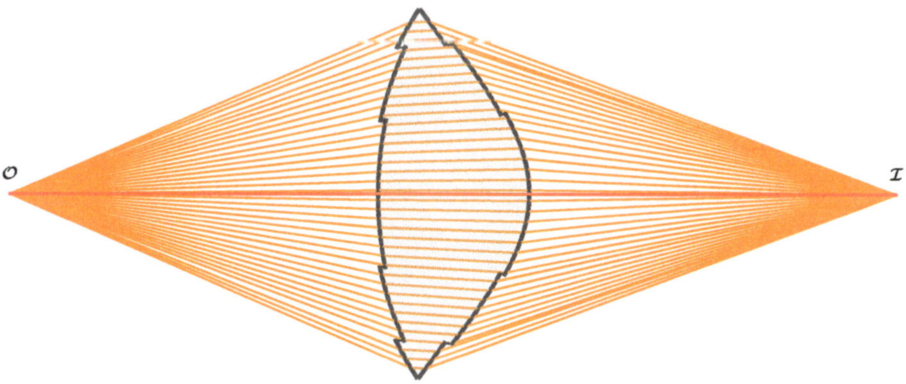


**Figure 6.3.** An on-axis Fresnel lens computed using equation (4.36). The number of segments of the first surface is equal to the number of segments of the second surface.

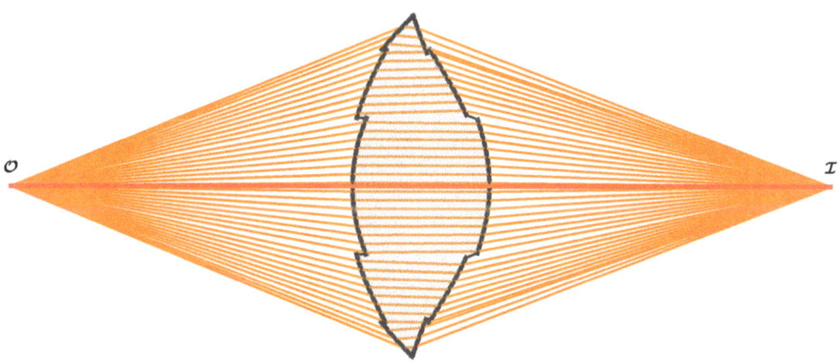


**Figure 6.4.** Another on-axis Fresnel lens computed using equation (4.36). The singlet is on-axis stigmatic only in the segments where the first and second surfaces are continuous.

Since there is a one-to-one relation in both surfaces the rays should not cross each other, because the pair of points $(p, q) \in R_a \times Z_a$ have a unique correspondence pair $(p', q') \in R_b \times Z_b$. The pair of points $(p', q')$ are positioned such that the optical path of the ray that touches them is equal to the path of the axial ray. If two or more rays pass through $(p', q')$, what experimentally happens is that $R_b$, $Z_b$ overlaps itself.

In figure 6.5, the first surface is a cosine function with high frequency that causes the rays to cross each other inside the lens; therefore, there is an overlap on the second surface and homeomorphism is lost.

In figures 6.6 and 6.7, there is a comparison between the surfaces that don't/do preserve homeomorphism between the surfaces.

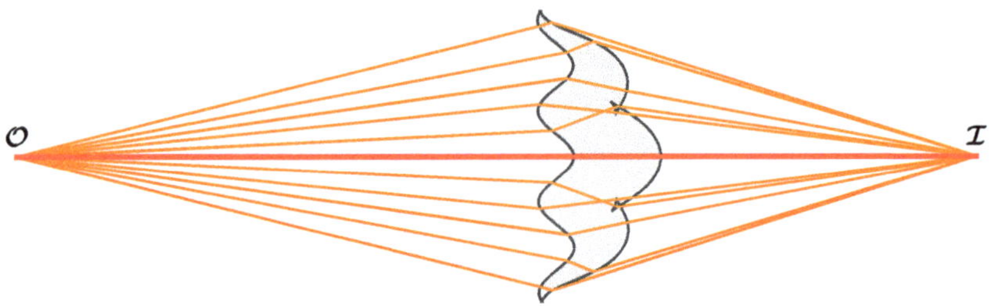


**Figure 6.5.** The first surface is a cosine function with high frequency that causes the rays to cross each other inside the lens; therefore, there is an overlap on the second surface and homeomorphism is lost.

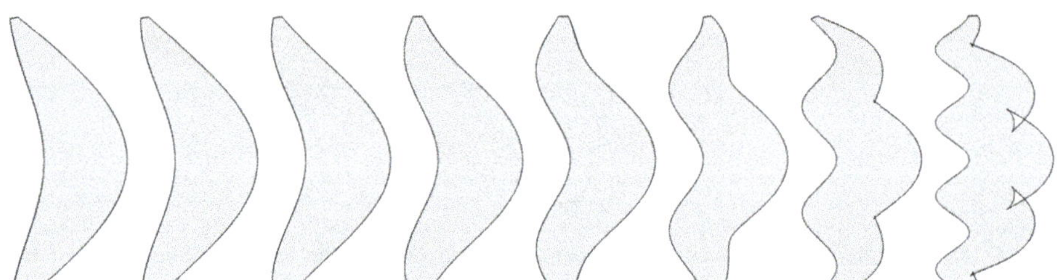

**Figure 6.6.** We have a stigmatic lens such that its first surface is a cosine. From left to right, the frequency of the cosine function is increased until the second surface overlaps.

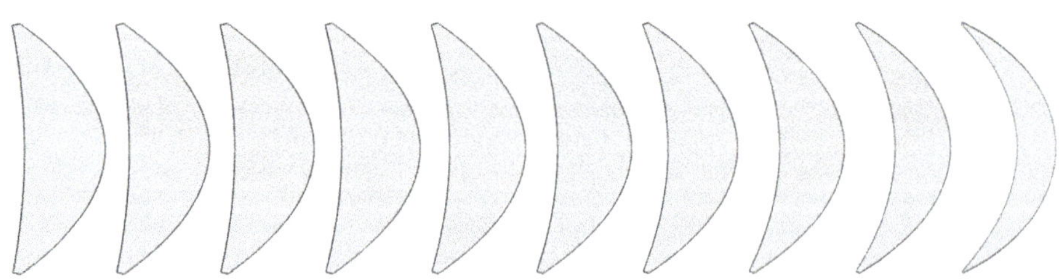

**Figure 6.7.** The curvature of the first surface is increased, but there is no overlapping on the second surface because the rays inside the lens do not cross each other.

## 6.4 Conclusions

As a conclusion in this chapter, we express the principal topological properties of lenses free of spherical aberration. The main features are that the first and second surfaces are topological Hausdorff spaces, and they are homeomorphic to each other. These properties illustrate the nature of this unique kind of lenses which are analogous to conic sections.

## Further reading


Acuña R G G and Gutiérrez-Vega J C 2019 General formula of the refractive telescope design free spherical aberration *Novel Optical Systems, Methods, and Applications XXII* vol 11105 (Bellingham, WA: International Society for Optics and Photonics)

Armstrong M A 2013 *Basic Topology* (Berlin: Springer Science & Business Media)

Baker C W 1991 *Introduction to Topology* (Malabar, FL: Krieger)

Born M and Wolf E 2013 *Principles of Optics: Electromagnetic Theory of Propagation Interference and Diffraction of Light* (Amsterdam: Elsevier)

Bredon G E 2013 *Topology and Geometry* vol 139 (Berlin: Springer Science & Business Media)

Chaves J 2016 *Introduction to Nonimaging Optics* 2nd edn (Boca Raton, FL: CRC Press)

Coxeter H S M 1998 *Non-Euclidean Geometry* (Cambridge: Cambridge University Press)

González-Acuña R G and Chaparro-Romo H A 2018 General formula for bi aspheric singlet lens design free of spherical aberration *Appl. Opt.* **57** 9341–5

González-Acuña R G and Guitiérrez-Vega J C 2018 Generalization of the axicon shape: the gaxicon *J. Opt. Soc. Am.* A **35** 1915–8

González-Acuña R G and Gutiérrez-Vega J C 2019a Analytic formulation of a refractive-reflective telescope free of spherical aberration *Opt. Eng.* **58** 085105

González-Acuña R G and Gutiérrez-Vega J C 2019b General formula for aspheric collimator lens design free of spherical aberration *Current Developments in Lens Design and Optical Engineering XX* vol 11104 (Bellingham, WA: International Society for Optics and Photonics)

González-Acuña R G, Avendaño-Alejo M and Gutiérrez-Vega J C 2019a Singlet lens for generating aberration-free patterns on deformed surfaces *J. Opt. Soc. Am.* A **36** 925–9

González-Acuña R G, Chaparro-Romo H A and Gutiérrez-Vega J C 2019b General formula to design freeform singlet free of spherical aberration and astigmatism *Appl. Opt.* **58** 1010–5

Kelley J L 2017 *General Topology* (Mineola, NY: Courier Dover Publications)

Kingslake R and Johnson R B 2009 *Lens Design Fundamentals* (New York: Academic)

Kreyszig E 1978 *Introductory Functional Analysis with Applications* vol 1 (New York: Wiley)

Kuratowski K 2014 *Topology* vol 1 (Amsterdam: Elsevier)

Luneburg R K and Herzberger M 1964 *Mathematical Theory of Optics* (Berkeley, CA: University of California Press)

Romano A and Cavaliere R 2016 *Geometric Optics: Theory and Design of Astronomical Optical Systems Using Mathematica®* (Basel: Birkhäuser)

Stavroudis O N 2006 *The Mathematics of Geometrical and Physical Optics: The k-Function and Its Ramifications* (New York: Wiley)

Sun H 2016 *Lens Design: A Practical Guide* (Boca Raton, FL: CRC Press)

Wallace A H 2006 *Differential Topology: First Steps* (North Chelmsford, MA: Courier Corporation)

**IOP** Publishing

Analytical Lens Design (Second Edition)

**Rafael G González-Acuña, Héctor A Chaparro-Romo and Julio C Gutiérrez-Vega**

# Chapter 7

# The gaxicon

We generalize the shape of the standard axicon by analytically finding the function of the output surface when the input surface is not flat but an arbitrary continuous function that possesses rotational symmetry. Illustrative examples are examined and evaluated using ray tracing techniques without the paraxial approximation.

## 7.1 Introduction

In 1954 J H McLeod invented the term axicon, when he was studying the optical characteristics of translucent cones. Today, an axicon is typically linked to a rotationally symmetric prism with a flat input surface and a linear conical output surface. The axicons are used to project a divergent circular ring.

In this chapter, we theorize the shape of the axicon by defining the output surface when the input surface is not flat but an optional continuous function given by the designer. To identify our generalized axicon from the standard axicon, we will name it *gaxicon*. The purpose is to design an optical component with the same optical transmittance and answer of an axicon by maintaining the versatility of forming the input surface.

The study is general, and it is not limited to the paraxial approach, just like all the investigations reported during a long treatise.

## 7.2 Geometrical model

For the analysis, we assume that the gaxicon has constant refraction index $n$ and axial thickness $t$. Also, we assume that the gaxicon is surrounded by air. The user gives the shape of the input surface ($z_a$, $r_a$). Notice that the subscript $a$ refers to the elements of the first input surface. The shape of the output ($z_b$, $r_b$) surface is unknown. Please observe that the subscript $b$ refers to the second surface. As usual,

doi:10.1088/978-0-7503-5774-6ch7

we locate the origin of the Cartesian coordinate system at the center of the first surface; thus, $z_a(0) = 0$. See figure 7.1.

We begin the analysis with the refraction suffered at the first surface. In chapter 4 we get the refracted vector inside the lens $\vec{\mathbf{v}}_2$ for a ray striking the input surface $(r_a, z_a)$ a point with a distance $t_a$ from the first surface,

$$\vec{\mathbf{v}}_2 = \frac{r_b - r_a}{\vartheta}\vec{\mathbf{e}}_1 + \frac{z_b - z_a}{\vartheta}\vec{\mathbf{e}}_2 = \wp_z\vec{\mathbf{e}}_1 + \wp_z\vec{\mathbf{e}}_2, \tag{7.1}$$

where $\overrightarrow{\mathbf{e}}_1$ is for the $r$ direction and $\overrightarrow{\mathbf{e}}_2$ is for the $z$ direction, and,

$$\vartheta \equiv \sqrt{(z_b - z_a)^2 + (r_b - r_a)^2}. \tag{4.19}$$

The cosine directors are given by equations (4.17) and (4.18),

$$\begin{aligned}\wp_r \equiv &\frac{(z_a - t_a)z_a{}' + r_a}{-\mathrm{sgn}(t_a)n\sqrt{r_a^2 + (t_a - z_a)^2}\,(1 + z_a{}'^2)} \\ &- z_a{}'\frac{\sqrt{1 - \dfrac{[r_a + (z_a - t_a)z_a{}']^2}{n^2[r_a^2 + (t_a - z_a)^2](1 + z_a{}'^2)}}}{\sqrt{1 + z_a{}'^2}},\end{aligned} \tag{4.17}$$

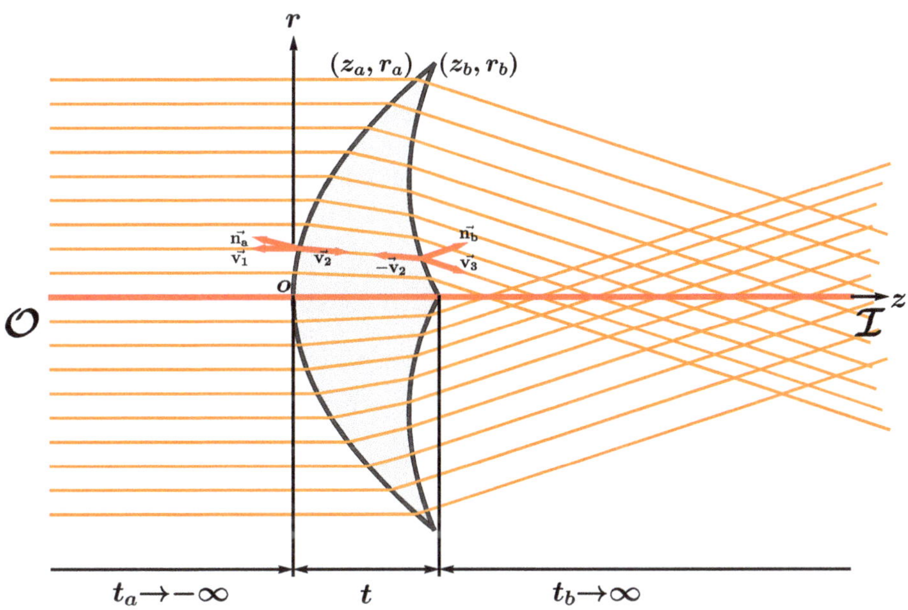


**Figure 7.1.** Meridional half-plane of the gaxicon, showing the variation of the incoming rays by the surfaces and their optical paths crossed. Also shown is the notation for the unit vectors of the incoming and emerging rays.

and,

$$\wp_z \equiv \frac{[r_a + (z_a - t_a)z_a']z_a'}{-\text{sgn}(t_a)n\sqrt{r_a^2 + (t_a - z_a)^2}(1 + z_a'^2)} + \frac{\sqrt{1 - \frac{[r_a + (z_a - t_a)z_a']^2}{n^2[r_a^2 + (t_a - z_a)^2](1 + z_a'^2)}}}{\sqrt{1 + z_a'^2}}. \tag{4.18}$$

Therefore, separating the components, we get,

$$\frac{z_b - z_a}{\vartheta} = \wp_z, \tag{7.2}$$

and,

$$\frac{r_b - r_a}{\vartheta} = \wp_r. \tag{7.3}$$

Then, we can directly solve for $z_b$ and $r_b$,

$$z_b = z_a + \vartheta\wp_z, \tag{7.4}$$

and,

$$r_b = r_a + \vartheta\wp_r. \tag{7.5}$$

Equations (7.4) and (7.5) have the structure of the solution. What we need is to get $\vartheta$ such that the singlet behaves like an axicon. For that we need an additional expression that relates $\vartheta$ with the other parameters.

An additional expression can be obtained using the Fermat principle. For derivation purposes, let's assume that the point object is located at $(t_a, 0)$ and the point image is located $(t_b + t, h_b)$, where $t$ is the central thickness of the gaxicon. We can equal the OPL of a reference ray with the OPL of a non-reference ray. If we choose the reference ray to be the axial ray, we have the OPL of the reference ray as,

$$-t_a + nt + \sqrt{t_b^2 + h_b^2} = \text{const.}, \tag{7.6}$$

because it starts and moves a distance $t_a$. Then, inside the gaxicon, it moves a distance $t$ and finally from the second surface to the point image it moves a distance $\sqrt{t_b^2 + h_b^2}$.

The OPL of the non-reference ray is given by,

$$\sqrt{r_a^2 + (z_a - t_a)^2} + n\sqrt{(r_b - r_a)^2 + (z_b - z_a)^2} + \sqrt{(r_b - h_b)^2 + (z_b - t - t_b)^2} = \text{const.} \tag{7.7}$$

The OPL of the non-reference ray from the point object to the first surface is $\sqrt{r_a^2 + (z_a - t_a)^2}$. Inside the lens the OPL is $n\sqrt{(r_b - r_a)^2 + (z_b - z_a)^2}$ and finally from the second surface to the image point the non-reference ray needs to travel a distance $\sqrt{(r_b - h_b)^2 + (z_b - t - t_b)^2}$.

Therefore, if we compare the OPL of the reference ray with the OPL of the non-reference ray we have,

$$\begin{aligned} &\sqrt{r_a^2 + (z_a - t_a)^2} + n\sqrt{(r_b - r_a)^2 + (z_b - z_a)^2} + \sqrt{(r_b - h_b)^2 + (z_b - t - t_b)^2} \\ &= -t_a + nt + \sqrt{t_b^2 + h_b^2}, \end{aligned} \tag{7.8}$$

replacing $\vartheta \equiv \sqrt{(r_b - r_a)^2 + (z_b - z_a)^2}$,

$$\begin{aligned} &\sqrt{r_a^2 + (z_a - t_a)^2} + n\vartheta + \sqrt{(r_b + h_b)^2 + (z_b - t - t_b)^2} \\ &+ t_a - nt - \sqrt{t_b^2 + h_b^2} = 0. \end{aligned} \tag{7.9}$$

Let's compute the limit $h_b \to wt_b$, where $w \equiv \tan\beta = h_b/t_b \in (-\infty, \infty)$ is the slope of the output rays. Negative/positive values of $w$ correspond to convergent/divergent conical wavefronts, respectively.

$$\begin{aligned} &n\vartheta + \sqrt{r_a^2 + (z_a - t_a)^2} + t_a + \sqrt{(r_b - wt_b)^2 + (-t_b + z_b - t)^2} \\ &- \sqrt{w^2 t_b^2 + t_b^2} - nt = 0. \end{aligned} \tag{7.10}$$

Computing the limit $t_b \to \infty$, and the image point to tend towards infinity but along the straight line defined by the angle $\beta$,

$$\begin{aligned} &n\left[\left(\lim_{t_b\to\infty} \vartheta\right) - t\right] + \sqrt{r_a^2 + (t_a - z_a)^2} + t_a - \frac{w\left(\lim_{t_b\to\infty} r_b\right) + \left(\lim_{t_b\to\infty} z_b\right)}{\sqrt{w^2 + 1}} \\ &+ \frac{t}{\sqrt{w^2 + 1}} = 0. \end{aligned} \tag{7.11}$$

Now to recover the parallelism of the input and output rays, we will let the object point tend to minus infinity, $t_a \to -\infty$,

$$\begin{aligned} &\sqrt{w^2 + 1}\left\{n\left[\lim_{t_a\to-\infty}\left(\lim_{t_b\to\infty} \vartheta\right)\right] + z_a\right\} - w\left[\lim_{t_a\to-\infty}\left(\lim_{t_b\to\infty} r_b\right)\right] \\ &- \left[\lim_{t_a\to-\infty}\left(\lim_{t_b\to\infty} z_b\right)\right] - nt\sqrt{w^2 + 1} + t = 0. \end{aligned} \tag{7.12}$$

It is important to note how the limits of $t_b \to \infty$ and $t_a \to -\infty$ affect $z_b$ and $r_b$,

$$\left[\lim_{t_a\to-\infty}\left(\lim_{t_b\to\infty} z_b\right)\right] = z_a + \left[\lim_{t_a\to-\infty}\left(\lim_{t_b\to\infty} \vartheta\right)\right]\left(\lim_{t_a\to-\infty} \wp_z\right), \tag{7.13}$$

and,

$$\left[\lim_{t_a\to-\infty}\left(\lim_{t_b\to\infty} r_b\right)\right] = r_a + \left[\lim_{t_a\to-\infty}\left(\lim_{t_b\to\infty} \vartheta\right)\right]\left(\lim_{t_a\to-\infty} \wp_r\right). \tag{7.14}$$

Observe that the limit $t_b \to \infty$ does not affect the cosine directors.

Thus, if we replace, $\left[\lim_{t_a\to-\infty}\left(\lim_{t_b\to\infty} z_b\right)\right]$ and $\left[\lim_{t_a\to-\infty}\left(\lim_{t_b\to\infty} r_b\right)\right]$ in equation (7.12) we have,

$$\begin{aligned}&\sqrt{w^2+1}\left\{n\left[\lim_{t_a\to-\infty}\left(\lim_{t_b\to\infty} \vartheta\right)\right] + z_a\right\}\\ &- w\left\{r_a + \left[\lim_{t_a\to-\infty}\left(\lim_{t_b\to\infty} \vartheta\right)\right]\left(\lim_{t_a\to-\infty} \wp_r\right)\right\}\\ &-\left\{z_a + \left[\lim_{t_a\to-\infty}\left(\lim_{t_b\to\infty} \vartheta\right)\right]\left(\lim_{t_a\to-\infty} \wp_z\right)\right\} - nt\sqrt{w^2+1} + t = 0.\end{aligned} \tag{7.15}$$

Now, we can directly solve for $\left[\lim_{t_a\to-\infty}\left(\lim_{t_b\to\infty} \vartheta\right)\right]$,

$$\left[\lim_{t_a\to-\infty}\left(\lim_{t_b\to\infty} \vartheta\right)\right] = \frac{wr_a - \left(\sqrt{w^2+1} - 1\right)z_a + t\left(n\sqrt{w^2+1} - 1\right)}{n\sqrt{w^2+1} - w\left(\lim_{t_a\to-\infty} \wp_r\right) - \left(\lim_{t_a\to-\infty} \wp_z\right)}. \tag{7.16}$$

Therefore, we can finally write the second surface of the gaxicon as,

$$\boxed{\begin{cases}\left[\lim_{t_a\to-\infty}\left(\lim_{t_b\to\infty} z_b\right)\right] = z_a + \left[\lim_{t_a\to-\infty}\left(\lim_{t_b\to\infty} \vartheta\right)\right]\left(\lim_{t_a\to-\infty} \wp_z\right),\\ \left[\lim_{t_a\to-\infty}\left(\lim_{t_b\to\infty} r_b\right)\right] = r_a + \left[\lim_{t_a\to-\infty}\left(\lim_{t_b\to\infty} \vartheta\right)\right]\left(\lim_{t_a\to-\infty} \wp_r\right).\end{cases}} \tag{7.17}$$

Notice that we already computed the cosine directors when $t_a \to -\infty$, with equations (4.37) and (4.38),

$$\lim_{t_a\to-\infty} \wp_r \equiv -\frac{z_a'\left(n\sqrt{(z_a')^2+1}\sqrt{\dfrac{(n^2-1)(z_a')^2+n^2}{n^2((z_a')^2+1)}} - 1\right)}{n[(z_a')^2+1]}, \tag{4.37}$$

$$\lim_{t_a \to -\infty} \wp_z \equiv \frac{\sqrt{\dfrac{(n^2 - 1)(z_a')^2 + n^2}{n^2[(z_a')^2 + 1]}}}{\sqrt{(z_a')^2 + 1}} + \frac{(z_a')^2}{n(z_a')^2 + n}. \tag{4.38}$$

Equation (7.17) is the most important result of this chapter. It describes the shape of the second surface analytically in terms of the function of its input surface and its characteristic output angle $\beta$. These equations may look cumbersome, but it is quite relevant that they could be expressed in close-form for arbitrary input functions $z_a(r_a)$.

We emphasize that a necessary condition for the validity of equation (7.17) is that the slope of the input surface vanishes at the optical axis, i.e. $z_a'(0) = 0$. In the following section, we will illustrate some relevant cases.

Also note that if we take $w = 0$, we recover the $\vartheta$ of the single-lens telescope,

$$\left[\lim_{t_a \to -\infty}\left(\lim_{t_b \to \infty} \vartheta\right)\right]\Bigg|_{w=0} = \frac{(n - 1)t}{n - \left(\lim\limits_{t_a \to -\infty} \wp_z\right)}. \tag{4.51}$$

## 7.3 Gallery of axicons

In this section, we present a gallery of axicons with different shapes. The figures of the gallery are 7.2–7.4. The design parameters of each sample of the gallery are displayed in its corresponding image with its ray tracing.

We have verified that equation (7.17) reduce to the expected linear functions

$$r_b = r_a, \qquad \text{and} \qquad z_b(r_a) = t - \left(\frac{\sin|\beta|}{n - \cos|\beta|}\right) r_a, \tag{7.18}$$

of the traditional axicon when $z_a(r_a) = 0$ as can be seen in the example of figure 7.2.

## 7.4 Conclusions

We theorized the output surface of a traditional axicon to get the behavior of an axicon when the input surface is not necessarily flat. We examined the model for a variety of input forms and confirmed its functionality with ray tracing techniques avoiding paraxial approaches.

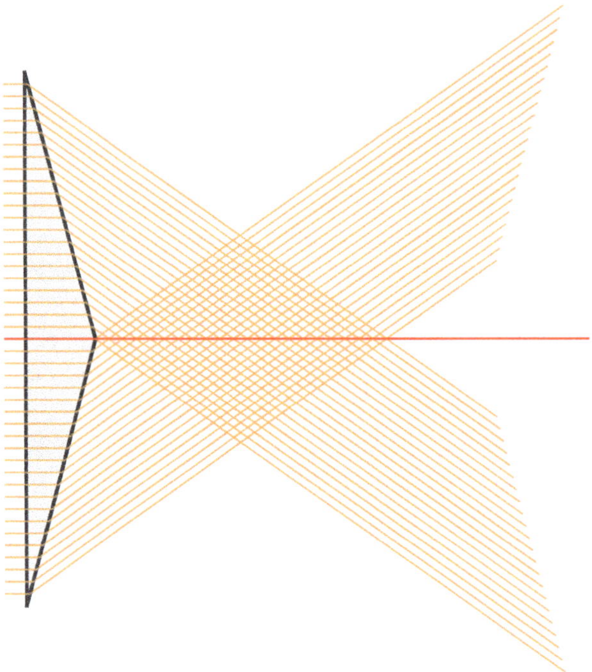

**Figure 7.2.** When the first surface $z_a(r_a) = 0$, we get the traditional axicon using equation (7.17).

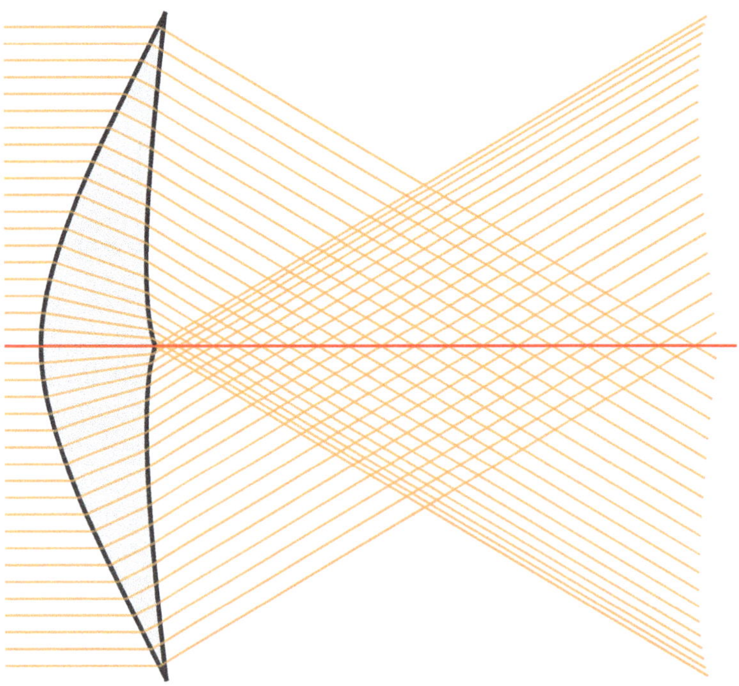

**Figure 7.3.** Design specifications: $n = 1.5$, $t = 50$ mm, $w = -0.33$, $z_a = -30 + \sqrt{r_a^2 + 30^2}$ and $z_b =$ equation (7.17).

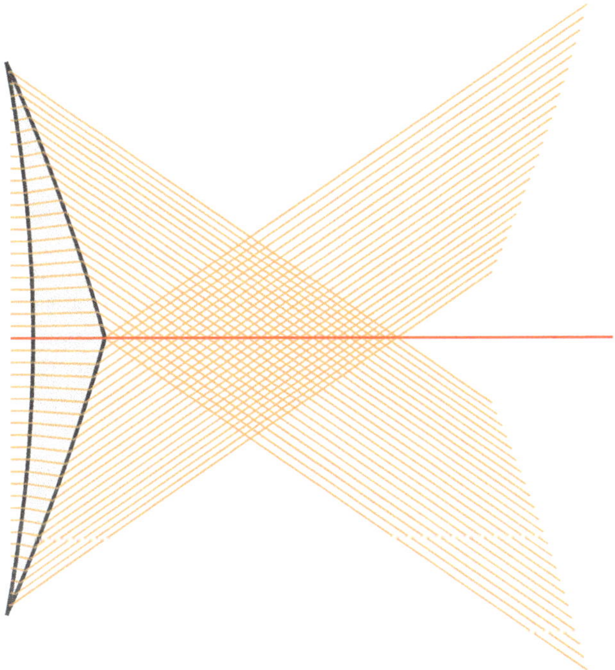

**Figure 7.4.** Design specifications: $n = 1.5$, $t = 50$ mm, $w = -0.33$, $z_a = 220 - \sqrt{r_a^2 + 220^2}$ and $z_b$ = equation (7.17).

# Further reading


Acuña R G G and Gutiérrez-Vega J C 2019 General formula of the refractive telescope design free spherical aberration *Novel Optical Systems Methods, and Applications XXII* vol 11105 (Bellingham, WA: International Society for Optics and Photonics)

Bakken G S 1974 The parabolic axicon *Appl. Opt.* **13** 1291–2

Born M and Wolf E 2013 *Principles of Optics: Electromagnetic Theory of Propagation Interference and Diffraction of Light* (Amsterdam: Elsevier)

Brzobohatý O, Čižmár T and Zemánek P 2008 High quality quasi-Bessel beam generated by round-tip axicon *Opt. Express* **16** 12688–700

Burvall A, Goncharov A and Dainty C 2005 Telephoto axicon *Optical Design and Engineering II* vol 5962 (Bellingham, WA: International Society for Optics and Photonics) p 596213

Chaves J 2016 *Introduction to Nonimaging Optics* 2nd edn (Boca Raton, FL: CRC Press)

Chebbi B and Golub I 2014 Development of spot size and lateral intensity distribution generated by exponential, logarithmic, and linear axicons *J. Opt. Soc. Am.* A **31** 2447–52

Edmonds W R 1974 Imaging properties of a conic axicon *Appl. Opt.* **13** 1762–4

Golub I 2006 Fresnel axicon *Opt. Lett.* **31** 1890–2

Golub I and Mirtchev T 2009 Absorption-free beam generated by a phase-engineered optical element *Opt. Lett.* **34** 1528–30

Golub I, Chebbi B, Shaw D and Nowacki D 2010 Characterization of a refractive logarithmic axicon *Opt. Lett.* **35** 2828–30

Golub I, Mirtchev T, Nuttall J and Shaw D 2012 The taming of absorption: generating a constant intensity beam in a lossy medium *Opt. Lett.* **37** 2556–8
González-Acuña R G and Chaparro-Romo H A 2018 General formula for bi-aspheric singlet lens design free of spherical aberration *Appl. Opt.* **57** 9341–5
González-Acuña R G and Guitiérrez-Vega J C 2018 Generalization of the axicon shape: the gaxicon *J. Opt. Soc. Am.* A **35** 1915–8
González-Acuña R G and Gutiérrez-Vega J C 2019a Analytic formulation of a refractive-reflective telescope free of spherical aberration *Opt. Eng.* **58** 085105
González-Acuña R G and Gutiérrez-Vega J C 2019b General formula for aspheric collimator lens design free of spherical aberration *Current Developments in Lens Design and Optical Engineering XX* (Bellingham, WA: International Society for Optics and Photonics)
González-Acuña R G, Avendaño-Alejo M and Gutiérrez-Vega J C 2019a Singlet lens for generating aberration-free patterns on deformed surfaces *J. Opt. Soc. Am.* A **36** 925–9
González-Acuña R G, Chaparro-Romo H A and Gutiérrez-Vega J C 2019b General formula to design freeform singlet free of spherical aberration and astigmatism *Appl. Opt.* **58** 1010–5
Gutiérrez-Vega J C, Rodrıguez-Masegosa R and Chávez-Cerda S 2003 Bessel-Gauss resonator with spherical output mirror: geometrical- and wave-optics analysis *J. Opt. Soc. Am.* A **20** 2113–22
Jaroszewicz Z, Burvall A and Friberg A T 2005 Axicon–the most important optical element *Opt. Photonics News* **16** 34–9
Kingslake R and Johnson R B 2009 *Lens Design Fundamentals* (New York: Academic)
Kotlyar V V and Kovalev A A 2006 Annular Radon transform and axicon transform *Opt. Eng.* **45** 078201
Luneburg R K and Herzberger M 1964 *Mathematical Theory of Optics* (Berkeley, CA: University of California Press)
Marchand E W 1990 Axicon gradient lenses *Appl. Opt.* **29** 4001–2
McGloin D, Garcés-Chávez V and Dholakia K 2003 Interfering Bessel beams for optical micromanipulation *Opt. Lett.* **28** 657–9
McKell C and Bonin K D 2018 Optical corral using a standing-wave Bessel beam *J. Opt. Soc. Am.* B **35** 1910–20
McLeod J H 1954 The axicon: a new type of optical element *J. Opt. Soc. Am.* **44** 592–7
Nagai S and Iizuka K 1982 A practical ultrasound axicon for non-destructive testing *Ultrasonics* **20** 265–70
Proust J, Martin J, Gérard D, Bijeon J L and Plain J 2012 Plasmonic axicon micro-lenses for chemical sensing *Conf. Lasers and Electro-Optics (CLEO), 2012* (Piscataway, NJ: IEEE) pp 1–2
Rayces J L 1958 Formation of axicon images *J. Opt. Soc. Am.* **48** 576-1–8
Sochacki J, Bara S, Jaroszewicz Z and Kołodziejczyk A 1992 Phase retardation of the uniform-intensity axilens *Opt. Lett.* **17** 7–9
Stavroudis O N 2006 *The Mathematics of Geometrical and Physical Optics: The k-Function and Its Ramifications* (New York: Wiley)
Steel W H 1979 Axicon means axial image *Appl. Opt.* **18** 2089–9
Sun H 2016 *Lens Design: A Practical Guide* (Boca Raton, FL: CRC Press)
Thaning A, Friberg A T, Popov S Y and Jaroszewicz Z 2002 Design of diffractive axicons producing uniform line images in Gaussian Schell-model illumination *J. Opt. Soc. Am.* A **19** 491–6

**Rafael G González-Acuña, Héctor A Chaparro-Romo and Julio C Gutiérrez-Vega**

# Chapter 8

# On-axis spherochromatic singlet

We derive the analytic formula of the output surface of a spherochromatic lens. The spherical aberration is small enough to consider the lens as diffracted limited. We test the singlet lens using ray-tracing methods and we find satisfactory results for three different wavelengths.

## 8.1 Introduction

A spherochromatic singlet is a lens that is designed to decrease the spherical and chromatic aberration. Commonly, the achromatic lenses are built with two or three lenses paired together. This paradigm was established in the eighteenth century with Isaac Newton's claim that eliminating chromatic aberration with a singlet lens was impossible. Primarily because of this view, Newton concentrated on creating telescopes with mirrors instead of lenses.

In 1729, Chester Moore Hall developed the first achromatic doublet. Later, in 1758, John Dollond enrolled the first patent of a doublet lens. The same standard is still present today, the doublets and triplets are commonly the essential components to diminish the chromatic aberration in an optical system. The conventional procedure consists of numerically optimizing the system parameters to minimize the spherochromatic aberration. These numerical approaches are beneficial for particular applications, but they do not provide mathematically rigorous solutions.

In this chapter, we determine a closed-form analytic equation to design the surface of a spherochromatic singlet lens. In the process of deriving the expression, we introduce a procedure for design singlets free of spherical and chromatic aberration for certain circumstances that during the chapter, we present in detail.

## 8.2 Mathematical model

The geometry of the singlet is presented in figure 8.1. We assume that the singlet is rotationally symmetric with its axial thickness $t$ surrounded by air and its first surface $(z_a, r_a)$ flat, therefore, $z_a = 0$. Notice that the subscript $a$ refers to the elements

doi:10.1088/978-0-7503-5774-6ch8

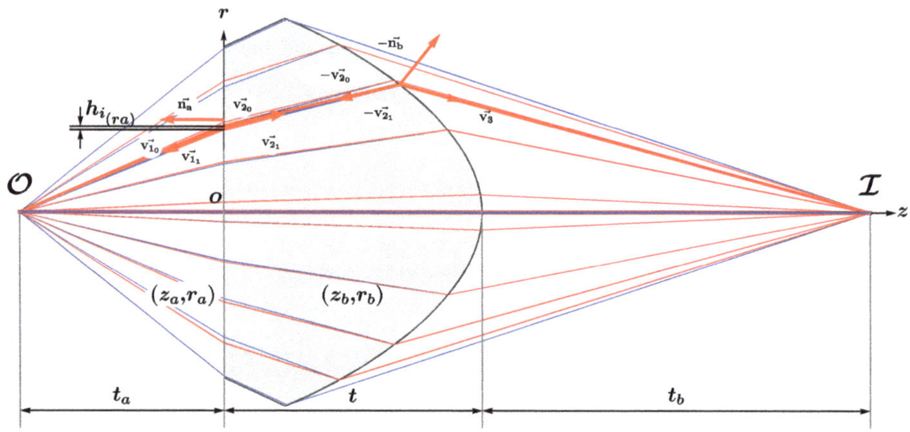


**Figure 8.1.** Meridional half-plane of the spherochromatic singlet, illustrating the deviation of the incoming rays by the surfaces and their optical paths traversed.

on the first surface. We placed the origin of the Cartesian coordinate system at the center of the first lens. The second surface is described by the unknown function $(z_b, r_b)$. Observe that the subscript $b$ refers to the coordinates at the second surface.

A polychromatic source point object in a steady-state is placed on-axis at $z = t_a$ from the input surface. The point image is placed behind the lens at an axial distance $t_b$ from the output surface, as shown in figure 8.1. The material dispersion of the glass is described by the variation of the refractive index versus the wavelength $n(\lambda)$. Although our analysis is general for many wavelengths, we will show it for three wavelengths $n(\lambda_i) = n_i$, for $i \in \{1, 2, 3\}$.

The intention is to discover the function $(z_b, r_b)$ of the output surface to eliminate the spherical and chromatic aberration at the point image. This indicates that rays rising from the source point with different angles and wavelengths converge precisely on the point image after the lens has refracted them. Note that the axial rays are not affected by the lens independently of their wavelengths. Therefore, axial rays are free of spherical and chromatic aberration. We will choose the axial ray as the reference ray.

We will work with three different wavelengths $(\lambda_1, \lambda_2, \lambda_3)$, consider three rays with different incidences on the first surface. Each incidence corresponds to a wavelength.

For clarity, in figure 8.1, we show only two rays. The first ray $(\lambda_1)$ strikes the lens at the point $(0, r_a)$ and it is going to be considered as the primary ray (red). A second ray $(\lambda_2)$ or $(\lambda_3)$ is selected such that it strikes the input surface at a radial separation $h_2(r_a)$ from the first ray (blue). It emerges from the lens just at the same point as the red ray does and with the same direction after the refraction suffered in the second surface. In this way, both rays arrive at the image point with the same propagation angle. The distance $h_2(r_a)$ is *a priori* unknown and should be determined in the process. In general, for a ray with wavelength $\lambda_i$, the separation between its input point on the first surface concerning the primary ray $(\lambda_1)$ will be denoted as $h_i(r_a)$, thus,

$$\boxed{h_1 \equiv 0.} \tag{8.1}$$

This study hypothesizes that there must be an analytic function $(z_b, r_b)$ for the output surface that guarantees that two rays with different angles and wavelengths complete the requirement of reaching the image point with the same direction. If this is true for two rays with arbitrary angles and wavelengths, then by extension it is going to be right for three, four, and more rays with arbitrary angles and wavelengths.

To begin the analysis, we will first study the refraction on the first surface of the singlet. Using Snell's law in two dimensions in geometric algebra we have,

$$\boxed{\begin{array}{c}\vec{\mathbf{v}}_{2,i} = \dfrac{1}{n_i}[\vec{\mathbf{v}}_{1,i} - (\vec{\mathbf{n}}_a \cdot \vec{\mathbf{v}}_{1,i})\vec{\mathbf{n}}_a] - \vec{\mathbf{n}}_a\sqrt{1 + \dfrac{1}{n^2}(\vec{\mathbf{n}}_a \wedge \vec{\mathbf{v}}_{1,i})^2} \\ \text{for , } \vec{\mathbf{v}}_{2,i}, \vec{\mathbf{v}}_{1,i}, \vec{\mathbf{n}}_a \in \mathbb{R}^2, \ i \in \{1, 2, 3\},\end{array}} \tag{8.2}$$

where $\vec{\mathbf{v}}_{1,i}$ is the unit vector of the incident ray, $\vec{\mathbf{v}}_{2,i}$ is unit vector of the refracted ray and finally $\vec{\mathbf{n}}_a$ is the normal vector of the first surface, see figure 8.1.

The unit vectors at the first surface are defined by,

$$\vec{\mathbf{v}}_{1,i} = \frac{(r_a + h_i)\vec{\mathbf{e}}_1 - t_a\vec{\mathbf{e}}_2}{\sqrt{(r_a + h_i)^2 + t_a^2}}, \qquad \vec{\mathbf{v}}_{2,i} = \frac{(r_b - r_a - h_i)\vec{\mathbf{e}}_1 + z_b\vec{\mathbf{e}}_2}{\sqrt{(r_b - r_a - h_i)^2 + z_b^2}}, \qquad \vec{\mathbf{n}}_a = -\vec{\mathbf{e}}_2, \tag{8.3}$$

where $\vec{\mathbf{e}}_1$ is for the $r$ direction and $\vec{\mathbf{e}}_2$ is for the $z$ direction. We need to replace the unit vectors in equation (8.2). Let's start with, $(\vec{\mathbf{n}}_a \wedge \vec{\mathbf{v}}_{1,i})^2$,

$$(\vec{\mathbf{n}}_a \wedge \vec{\mathbf{v}}_{1,i}) = \frac{r_a + h_i}{\sqrt{(r_a + h_i)^2 + t_a^2}}(\vec{\mathbf{e}}_1 \wedge \vec{\mathbf{e}}_2). \tag{8.4}$$

The next step is to square it, please observe that $(\vec{\mathbf{e}}_1 \wedge \vec{\mathbf{e}}_2)^2 = -1$,

$$(\vec{\mathbf{n}}_a \wedge \vec{\mathbf{v}}_{1,i})^2 = -\frac{(r_a + h_i)^2}{(r_a + h_i)^2 + t_a^2}. \tag{8.5}$$

Then, the terms $-\vec{\mathbf{n}}_a\sqrt{1 + \frac{1}{n^2}(\vec{\mathbf{n}}_a \wedge \vec{\mathbf{v}}_{1,i})^2}$ can be expressed as,

$$-\vec{\mathbf{n}}_a\sqrt{1 + \frac{1}{n^2}(\vec{\mathbf{n}}_a \wedge \vec{\mathbf{v}}_{1,i})^2} = \vec{\mathbf{e}}_2\sqrt{1 - \frac{(r_a + h_i)^2}{n_i^2[(r_a + h_i)^2 + t_a^2]}}. \tag{8.6}$$

If we replace the unit vectors in, $\frac{1}{n_i}[\vec{\mathbf{v}}_1 - (\vec{\mathbf{n}}_a \cdot \vec{\mathbf{v}}_1)\vec{\mathbf{n}}_a]$, we have,

$$\frac{1}{n_i}[\vec{\mathbf{v}}_1 - (\vec{\mathbf{n}}_a \cdot \vec{\mathbf{v}}_1)\vec{\mathbf{n}}_a] = \frac{(r_a + h_i)\vec{\mathbf{e}}_1 - t_a\vec{\mathbf{e}}_2}{n_i\sqrt{(r_a + h_i)^2 + t_a^2}} + \frac{t_a\vec{\mathbf{e}}_2}{n_i\sqrt{(r_a + h_i)^2 + t_a^2}} \tag{8.7}$$

$$= \frac{(r_a + h_i)\vec{\mathbf{e}}_1}{n_i\sqrt{(r_a + h_i)^2 + t_a^2}}. \tag{8.8}$$

Thus, $\frac{1}{n_i}[\vec{\mathbf{v}}_{1,i} - (\vec{\mathbf{n}}_a \cdot \vec{\mathbf{v}}_{1,i})\vec{\mathbf{n}}_a] - \vec{\mathbf{n}}_a\sqrt{1 + \frac{1}{n^2}(\vec{\mathbf{n}}_a \wedge \vec{\mathbf{v}}_{1,i})^2}$ is given by,

$$\vec{\mathbf{v}}_2 = \frac{(r_a + h_i)\vec{\mathbf{e}}_1}{n_i\sqrt{(r_a + h_i)^2 + t_a^2}} + \vec{\mathbf{e}}_2\sqrt{1 - \frac{(r_a + h_i)^2}{n_i^2[(r_a + h_i)^2 + t_a^2]}}. \tag{8.9}$$

Let's separate the coordinates, $\vec{\mathbf{e}}_2$ and $\vec{\mathbf{e}}_2$,

$$\frac{(r_b - r_a - h_i)}{\sqrt{(r_b - r_a - h_i)^2 + z_b^2}} = \frac{(r_a + h_i)}{n_i\sqrt{(r_a + h_i)^2 + t_a^2}}, \tag{8.10}$$

and,

$$\frac{z_b}{\sqrt{(r_b - r_a - h_i)^2 + z_b^2}} = \sqrt{1 - \frac{(r_a + h_i)^2}{n_i^2[(r_a + h_i)^2 + t_a^2]}}. \tag{8.11}$$

In the left side of equations (8.10) and (8.11) are the unknowns $z_b$, $r_b$ and in the right side are the cosine director $\wp_z$ and $\wp_r$, $\wp_r^2 + \wp_z^2 = 1$. Thus,

$$\wp_{r,i} \equiv \frac{(r_a + h_i)}{n_i\sqrt{(r_a + h_i)^2 + t_a^2}}, \tag{8.12}$$

and,

$$\wp_{z,i} \equiv \sqrt{1 - \frac{(r_a + h_i)^2}{n_i^2[(r_a + h_i)^2 + t_a^2]}}. \tag{8.13}$$

Now, let's work with the first wavelength $i = 1$, the cosine directors of the refracted ray inside the lens are,

$$\frac{(r_b - r_a)}{\sqrt{(r_b - r_a)^2 + z_b^2}} = \frac{r_a}{n_1\sqrt{r_a^2 + t_a^2}} = \wp_{r,1}, \tag{8.14}$$

and,

$$\frac{z_b}{\sqrt{(r_b - r_a)^2 + z_b^2}} = \sqrt{1 - \frac{r_a^2}{n_1^2[r_a^2 + t_a^2]}} = \wp_{z,1}, \tag{8.15}$$

dividing the last two expressions, we get

$$\frac{r_b - r_a}{z_b} = \frac{\wp_{r,1}}{\wp_{z,1}}. \tag{8.16}$$

We can solve for $r_b$,

$$r_b = r_a + \frac{\wp_{r,1} z_b}{\wp_{z,1}}. \tag{8.17}$$

Now, let's focus on the other wavelengths, if we divide equation (8.10) and (8.11) we get,

$$\frac{r_b - r_a - h_i}{z_b} = \frac{r_a + h_i}{\left(n_i\sqrt{(r_a + h_i)^2 + t_a^2}\right)\sqrt{1 - \frac{(r_a + h_i)^2}{n_i^2[(r_a + h_i)^2 + t_a^2]}}}. \tag{8.18}$$

The last equation is nonlinear and we cannot solve for $h_i$. Fortunately, we can solve for $h_i$ under the following approximation,

$$\boxed{\frac{h_i}{z_b} \approx 0.} \tag{8.19}$$

This approximation is physically valid by virtue of the small variation of the refraction index with respect to the wavelength. With this approximation the numerator in the left-hand side of equation (8.18) reduces to $(r_b - r_a)$,

$$\frac{r_b - r_a}{z_b} = \frac{r_a + h_i}{\left(n_i\sqrt{(r_a + h_i)^2 + t_a^2}\right)\sqrt{1 - \frac{(r_a + h_i)^2}{n_i^2[(r_a + h_i)^2 + t_a^2]}}}. \tag{8.20}$$

The resulting equation can be solved analytically by replacing $r_b = r_a + \frac{\wp_{r,1} z_b}{\wp_{z,1}}$ on the left side,

$$\frac{\wp_{r,1}}{\wp_{z,1}} = \frac{r_a + h_i}{\left(n_i\sqrt{(r_a + h_i)^2 + t_a^2}\right)\sqrt{1 - \frac{(r_a + h_i)^2}{n_i^2[(r_a + h_i)^2 + t_a^2]}}}, \tag{8.21}$$

So, solving for $h_i$,

$$h_i = \frac{n_i^2 \wp_{r,1}^2 t_a^2}{\sqrt{n_i^2 \wp_{r,1}^2 t_a^2\left[\wp_{z,1}^2 - (n_i^2 - 1)\wp_{r,1}^2\right]}} - r_a. \tag{8.22}$$

Therefore, we know how the heights $h_i$ and $r_b$ are in terms of $z_b$. Then, let's use Fermat's principle to solve for $z_b$.

Consider that the point object emits two rays with the same wavelength $\lambda_i$, the first one travels along with the optical axis $z$, and the second one strikes the input surface of the singlet at $(0, r_a + h_i)$. The first ray, the axial ray, is considered our reference ray. The second ray is the non-reference ray. Both rays pass through the singlet and meet again at the image point placed at $z = t + t_b$. The Fermat principle predicts that the optical path length (OPL) of the reference ray and the OPL of the non-reference ray are equal. Thus,

$$\boxed{\begin{aligned} -t_a + n_i t + t_b = \sqrt{(r_a + h_i)^2 + t_a^2} + n_i\sqrt{z_b^2 + (r_b - r_a - h_i)^2} \\ + \sqrt{r_b^2 + (z_b - t - t_b)^2}, \end{aligned}} \tag{8.23}$$

for each wavelength $i \in \{1, 2, 3\}$, and where $h_1 = 0$. Therefore, for the first wavelength,

$$-t_a + t_b + n_1 t = n_1\sqrt{(r_b - r_a)^2 + z_b^2} + \sqrt{r_a^2 + t_a^2} + \sqrt{r_b^2 + (-t_b + z_b - t)^2}. \quad (8.24)$$

For the second wavelength, we have,

$$\begin{aligned} -t_a + t_b + n_2 t = n_2\sqrt{(-r_a - h_2 + r_b)^2 + z_b^2} + \sqrt{(r_a + h_2)^2 + t_a^2} \\ + \sqrt{r_b^2 + (-t_b + z_b - t)^2}. \end{aligned} \quad (8.25)$$

Finally for the third wavelength,

$$\begin{aligned} -t_a + t_b + n_3 t = n_3\sqrt{(-r_a + r_b - h_3)^2 + z_b^2} + \sqrt{(r_a + h_3)^2 + t_a^2} \\ + \sqrt{r_b^2 + (-t_b + z_b - t)^2}. \end{aligned} \quad (8.26)$$

Then, let's sum over the optical paths,

$$\begin{aligned} -3t_a + 3t_b + (n_1 + n_2 + n_3)t &= n_1\sqrt{(r_b - r_a)^2 + z_b^2} + \sqrt{r_a^2 + t_a^2} \\ &+ 3\sqrt{r_b^2 + (z_b - t_b - t)^2} \\ &+ n_2\sqrt{(r_b - r_a - h_2)^2 + z_b^2} + \sqrt{(r_a + h_2)^2 + t_a^2} \\ &+ n_3\sqrt{(r_b - r_a - h_3)^2 + z_b^2} + \sqrt{(r_a + h_3)^2 + t_a^2}. \end{aligned} \quad (8.27)$$

The last expression has several terms that do not depend on the unknowns, let's put all of them in a single parameter given by,

$$\begin{aligned} f \equiv 3t_b + (n_1 + n_2 + n_3)t - \sqrt{(r_a + h_2)^2 + t_a^2} \\ - \sqrt{(r_a + h_3)^2 + t_a^2} - \sqrt{r_a^2 + t_a^2} - 3t_a. \end{aligned} \quad (8.28)$$

Replacing the parameter $f$ on equation (8.27),

$$\begin{aligned} f = 3\sqrt{r_b^2 + (z_b - t_b - t)^2} + n_1\sqrt{(r_b - r_a)^2 + z_b^2} \\ + n_2\sqrt{(r_b - r_a - h_2)^2 + z_b^2} + n_3\sqrt{(r_b - r_a - h_3)^2 + z_b^2}. \end{aligned} \quad (8.29)$$

Now, from Snell's law at the first surface we found the following relations,

$$r_b = r_a + \frac{\wp_{r,1} z_b}{\wp_{z,1}}, \quad \text{and} \quad \frac{z_b}{\wp_{z,i}} = \sqrt{(r_b - r_a - h_i)^2 + z_b^2}. \quad (8.30)$$

We can replace them in equation (8.29) to simplify it,

$$f = 3\sqrt{\left(r_a + \frac{\wp_{r,1}z_b}{\wp_{z,1}}\right)^2 + (t_b - z_b + t)^2} + \frac{n_1 z_b}{\wp_{z,1}} + \frac{n_2 z_b}{\wp_{z,2}} + \frac{n_3 z_b}{\wp_{z,3}}. \tag{8.31}$$

The last terms of the previous equation can be grouped into a sum, with running index $i \in \{1, 2, 3\}$,

$$f = 3\sqrt{\left(r_a + \frac{\wp_{r,1}z_b}{\wp_{z,1}}\right)^2 + (t_b - z_b + t)^2} + \sum_{i=1}^{3}\frac{n_i z_b}{\wp_{z,i}}. \tag{8.32}$$

We can extract $z_b$ from the sum since it does not have the running index $i \in \{1, 2, 3\}$,

$$f = 3\sqrt{\left(r_a + \frac{\wp_{r,1}z_b}{\wp_{z,1}}\right)^2 + (t_b - z_b + t)^2} + z_b\sum_{i=1}^{3}\frac{n_i}{\wp_{z,i}}. \tag{8.33}$$

Elevating to the power of two, in both sides,

$$\frac{1}{9}\left(f - z_b\sum_{i=1}^{3}\frac{n_i}{\wp_{z,i}}\right)^2 = \left(r_a + \frac{\wp_{r,1}z_b}{\wp_{z,1}}\right)^2 + (t_b + t - z_b)^2. \tag{8.34}$$

Expanding,

$$\begin{aligned}\frac{1}{9}f^2 - \frac{2}{9}\left(f\sum_{i=1}^{3}\frac{n_i}{\wp_{z,i}}\right)z_b + \frac{1}{9}\left(\sum_{i=1}^{3}\frac{n_i}{\wp_{z,i}}\right)^2 z_b^2 = r_a^2 + 2\frac{\wp_{r,1}r_a}{\wp_{z,1}}z_b + \left(\frac{\wp_{r,1}}{\wp_{z,1}}\right)^2 z_b^2 \\ + (t_b + t)^2 - 2(t_b + t)z_b + z_b^2.\end{aligned} \tag{8.35}$$

Collecting for $z_b$, and $z_b^2$,

$$\left[1 - \frac{1}{9}\left(\sum_{i=1}^{3}\frac{n_i}{\wp_{z,i}}\right)^2 + \left(\frac{\wp_{r,1}}{\wp_{z,1}}\right)^2\right]z_b^2 + 2\left[\frac{\wp_{r,1}r_a}{\wp_{z,1}} - (t_b + t) + \frac{1}{9}\left(f\sum_{i=1}^{3}\frac{n_i}{\wp_{z,i}}\right)\right]z_b \tag{8.36}$$

$$+\left[r_a^2 - \frac{1}{9}f^2 + (t_b + t)^2\right] = 0. \tag{8.37}$$

Multiplying both sides by $4\left[1 - \frac{1}{9}\left(\sum_{i=1}^{3}\frac{n_i}{\wp_{z,i}}\right)^2 + \left(\frac{\wp_{r,1}}{\wp_{z,1}}\right)^2\right]$,

$$\begin{aligned}
&4\left[1-\frac{1}{9}\left(\sum_{i=1}^{3}\frac{n_i}{\wp_{z,i}}\right)^2+\left(\frac{\wp_{r,1}}{\wp_{z,1}}\right)^2\right]^2 z_b^2\\
&+4\left[1-\frac{1}{9}\left(\sum_{i=1}^{3}\frac{n_i}{\wp_{z,i}}\right)^2+\left(\frac{\wp_{r,1}}{\wp_{z,1}}\right)^2\right]\left[r_a^2-\frac{1}{9}f^2+(t_b+t)^2\right]\\
&+8\left[1-\frac{1}{9}\left(\sum_{i=1}^{3}\frac{n_i}{\wp_{z,i}}\right)^2+\left(\frac{\wp_{r,1}}{\wp_{z,1}}\right)^2\right]\\
&\left[\frac{\wp_{r,1}r_a}{\wp_{z,1}}-(t_b+t)+\frac{1}{9}\left(f\sum_{i=1}^{3}\frac{n_i}{\wp_{z,i}}\right)\right]z_b=0.
\end{aligned} \tag{8.38}$$

Adding up $4\left[\frac{\wp_{r,1}r_a}{\wp_{z,1}}-(t_b+t)+\frac{1}{9}\left(f\sum_{i=1}^{3}\frac{n_i}{\wp_{z,i}}\right)\right]^2$ in both sides,

$$\begin{aligned}
&4\left[1-\frac{1}{9}\left(\sum_{i=1}^{3}\frac{n_i}{\wp_{z,i}}\right)^2+\left(\frac{\wp_{r,1}}{\wp_{z,1}}\right)^2\right]^2 z_b^2+4\left[\frac{\wp_{r,1}r_a}{\wp_{z,1}}-(t_b+t)+\frac{1}{9}\left(f\sum_{i=1}^{3}\frac{n_i}{\wp_{z,i}}\right)\right]^2\\
&+4\left[1-\frac{1}{9}\left(\sum_{i=1}^{3}\frac{n_i}{\wp_{z,i}}\right)^2+\left(\frac{\wp_{r,1}}{\wp_{z,1}}\right)^2\right]\left[r_a^2-\frac{1}{9}f^2+(t_b+t)^2\right]\\
&+8\left[1-\frac{1}{9}\left(\sum_{i=1}^{3}\frac{n_i}{\wp_{z,i}}\right)^2+\left(\frac{\wp_{r,1}}{\wp_{z,1}}\right)^2\right]\left[\frac{\wp_{r,1}r_a}{\wp_{z,1}}-(t_b+t)+\frac{1}{9}\left(f\sum_{i=1}^{3}\frac{n_i}{\wp_{z,i}}\right)\right]z_b\\
&=4\left[\frac{\wp_{r,1}r_a}{\wp_{z,1}}-(t_b+t)+\frac{1}{9}\left(f\sum_{i=1}^{3}\frac{n_i}{\wp_{z,i}}\right)\right]^2.
\end{aligned} \tag{8.39}$$

Completing the square binomial,

$$\begin{aligned}
&\left\{2\left[1-\frac{1}{9}\left(\sum_{i=1}^{3}\frac{n_i}{\wp_{z,i}}\right)^2+\left(\frac{\wp_{r,1}}{\wp_{z,1}}\right)^2\right]z_b+2\left[\frac{\wp_{r,1}r_a}{\wp_{z,1}}-(t_b+t)+\frac{1}{9}\left(f\sum_{i=1}^{3}\frac{n_i}{\wp_{z,i}}\right)\right]\right\}^2\\
&=4\left[\frac{\wp_{r,1}r_a}{\wp_{z,1}}-(t_b+t)+\frac{1}{9}\left(f\sum_{i=1}^{3}\frac{n_i}{\wp_{z,i}}\right)\right]^2\\
&-4\left[1-\frac{1}{9}\left(\sum_{i=1}^{3}\frac{n_i}{\wp_{z,i}}\right)^2+\left(\frac{\wp_{r,1}}{\wp_{z,1}}\right)^2\right]\left[r_a^2-\frac{1}{9}f^2+(t_b+t)^2\right].
\end{aligned} \tag{8.40}$$

Applying the square root in both sides, and solving for $z_b$,

$$\boxed{\begin{aligned} z_b = &-\frac{\dfrac{\wp_{l,l}r_a}{\wp_{z,1}} - (t_b + t) + \dfrac{1}{9}\left(f\sum_{i=1}^{3}\dfrac{n_i}{\wp_{z,i}}\right)}{1 - \dfrac{1}{9}\left(\sum_{i=1}^{3}\dfrac{n_i}{\wp_{z,i}}\right)^2 + \left(\dfrac{\wp_{r,1}}{\wp_{z,1}}\right)^2} \\ &\pm \frac{\sqrt{\begin{array}{l}\left[\dfrac{\wp_{r,1}r_a}{\wp_{z,1}} - (t_b + t) + \dfrac{1}{9}\left(f\sum_{i=1}^{3}\dfrac{n_i}{\wp_{z,i}}\right)\right]^2 \\ -\left[1 - \dfrac{1}{9}\left(\sum_{i=1}^{3}\dfrac{n_i}{\wp_{z,i}}\right)^2 + \left(\dfrac{\wp_{r,1}}{\wp_{z,1}}\right)^2\right]\left[r_a^2 - \dfrac{1}{9}f^2 + (t_b + t)^2\right]\end{array}}}{1 - \dfrac{1}{9}\left(\sum_{i=1}^{3}\dfrac{n_i}{\wp_{z,i}}\right)^2 + \left(\dfrac{\wp_{r,1}}{\wp_{z,1}}\right)^2}. \end{aligned}} \tag{8.41}$$

and,

$$r_b = r_a + \frac{\wp_{r,1}z_b}{\wp_{z,1}}. \tag{8.42}$$

The simplified analytic solution equation (8.41) constitutes the most important result of this chapter.

Equation (8.41) expresses in an analytic closed-form the shape of the output surface of a dispersive plane-aspheric singlet free of spherical and chromatic aberration for the previously defined wavelengths ($\lambda_1$, $\lambda_2$, $\lambda_3$). The surface is described in parametric form with the functions $z_b(r_a)$ and $r_b(r_a)$, where $r_a$ plays the role of an independent radial variable.

Equation (8.41) describes the second surface of singlet under the approximation $h_i/z_b$ which is physically valid for typical values in practical applications and as long as $z_b > > h_i$. Therefore, in the borders of the lens, it may fail.

We need to mention that equation (8.41) is valid if and only if all the refraction indices are positive or negative for each wavelength. So, the relations may be applied to devices built with metamaterials. Equation (8.41) will fail if some of the refraction indexes are positive, and others are negative.

We have restricted the analysis to three wavelengths, but the method can be straightforwardly extended to more wavelengths. So, if we generalize equation (8.41) for an arbitrary running index $i \in \{1, 2, 3, 4, \dots, N\}$ where $N \in \mathbb{N}$, we get the following expression,

$$z_b = -\frac{\frac{\wp_{r,1} r_a}{\wp_{z,1}} - (t_b + t) + \frac{1}{N^2}\left(f \sum_{i=1}^{N} \frac{n_i}{\wp_{z,i}}\right)}{1 - \frac{1}{N^2}\left(\sum_{i=1}^{N} \frac{n_i}{\wp_{z,i}}\right)^2 + \left(\frac{\wp_{r,1}}{\wp_{z,1}}\right)^2} \pm \frac{\sqrt{\left[\frac{\wp_{r,1} r_a}{\wp_{z,1}} - (t_b + t) + \frac{1}{N^2}\left(f \sum_{i=1}^{N} \frac{n_i}{\wp_{z,i}}\right)\right]^2 - \left[1 - \frac{1}{N^2}\left(\sum_{i=1}^{N} \frac{n_i}{\wp_{z,i}}\right)^2 + \left(\frac{\wp_{r,1}}{\wp_{z,1}}\right)^2\right]\left[r_a^2 - \frac{1}{N^2} f^2 + (t_b + t)^2\right]}}{1 - \frac{1}{N^2}\left(\sum_{i=1}^{N} \frac{n_i}{\wp_{z,i}}\right)^2 + \left(\frac{\wp_{r,1}}{\wp_{z,1}}\right)^2}. \tag{8.43}$$

and,

$$r_b = r_a + \frac{\wp_{r,1} z_b}{\wp_{z,1}}. \tag{8.44}$$

The generalization works if and only if there is a small variation of the refraction index with respect to the wavelength.

## 8.3 Illustrative examples

The following gallery presents spherochromatic singlets. To test the model, we consider spherochromatic singlets built with BK7 glass with refractive indices $n_1 = 1.5168$, $n_2 = 1.5224$, and $n_3 = 1.5143$, corresponding to the wavelengths $\lambda_1 = 587.6$ nm, $\lambda_2 = 486.1$ nm, and $\lambda_3 = 656.3$ nm, respectively. The dashed lines, the continuous lines and the dotted lines correspond to $n_1$, $n_2$ and $n_3$, respectively.

To have a numeric way to test the accuracy of the singlets, we have computed the percentage similitude of the optical paths (PSOP). The PSOP measures how similar is the optical path of a non-axial ray with the optical path of the axial ray. Thus,

$$\text{PSOP} = \left(1 - \left|\frac{\text{Left side of equation (8.23)} - \text{Right side of equation (8.23)}}{\text{Left side of equation (8.23)}}\right|\right) \times 100\%, \tag{8.45}$$

where the term inside the absolute value is the normalized optical path difference (NOPD) between an axial ray and a non-axial ray. The PSOP of a singlet of chapter 4, designed by equation (4.36), is 100%. The figures of the gallery are 8.2, 8.4 and 8.6 and, their respective PSOP are 8.3, 8.5 and 8.7.

## 8.4 Spherochromatic collimator

The next problem we are interested in is the design of a spherochromatic collimator. A spherochromatic collimator is a lens such that it collimates the light for several wavelengths. We can design a spherochromatic collimator for several wavelengths, but here we will design it for three wavelengths. We need to implement the limit when $t_b \to \infty$ in equations (8.24) and (8.25) and (8.26).

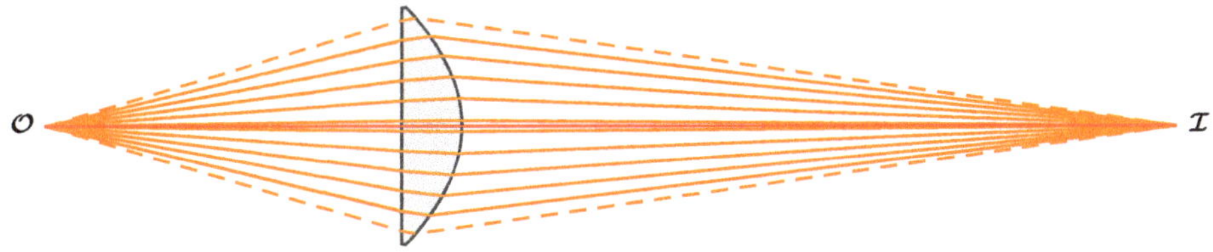


**Figure 8.2.** Design specifications: $n_1 = 1.5168$, $n_2 = 1.5224$, $n_3 = 1.5143$, $t_a = -30$ mm, $t = 5$ mm, $t_b = 60$, $z_a = 0$ and $z_b$= equation (8.41).

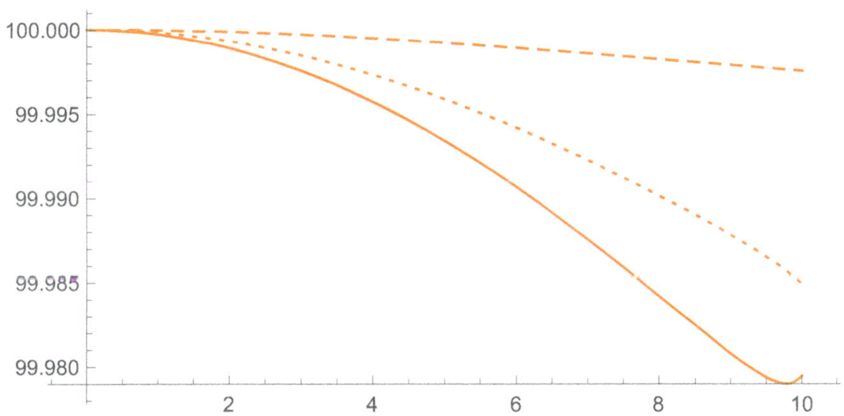


**Figure 8.3.** Percentage similitude of the optical paths (PSOP) of figure 8.2. The dashed lines, the continuous lines and the dotted lines correspond to $n_1$, $n_2$ and $n_3$, respectively.

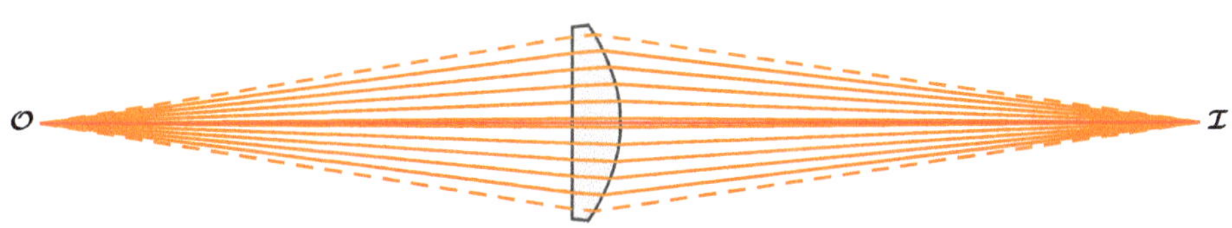


**Figure 8.4.** Design specifications: $n_1 = 1.5168$, $n_2 = 1.5224$, $n_3 = 1.5143$, $t_a = -55$ mm, $t = 5$ mm, $t_b = 60$, $z_a = 0$ and $z_b$= equation (8.41).

Therefore, we recall equation (8.24),

$$-t_a + t_b + n_1 t = n_1\sqrt{(r_b - r_a)^2 + z_b^2} + \sqrt{r_a^2 + t_a^2} + \sqrt{r_b^2 + (-t_b + z_b - t)^2}, \quad (8.24)$$

and we apply the limit when $t_b \to \infty$ over it,

$$-\sqrt{r_a^2 + t_a^2} - t_a + n_1 t - t = n_1\sqrt{\left\{r_a - \left[\lim_{t_b \to \infty}(r_b)\right]\right\}^2 + \left[\lim_{t_b \to \infty}(z_b)\right]^2} - \left[\lim_{t_b \to \infty}(z_b)\right]. \quad (8.46)$$

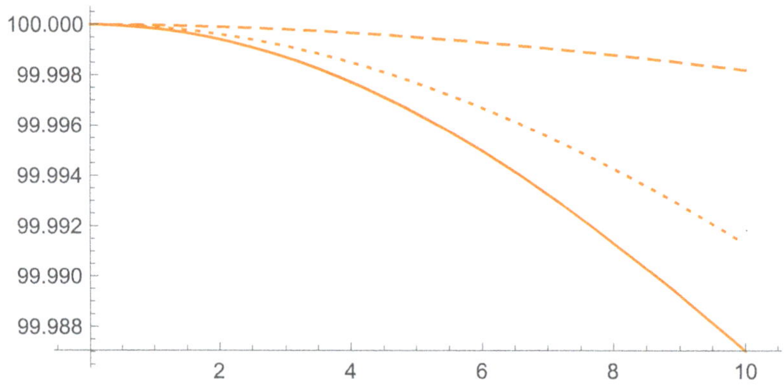


**Figure 8.5.** PSOP figure 8.4. The dashed lines, the continuous lines and the dotted lines correspond to $\lambda_1$, $\lambda_2$ and $\lambda_3$, respectively.

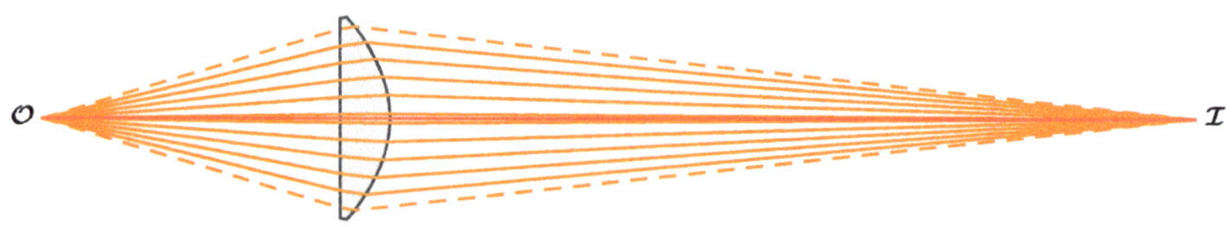


**Figure 8.6.** Design specifications: $n_1 = 1.5168$, $n_2 = 1.5224$, $n_3 = 1.5143$, $t_a = -30$ mm, $t = 5$ mm, $t_b = 80$, $z_a = 0$ and $z_b$= equation (8.41).

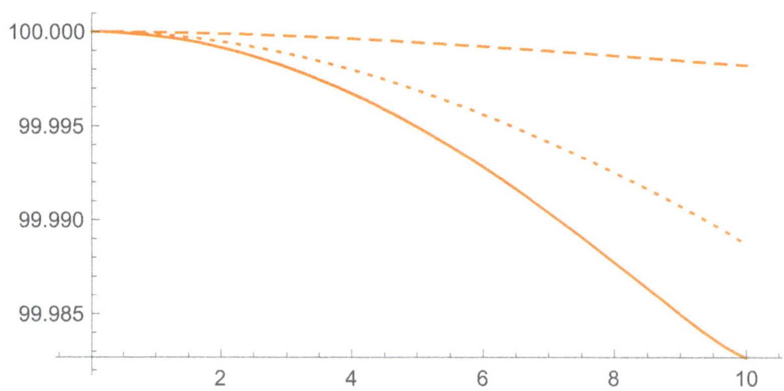


**Figure 8.7.** PSOP figure 8.6.

For the second wavelength, we have,

$$\begin{aligned}-t_a + t_b + n_2 t = n_2\sqrt{(-r_a - h_2 + r_b)^2 + z_b^2} \\ + \sqrt{(r_a + h_2)^2 + t_a^2} + \sqrt{r_b^2 + (-t_b + z_b - t)^2}.\end{aligned} \tag{8.25}$$

Let's apply the limit when $t_b \to \infty$ in equation (8.25),

$$\begin{aligned}&-\sqrt{(r_a + h_2)^2 + t_a^2} - t_a + n_2 t - t \\ &= n_2\sqrt{\left\{r_a - \left[\lim_{t_b \to \infty}(r_b)\right] + h_2\right\}^2 + \left[\lim_{t_b \to \infty}(z_b)\right]^2} \\ &-\left[\lim_{t_b \to \infty}(z_b)\right].\end{aligned} \tag{8.47}$$

Finally for the third wavelength, we have,

$$\begin{aligned}-t_a + t_b + n_3 t = n_3\sqrt{(-r_a + r_b - h_3)^2 + z_b^2} \\ + \sqrt{(r_a + h_3)^2 + t_a^2} + \sqrt{r_b^2 + (-t_b + z_b - t)^2}.\end{aligned} \tag{8.26}$$

If we apply the limit when $t_b \to \infty$ in the last expression we get,

$$\begin{aligned}&-\sqrt{(r_a + h_3)^2 + t_a^2} - t_a + n_3 t - t \\ &= n_3\sqrt{\left\{r_a - \left[\lim_{t_b \to \infty}(r_b)\right] + h_3\right\}^2 + \left[\lim_{t_b \to \infty}(z_b)\right]^2} \\ &-\left[\lim_{t_b \to \infty}(z_b)\right].\end{aligned} \tag{8.48}$$

Therefore, when we evaluated the limit when $t_b \to \infty$ on the Fermat principle, we get,

$$\boxed{\begin{aligned}&-\sqrt{(r_a + h_i)^2 + t_a^2} - t_a + n_i t - t \\ &= n_i\sqrt{\left\{r_a - \left[\lim_{t_b \to \infty}(r_b)\right] + h_i\right\}^2 + \left[\lim_{t_b \to \infty}(z_b)\right]^2} \\ &-\left[\lim_{t_b \to \infty}(z_b)\right],\end{aligned}} \tag{8.49}$$

for $i \in \{1, 2, 3\}$.

The next step is to sum equations (8.46), (8.47) and (8.48),

$$
\begin{aligned}
&- \sqrt{r_a^2 + t_a^2} - \sqrt{(r_a + h_2)^2 + t_a^2} - \sqrt{(r_a + h_3)^2 + t_a^2} \\
&\qquad - 3t_a + (n_1 + n_2 + n_3)t - 3t \\
&= n_1\sqrt{\left\{r_a - \left[\lim_{t_b \to \infty}(r_b)\right]\right\}^2 + \left[\lim_{t_b \to \infty}(z_b)\right]^2} - \left[\lim_{t_b \to \infty}(z_b)\right] \\
&+ n_2\sqrt{\left\{r_a - \left[\lim_{t_b \to \infty}(r_b)\right] + h_2\right\}^2 + \left[\lim_{t_b \to \infty}(z_b)\right]^2} - \left[\lim_{t_b \to \infty}(z_b)\right] \\
&+ n_3\sqrt{\left\{r_a - \left[\lim_{t_b \to \infty}(r_b)\right] + h_3\right\}^2 + \left[\lim_{t_b \to \infty}(z_b)\right]^2} \\
&- \left[\lim_{t_b \to \infty}(z_b)\right].
\end{aligned}
\tag{8.50}
$$

To clean up this mess, let's define the following parameter $f$,

$$
\begin{aligned}
f \equiv &- \sqrt{(r_a + h_2)^2 + t_a^2} - \sqrt{(r_a + h_3)^2 + t_a^2} \\
&- \sqrt{r_a^2 + t_a^2} - 3t_a + (n_1 + n_2 + n_3)t - 3t.
\end{aligned}
\tag{8.51}
$$

Replacing $f$ in equation (8.50),

$$
\begin{aligned}
f = &\, n_3\sqrt{\left\{r_a - \left[\lim_{t_b \to \infty}(r_b)\right] + h_3\right\}^2 + \left[\lim_{t_b \to \infty}(z_b)\right]^2} \\
&+ n_1\sqrt{\left\{r_a - \left[\lim_{t_b \to \infty}(r_b)\right]\right\}^2 + \left[\lim_{t_b \to \infty}(z_b)\right]^2} \\
&+ n_2\sqrt{\left\{r_a - \left[\lim_{t_b \to \infty}(r_b)\right] + h_2\right\}^2 + \left[\lim_{t_b \to \infty}(z_b)\right]^2} \\
&- 3\left[\lim_{t_b \to \infty}(z_b)\right].
\end{aligned}
\tag{8.52}
$$

Now from Snell's law at the first surface we learned the following relation,

$$
\frac{\left[\lim_{t_b \to \infty}(z_b)\right]}{\wp_{z,i}} = \sqrt{\left\{r_a - \left[\lim_{t_b \to \infty}(r_b)\right] + h_i\right\}^2 + \left[\lim_{t_b \to \infty}(z_b)\right]^2}.
\tag{8.53}
$$

Replacing this last relation in equation (8.52),

$$
\begin{aligned}
f = &\frac{n_1}{\wp_{z,1}}\left[\lim_{t_b \to \infty}(z_b)\right] + \frac{n_2}{\wp_{z,2}}\left[\lim_{t_b \to \infty}(z_b)\right] + \frac{n_3}{\wp_{z,3}}\left[\lim_{t_b \to \infty}(z_b)\right] \\
&- 3\left[\lim_{t_b \to \infty}(z_b)\right].
\end{aligned}
\tag{8.54}
$$

Finally we can solve for $\left[\lim_{t_b\to\infty}(z_b)\right]$,

$$\boxed{\left[\lim_{t_b\to\infty}(z_b)\right] = -\frac{f\wp_{z,1}\wp_{z,2}\wp_{z,3}}{-n_3\wp_{z,1}\wp_{z,2} - n_1\wp_{z,3}\wp_{z,2} - n_2\wp_{z,1}\wp_{z,3} + 3\wp_{z,1}\wp_{z,3}\wp_{z,2}}}, \tag{8.55}$$

and,

$$\left[\lim_{t_b\to\infty}(r_b)\right] = r_a + \frac{\wp_{r,1}\left[\lim_{t_b\to\infty}(z_b)\right]}{\wp_{z,1}}. \tag{8.56}$$

The last equation expresses the shape of the second surface of the spherochromatic collimator.

## 8.5 Gallery of spherochromatic collimators

In this section, we present a gallery of spherochromatic collimators. All the singlets are designed computing equation (8.55). In the figures are presented the input values and the ray-tracing. Again we take BK7 glass as the building material of the singlets. Thus, $n_1 = 1.5168$, $n_2 = 1.5224$, and $n_3 = 1.5143$, and the dashed lines, the continuous lines and the dotted lines correspond to $n_1$, $n_2$ and $n_3$, respectively.

To compute PSOP of the collimator we use equation (8.49). Thus,

$$\text{PSOP} = \left(1 - \left|\frac{\text{Left side of equation(8.49)} - \text{Right side of equation (8.49)}}{\text{Left side of equation (8.49)}}\right|\right) \times 100\%. \tag{8.57}$$

The PSOP of a collimator of chapter 4, using equation (4.49), is 100%. The figures of the gallery are 8.8 and 8.10, and their respective PSOPs are 8.9 and 8.11.

## 8.6 Discussion and conclusions

In conclusion, we have proposed an analytical method to develop the output surface of a lens-free of spherical and chromatic aberration. We obtained the exact nonlinear system of algebraic equations and gave a very precise simplified model that can be solved in closed form. The formulations can be applied for an arbitrary

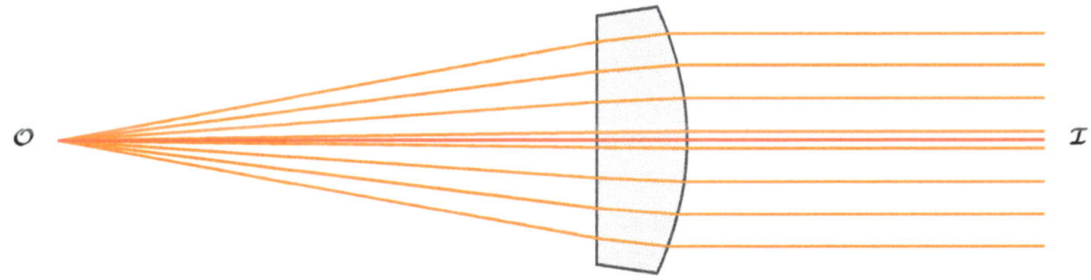


**Figure 8.8.** Design specifications: $n_1 = 1.5168$, $n_2 = 1.5224$, $n_3 = 1.5143$, $t_a = -30$ mm, $t = 5$ mm, $z_a = 0$ and $z_b$= equation (8.41).

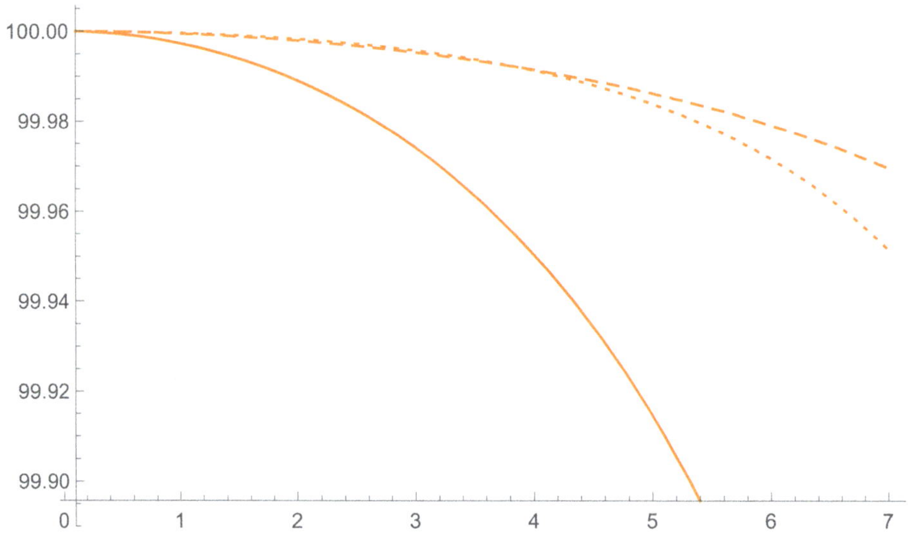


**Figure 8.9.** PSOP of figure 8.8.

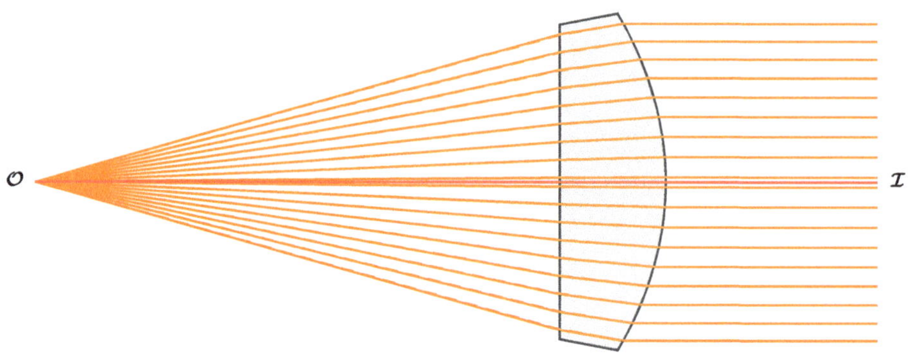


**Figure 8.10.** Design specifications: $n_1 = 1.5168$, $n_2 = 1.5224$, $n_3 = 1.5143$, $t_a = -50$ mm, $t = 10$ mm, $z_a = 0$ and $z_b$= equation (8.41).

number of wavelengths. In the process, we have refuted the old paradigm introduced by Newton, which says that the spherochromatic aberration cannot be diminished using a single lens.

It is necessary to observe that equation (8.41) fails when the refractive indices are distant, for example, $n_1 = 7$, $n_2 = 5$ and $n_3 = 2$. Fortunately, this situation is almost improbable to be found in practical purposes. Most of the optical materials have a small change of refraction index concerning the wavelength.

Ultimately, we want to mention that the enigma introduced and solved in this chapter is essentially the generalization of the problem discussed by Christian Huygens in the preface and sixth chapter of his magnum opus *Traité de la lumiére*. The original problem formulated in Huygens's view is how the form of glass (lens) must be such that it can collect the rays of one wavelength when one of the surfaces

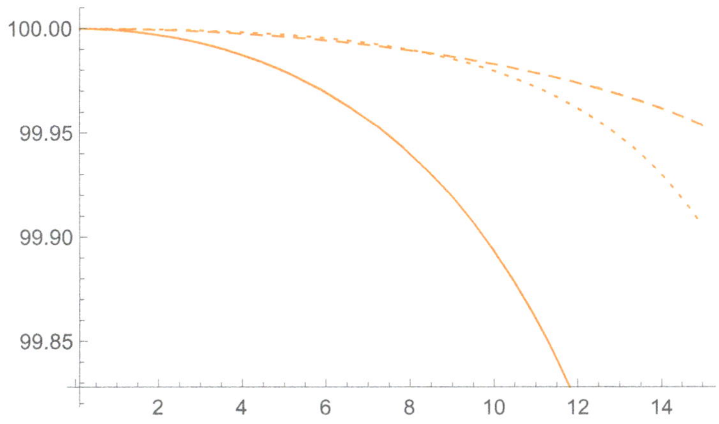


**Figure 8.11.** PSOP of figure 8.10.

is given. In chapter 4, we solved the original problem for one wavelength. Here in this chapter, we present a solution for the case when the first surface is flat and for three different wavelengths. If we take $n_1 = n_2 = n_3$, equation (8.41) converges to the solution proposed in chapter 4 when the first surface is flat, and the spot diagrams tend to be a single point.

## Further reading

Acuña R G G and Gutiérrez-Vega J C 2019 General formula of the refractive telescope design free spherical aberration *Novel Optical Systems, Methods, and Applications XXII* vol 11105 (Bellingham, WA: International Society for Optics and Photonics)

Born M and Wolf E 2013 *Principles of Optics: Electromagnetic Theory of Propagation Interference and Diffraction of Light* (Amsterdam: Elsevier)

Braunecker B, Hentschel R and Tiziani H J 2008 *Advanced Optics Using Aspherical Elements* vol 173 (Bellingham, WA: SPIE)

Chaves J 2016 *Introduction to Nonimaging Optics* 2nd edn (Boca Raton, FL: CRC Press)

Daumas M 1972 Scientific instruments of the seventeenth and eighteenth centuries and their makers *Scientific Instruments of the Seventeenth and Eighteenth Centuries and Their Makers* ed M Daumas (London: Batsford)

González-Acuña R G and Chaparro-Romo H A 2018 General formula for bi-aspheric singlet lens design free of spherical aberration *Appl. Opt.* **57** 9341–5

González-Acuña R G and Guitiérrez-Vega J C 2018 Generalization of the axicon shape: the gaxicon *J. Opt. Soc. Am.* A **35** 1915–8

González-Acuña R G and Gutiérrez-Vega J C 2019a Analytic formulation of a refractive-reflective telescope free of spherical aberration *Opt. Eng.* **58** 085105

González-Acuña R G and Gutiérrez-Vega J C 2019b General formula for aspheric collimator lens design free of spherical aberration *Current Developments in Lens Design and Optical Engineering XX* vol 11104 (Bellingham, WA: International Society for Optics and Photonics)

González-Acuña R G, Avendaño-Alejo M and Gutiérrez-Vega J C 2019a Singlet lens for generating aberration-free patterns on deformed surfaces *J. Opt. Soc. Am.* A **36** 925–9

González-Acuña R G, Chaparro-Romo H A and Gutiérrez-Vega J C 2019b General formula to design freeform singlet free of spherical aberration and astigmatism *Appl. Opt.* **58** 1010–5

Kingslake R and Johnson R B 2009 *Lens Design Fundamentals* (New York: Academic)

Luneburg R K and Herzberger M 1964 *Mathematical Theory of Optics* (Berkeley, CA: University of California Press)

McCray W P 2016 *Stargazer: The Life and Times of the Telescope* (Cambridge, MA: Da Capo)

Nagar J, Campbell S D and Werner D H 2018 Apochromatic singlets enabled by metasurface-augmented GRIN lenses *Optica* **5** 99–102

Newton I 1979 *Opticks, or, a Treatise of the Reflections, Refractions, Inflections & Colours of Light* (North Chelmsford, MA: Courier Corporation)

Schulz G 1983 Achromatic and sharp real imaging of a point by a single aspheric lens *Appl. Opt.* **22** 3242–8

Stavroudis O N 2006 *The Mathematics of Geometrical and Physical Optics: The k-Function and Its Ramifications* (New York: Wiley)

Sun H 2016 *Lens Design: A Practical Guide* (Boca Raton, FL: CRC Press)

Toomer G J 2012 *Diocles, On Burning Mirrors: The Arabic Translation of the Lost Greek Original* vol 1 (Berlin: Springer Science)

# Part III

## Stigmatic and astigmatic freeform singlets

**Rafael G González-Acuña, Héctor A Chaparro-Romo and Julio C Gutiérrez-Vega**

# Chapter 9

# On-axis stigmatic freeform lens

In this chapter, a precise closed-form formula for the design of on-axis stigmatic freeform lenses is presented. Given the equation of the freeform input surface, the method returns the equation of the second surface to adjust the spherical aberration. The derivation is based on the precise employment of the Fermat principle under the conventional geometrical optics approach.

## 9.1 Introduction

In the first part of the book is an introduction. For the second part, we just focused ourselves on three-dimensional radially symmetric surfaces. Thus, we worked on $\mathbb{R}^2$. It is time to move on, to the Euclidean space in $\mathbb{R}^3$. The lenses that we are going to design here do not have any particular symmetry. Therefore, we call them freeform singlets.

A freeform optical design has at least one freeform shape; a freeform shape is a surface such that it does not have symmetry. In chapter 4, we started with several quotes of renowned scientists interested in the problem of designing a stigmatic aspheric singlet. Particularly, Descartes described the problem in an interesting way; we recall the quotation of Descartes in his treatise *De la nature des lignes courbes*,

> *I might go farther and explain how,* ***if one surface of a lens is provided and is not entirely plane nor formed of conic sections or circles, the other surface can be so determined as to carry all the rays from a given point to a different point, also provided.*** *This is no more complicated than the problems I have just described: indeed. It is much easier since the way is now clear; I prefer, however, to leave this for others to struggle out, to the end that they may appreciate the more highly the discovery of those things here explained, through having themselves to face some challenges.*

doi:10.1088/978-0-7503-5774-6ch9

Descartes understood that this problem is not only on $\mathbb{R}^2$, therefore, in the following paragraph of *De la nature des lignes courbes* he extended it to $\mathbb{R}^3$,

> ***In all this analysis I have considered only curves that can be defined upon a plane surface, but my comments can easily be made to apply to all those curves which can be formulated of as generated by the regular movement of the points of a body in three-dimensional space.*** *This can be done by dropping perpendiculars from each point of the curve under thought upon two planes intersecting at right angles, for the ends of these perpendiculars will describe two other curves, one in each of the two planes, all points of which may be obtained in the way previously described, and all of which may be associated to those of a straight line common to the two planes; and by means of these, the points of the three-dimensional curve will be entirely obtained.*

Descartes visualized stigmatic freeform lenses over 350 years ago. In this chapter, we are going to design an on-axis stigmatic freeform lens.

## 9.2 Finite image-object

The problem is the following: given a freeform surface, $(x_a, y_a, z_a)$ how must the second surface $(x_b, y_b, z_b)$ of the freeform singlet be such that the lens is on-axis stigmatic. We place the freeform singlet at the origin of the Cartesian plane.

We assume that the lens is surrounded by air. Its central thickness is $t$ and its refraction index is $n$. The distance between the object and the first surface is $t_a$, and the distance between the second surface and the image is $t_b$. In the origin, the normal vector from the first and second surface must be parallel to the optical axis.

### 9.2.1 Fermat principle

To hunt down $(x_b, y_b, z_b)$, we need to compare the optical path length (OPL) on the reference ray with the OPL of any other ray. We choose the reference ray as the axial ray, the OPL of the reference ray is constant, and it is given by

$$-t_a + nt + t_b = \text{const.} \tag{9.1}$$

Then we compare the OPL reference ray with the OPL of any other ray,

$$\begin{aligned} -t_a + nt + t_b = &- \operatorname{sgn}(t_a)\sqrt{x_a^2 + y_a^2 + (z_a - t_a)^2} \\ &+ n\sqrt{(x_b - x_a)^2 + (y_b - y_a)^2 + (z_b - z_a)^2} \\ &+ \operatorname{sgn}(t_b)\sqrt{x_b^2 + y_b^2 + (z_b - t - t_b)^2}. \end{aligned} \tag{9.2}$$

Please note that we take account of whether the point object is real or virtual and also whether the point image is real or virtual. The generalization of the OPL of a non-axial ray from $\mathbb{R}^2$ to $\mathbb{R}^3$ is the following,

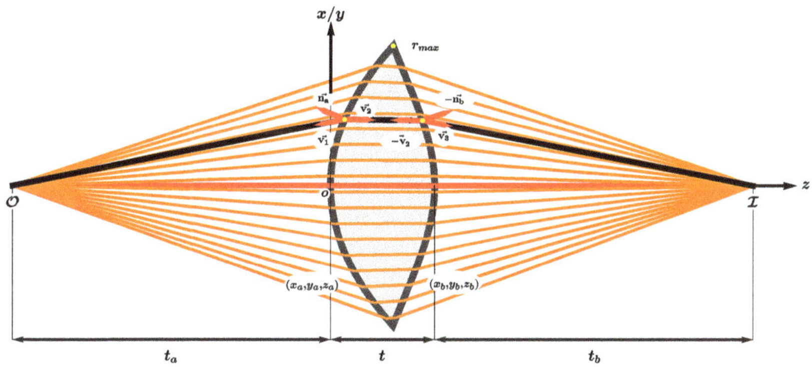


**Figure 9.1.** Diagram of an on-axis freeform stigmatic singlet. The first surface is given by $(x_a, y_a, z_a)$, and the second surface is given by $(x_b, y_b, z_b)$. The gap between the first surface and the object is $t_a$, the thickness at the center of the lens is $t$, and the gap between the second surface and the image is $t_b$. $\vec{\mathbf{v}}_1$ is unit vector of the incident ray, $\vec{\mathbf{v}}_2$ is the unit vector of the refracted ray inside the lens, $\vec{\mathbf{v}}_3$ is the output $\vec{\mathbf{n}}_a$ is the normal vector of the first surface and $\vec{\mathbf{n}}_b$ is the normal vector of the second surface.

$$r_a^2 \to x_a^2 + y_a^2, \tag{9.3}$$

$$(r_b - r_a)^2 \to (x_b - x_a)^2 + (y_b - y_a)^2. \tag{9.4}$$

See figure 9.1.

### 9.2.2 Snell's law

The second step to get $(x_b, y_b, z_b)$ is to compute Snell's law at the first surface of the freeform singlet, which is written as,

$$\boxed{\vec{\mathbf{v}}_2 = \frac{1}{n}[\vec{\mathbf{v}}_1 - (\vec{\mathbf{n}}_a \cdot \vec{\mathbf{v}}_1)\vec{\mathbf{n}}_a] - \vec{\mathbf{n}}_a\sqrt{1 - \frac{1}{n^2}(\vec{\mathbf{n}}_a \wedge \vec{\mathbf{v}}_1)^2} \quad \text{for } \vec{\mathbf{v}}_2, \vec{\mathbf{v}}_1, \vec{\mathbf{n}}_a \in \mathbb{R}^3,} \tag{9.5}$$

where $\vec{\mathbf{v}}_1$ is the unit vector of the incident ray; it travels from the object to the first surface. $\vec{\mathbf{v}}_2$ is the unit vector of the refracted ray; it moves inside the lens. Finally, $\vec{\mathbf{n}}_a$ is the normal vector of the first surface. Please see figure 9.1. The unit vectors of the refraction at the first surface are,

$$\begin{aligned}
\vec{\mathbf{v}}_1 &= \frac{x_a\vec{\mathbf{e}}_1 + y_a\vec{\mathbf{e}}_2 + (z_a - t_a)\vec{\mathbf{e}}_3}{\sqrt{x_a^2 + y_a^2 + (z_a - t_a)^2}}, \\
\vec{\mathbf{v}}_2 &= \frac{(x_b - x_a)\vec{\mathbf{e}}_1 + (y_b - y_a)\vec{\mathbf{e}}_2 + (z_b - z_a)\vec{\mathbf{e}}_3}{\sqrt{(x_b - x_a)^2 + (y_b - y_a)^2 + (z_b - z_a)^2}}, \\
\vec{\mathbf{n}}_a &= \frac{z_{a_x}\vec{\mathbf{e}}_1 + z_{a_y}\vec{\mathbf{e}}_2 - \vec{\mathbf{e}}_3}{\sqrt{z_{a_x}^2 + z_{a_y}^2 + 1}},
\end{aligned} \tag{9.6}$$

where, $z_{a_x} \equiv \partial_x z_a$ and $z_{a_y} \equiv \partial_y z_a$ are the partial derivatives of $z_a(x_a, y_a)$ with respect to $x_a$ and $y_a$.

We need to replace the unit vectors of equation (9.6) in equation (9.5). This procedure is simple but quite long. Let's begin with $(\vec{\mathbf{n}}_a \wedge \vec{\mathbf{v}}_1)$, which is inside the square root of equation (9.5),

$$\begin{aligned}(\vec{\mathbf{n}}_a \wedge \vec{\mathbf{v}}_1) = &\frac{(y_a z_{a_x} - x_a z_{a_y})(\vec{\mathbf{e}}_1 \wedge \vec{\mathbf{e}}_2)}{\sqrt{z_{a_x}^2 + z_{a_y}^2 + 1}\sqrt{(t_a - z_a)^2 + x_a^2 + y_a^2}} - \frac{\left[z_{a_x}(z_a - t_a) + x_a\right](\vec{\mathbf{e}}_3 \wedge \vec{\mathbf{e}}_1)}{\sqrt{z_{a_x}^2 + z_{a_y}^2 + 1}\sqrt{(t_a - z_a)^2 + x_a^2 + y_a^2}} \\ &+ \frac{\left[z_{a_y}(z_a - t_a) + y_a\right](\vec{\mathbf{e}}_2 \wedge \vec{\mathbf{e}}_3)}{\sqrt{z_{a_x}^2 + z_{a_y}^2 + 1}\sqrt{(t_a - z_a)^2 + x_a^2 + y_a^2}} \\ = &\frac{(y_a z_{a_x} - x_a z_{a_y})\vec{\mathbf{e}}_3}{\sqrt{z_{a_x}^2 + z_{a_y}^2 + 1}\sqrt{(t_a - z_a)^2 + x_a^2 + y_a^2}} - \frac{\left[z_{a_x}(z_a - t_a) + x_a\right]\vec{\mathbf{e}}_2}{\sqrt{z_{a_x}^2 + z_{a_y}^2 + 1}\sqrt{(t_a - z_a)^2 + x_a^2 + y_a^2}} \\ &+ \frac{\left[z_{a_y}(z_a - t_a) + y_a\right]\vec{\mathbf{e}}_1}{\sqrt{z_{a_x}^2 + z_{a_y}^2 + 1}\sqrt{(t_a - z_a)^2 + x_a^2 + y_a^2}}.\end{aligned} \tag{9.7}$$

Then, multiplying $(\vec{\mathbf{n}}_a \wedge \vec{\mathbf{v}}_1) \cdot (\vec{\mathbf{n}}_a \wedge \vec{\mathbf{v}}_1) = (\vec{\mathbf{n}}_a \wedge \vec{\mathbf{v}}_1)^2$,

$$(\vec{\mathbf{n}}_a \wedge \vec{\mathbf{v}}_1)^2 = \frac{(y_a z_{a_x} - x_a z_{a_y})^2 + [z_{a_x}(z_a - t_a) + x_a]^2 + \left(z_{a_y}(z_a - t_a) + y_a\right)^2}{\left(z_{a_x}^2 + z_{a_y}^2 + 1\right)\left[(t_a - z_a)^2 + x_a^2 + y_a^2\right]}. \tag{9.8}$$

Therefore, the term with the square root in equation (9.6) can be expressed as,

$$\begin{aligned}&\sqrt{1 - \frac{1}{n^2}(\vec{\mathbf{n}}_a \wedge \vec{\mathbf{v}}_1)^2} \\ &= \sqrt{1 - \frac{\left(y_a z_{a_x} - x_a z_{a_y}\right)^2 + [z_{a_x}(z_a - t_a) + x_a]^2 + \left[z_{a_y}(z_a - t_a) + y_a\right]^2}{n^2\left(z_{a_x}^2 + z_{a_y}^2 + 1\right)\left[(t_a - z_a)^2 + x_a^2 + y_a^2\right]}}.\end{aligned} \tag{9.9}$$

Now, let's focus on the term, $(\vec{\mathbf{n}}_a \cdot \vec{\mathbf{v}}_1)\vec{\mathbf{n}}_a$,

$$\begin{aligned}(\vec{\mathbf{n}}_a \cdot \vec{\mathbf{v}}_1)\vec{\mathbf{n}}_a &= \left[\frac{x_a z_{a_x} + y_a z_{a_y} - (z_a - t_a)}{\sqrt{(t_a - z_a)^2 + x_a^2 + y_a^2}\sqrt{z_{a_x}^2 + z_{a_y}^2 + 1}}\right]\frac{z_{a_x}\vec{\mathbf{e}}_1 + z_{a_y}\vec{\mathbf{e}}_2 - \vec{\mathbf{e}}_3}{\sqrt{z_{a_x}^2 + z_{a_y}^2 + 1}}, \\ &= \frac{[x_a z_{a_x} + y_a z_{a_y} - (z_a - t_a)]}{\sqrt{(t_a - z_a)^2 + x_a^2 + y_a^2}}\frac{z_{a_x}\vec{\mathbf{e}}_1 + z_{a_y}\vec{\mathbf{e}}_2 - \vec{\mathbf{e}}_3}{(z_{a_x}^2 + z_{a_y}^2 + 1)} \\ &= \frac{z_{a_x}[x_a z_{a_x} + y_a z_{a_y} - (z_a - t_a)]\vec{\mathbf{e}}_1}{\sqrt{(t_a - z_a)^2 + x_a^2 + y_a^2}(z_{a_x}^2 + z_{a_y}^2 + 1)} \\ &+ \frac{z_{a_y}[x_a z_{a_x} + y_a z_{a_y} - (z_a - t_a)]\vec{\mathbf{e}}_2}{\sqrt{(t_a - z_a)^2 + x_a^2 + y_a^2}(z_{a_x}^2 + z_{a_y}^2 + 1)} \\ &- \frac{[x_a z_{a_x} + y_a z_{a_y} - (z_a - t_a)]\vec{\mathbf{e}}_3}{\sqrt{(t_a - z_a)^2 + x_a^2 + y_a^2}(z_{a_x}^2 + z_{a_y}^2 + 1)}.\end{aligned} \tag{9.10}$$

Now, we can work with the term, $\vec{\mathbf{v}}_1 - (\vec{\mathbf{n}}_a \cdot \vec{\mathbf{v}}_1)\vec{\mathbf{n}}_a$,

$$\begin{aligned}\vec{\mathbf{v}}_1 - (\vec{\mathbf{n}}_a \cdot \vec{\mathbf{v}}_1)\vec{\mathbf{n}}_a = &-\frac{z_{a_x}[x_a z_{a_x} + y_a z_{a_y} - (z_a - t_a)]\vec{\mathbf{e}}_1}{\sqrt{(t_a - z_a)^2 + x_a^2 + y_a^2}\,(z_{a_x}^2 + z_{a_y}^2 + 1)} + \frac{x_a \vec{\mathbf{e}}_1}{\sqrt{x_a^2 + y_a^2 + (z_a - t_a)^2}} \\ &- \frac{z_{a_y}[x_a z_{a_x} + y_a z_{a_y} - (z_a - t_a)]\vec{\mathbf{e}}_2}{\sqrt{(t_a - z_a)^2 + x_a^2 + y_a^2}\,(z_{a_x}^2 + z_{a_y}^2 + 1)} + \frac{y_a \vec{\mathbf{e}}_2}{\sqrt{x_a^2 + y_a^2 + (z_a - t_a)^2}} \\ &+ \frac{[x_a z_{a_x} + y_a z_{a_y} - (z_a - t_a)]\vec{\mathbf{e}}_3}{\sqrt{(t_a - z_a)^2 + x_a^2 + y_a^2}\,(z_{a_x}^2 + z_{a_y}^2 + 1)} + \frac{(z_a - t_a)\vec{\mathbf{e}}_3}{\sqrt{x_a^2 + y_a^2 + (z_a - t_a)^2}}.\end{aligned} \tag{9.11}$$

Simplifying,

$$\begin{aligned}\vec{\mathbf{v}}_1 - (\vec{\mathbf{n}}_a \cdot \vec{\mathbf{v}}_1)\vec{\mathbf{n}}_a = &\frac{x_a\left(z_{a_y}^2 + 1\right) - z_{a_x}\left(y_a z_{a_y} + t_a - z_a\right)}{\left(z_{a_x}^2 + z_{a_y}^2 + 1\right)\sqrt{(t_a - z_a)^2 + x_a^2 + y_a^2}}\vec{\mathbf{e}}_1 \\ &+ \frac{y_a\left(z_{a_x}^2 + 1\right) - z_{a_y}(x_a z_{a_x} + t_a - z_a)}{\left(z_{a_x}^2 + z_{a_y}^2 + 1\right)\sqrt{(t_a - z_a)^2 + x_a^2 + y_a^2}}\vec{\mathbf{e}}_2 \\ &+ \frac{-(t_a - z_a)\left(z_{a_x}^2 + z_{a_y}^2\right) + x_a z_{a_x} + y_a z_{a_y}}{\left(z_{a_x}^2 + z_{a_y}^2 + 1\right)\sqrt{(t_a - z_a)^2 + x_a^2 + y_a^2}}\vec{\mathbf{e}}_3.\end{aligned} \tag{9.12}$$

It is time, we are in a position to replace equations (9.9) and (9.12) in equation (9.5), then we separate the components, we get the following three equations,

$$\begin{aligned}\frac{(x_b - x_a)}{\vartheta} = &\frac{x_a\left(z_{a_y}^2 + 1\right) - z_{a_x}\left(y_a z_{a_y} + t_a - z_a\right)}{-\operatorname{sgn}(t_a) n\left(z_{a_x}^2 + z_{a_y}^2 + 1\right)\sqrt{(t_a - z_a)^2 + x_a^2 + y_a^2}} \\ &- \frac{z_{a_x}\sqrt{1 - \dfrac{(y_a z_{a_x} - x_a z_{a_y})^2 + \left[z_{a_x}(z_a - t_a) + x_a\right]^2 + \left[z_{a_y}(z_a - t_a) + y_a\right]^2}{n^2\left(z_{a_x}^2 + z_{a_y}^2 + 1\right)\left[(t_a - z_a)^2 + x_a^2 + y_a^2\right]}}}{\sqrt{z_{a_x}^2 + z_{a_y}^2 + 1}},\end{aligned} \tag{9.13}$$

$$\begin{aligned}\frac{(y_b - y_a)}{\vartheta} = &\frac{y_a\left(z_{a_x}^2 + 1\right) - z_{a_y}(x_a z_{a_x} + t_a - z_a)}{-\operatorname{sgn}(t_a) n\left(z_{a_x}^2 + z_{a_y}^2 + 1\right)\sqrt{(t_a - z_a)^2 + x_a^2 + y_a^2}} \\ &- \frac{z_{a_y}\sqrt{1 - \dfrac{(y_a z_{a_x} - x_a z_{a_y})^2 + \left[z_{a_x}(z_a - t_a) + x_a\right]^2 + \left[z_{a_y}(z_a - t_a) + y_a\right]^2}{n^2\left(z_{a_x}^2 + z_{a_y}^2 + 1\right)\left[(t_a - z_a)^2 + x_a^2 + y_a^2\right]}}}{\sqrt{z_{a_x}^2 + z_{a_y}^2 + 1}},\end{aligned} \tag{9.14}$$

and,

$$\frac{(z_b - z_a)}{\vartheta} = \frac{-(t_a - z_a)\left(z_{a_x}^2 + z_{a_y}^2\right) + x_a z_{a_x} + y_a z_{a_y}}{-\text{sgn}(t_a)n\left(z_{a_x}^2 + z_{a_y}^2 + 1\right)\sqrt{(t_a - z_a)^2 + x_a^2 + y_a^2}} + \frac{\sqrt{1 - \frac{(y_a z_{a_x} - x_a z_{a_y})^2 + \left[z_{a_x}(z_a - t_a) + x_a\right]^2 + \left[z_{a_y}(z_a - t_a) + y_a\right]^2}{n^2\left(z_{a_x}^2 + z_{a_y}^2 + 1\right)\left[(t_a - z_a)^2 + x_a^2 + y_a^2\right]}}}{\sqrt{z_{a_x}^2 + z_{a_y}^2 + 1}}, \tag{9.15}$$

where,

$$\vartheta \equiv \sqrt{(x_b - x_a)^2 + (y_b - y_a)^2 + (z_b - z_a)^2}. \tag{9.16}$$

The right side of equations (9.13), (9.14) and (9.15) only depends on known parameters; in reality, on the right side are the cosine directors of the refracted ray.

$$\wp_x \equiv \frac{x_a\left(z_{a_y}^2 + 1\right) - z_{a_x}\left(y_a z_{a_y} + t_a - z_a\right)}{-\text{sgn}(t_a)n\left(z_{a_x}^2 + z_{a_y}^2 + 1\right)\sqrt{(t_a - z_a)^2 + x_a^2 + y_a^2}} - \frac{z_{a_x}\sqrt{1 - \frac{(y_a z_{a_x} - x_a z_{a_y})^2 + \left[z_{a_x}(z_a - t_a) + x_a\right]^2 + \left[z_{a_y}(z_a - t_a) + y_a\right]^2}{n^2\left(z_{a_x}^2 + z_{a_y}^2 + 1\right)\left[(t_a - z_a)^2 + x_a^2 + y_a^2\right]}}}{\sqrt{z_{a_x}^2 + z_{a_y}^2 + 1}}, \tag{9.17}$$

$$\wp_y \equiv \frac{y_a\left(z_{a_x}^2 + 1\right) - z_{a_y}(x_a z_{a_x} + t_a - z_a)}{-\text{sgn}(t_a)n\left(z_{a_x}^2 + z_{a_y}^2 + 1\right)\sqrt{(t_a - z_a)^2 + x_a^2 + y_a^2}} - \frac{z_{a_y}\sqrt{1 - \frac{(y_a z_{a_x} - x_a z_{a_y})^2 + \left[z_{a_x}(z_a - t_a) + x_a\right]^2 + \left[z_{a_y}(z_a - t_a) + y_a\right]^2}{n^2\left(z_{a_x}^2 + z_{a_y}^2 + 1\right)\left[(t_a - z_a)^2 + x_a^2 + y_a^2\right]}}}{\sqrt{z_{a_x}^2 + z_{a_y}^2 + 1}}, \tag{9.18}$$

and,

$$\wp_z \equiv \frac{-(t_a - z_a)\left(z_{a_x}^2 + z_{a_y}^2\right) + x_a z_{a_x} + y_a z_{a_y}}{-\text{sgn}(t_a)n\left(z_{a_x}^2 + z_{a_y}^2 + 1\right)\sqrt{(t_a - z_a)^2 + x_a^2 + y_a^2}} + \frac{\sqrt{1 - \frac{(y_a z_{a_x} - x_a z_{a_y})^2 + [z_{a_x}(z_a - t_a) + x_a]^2 + \left[z_{a_y}(z_a - t_a) + y_a\right]^2}{n^2\left(z_{a_x}^2 + z_{a_y}^2 + 1\right)\left[(t_a - z_a)^2 + x_a^2 + y_a^2\right]}}}{\sqrt{z_{a_x}^2 + z_{a_y}^2 + 1}}. \tag{9.19}$$

Thus, $\wp_x^2 + \wp_y^2 + \wp_z^2 = 1$. We can equal the cosine directors of equations (9.17), (9.18) and (9.19) with (9.13), (9.14) and (9.15), and we get,

$$\frac{x_b - x_a}{\vartheta} = \wp_x, \tag{9.20}$$

$$\frac{y_b - y_a}{\vartheta} = \wp_y, \tag{9.21}$$

and,

$$\frac{z_b - z_a}{\vartheta} = \wp_z. \tag{9.22}$$

Solving for $x_b$, $y_b$ and $z_b$, we get the structure of the second surface of the singlet.

$$\begin{cases} x_b = x_a + \vartheta\wp_x, \\ y_b = y_a + \vartheta\wp_y, \\ z_b = z_a + \vartheta\wp_z. \end{cases} \tag{9.23}$$

The next step is to find $\vartheta$ in terms of the known parameters. We mean, the first surface, the central thickness $t$, the refraction index $n$, the cosine directors $\wp_x$, $\wp_y$ and $\wp_z$ and the distances $t_a$ and $t_b$. For that we set a system of equations that involves Fermat's principle and equation (9.23).

### 9.2.3 Solution

In this section, we focus on solving for $\vartheta$. So, let's recall the OPL of the reference ray and the OPL of any other ray, equation (9.2),

$$\begin{aligned} -t_a + nt + t_b = &- \operatorname{sgn}(t_a)\sqrt{x_a^2 + y_a^2 + (z_a - t_a)^2} \\ &+ n\sqrt{(x_b - x_a)^2 + (y_b - y_a)^2 + (z_b - z_b)^2} \\ &+ \operatorname{sgn}(t_b)\sqrt{x_b^2 + y_b^2 + (z_b - t - t_b)^2}. \end{aligned} \tag{9.2}$$

Replacing equation (9.16) and (9.23) in the above equation, equation (9.2),

$$\begin{aligned} -t_a + nt + t_b \;=\; &- \operatorname{sgn}(t_a)\sqrt{x_a^2 + y_a^2 + (z_a - t_a)^2} + n\vartheta \\ &+ \operatorname{sgn}(t_b)\sqrt{(x_a + \vartheta\wp_x)^2 + (y_a + \vartheta\wp_y)^2 + (z_a + \vartheta\wp_z - t - t_b)^2}. \end{aligned} \tag{9.24}$$

To simplify the last equation we define the following parameters,

$$f \equiv -t_a + nt + t_b + \operatorname{sgn}(t_a)\sqrt{x_a^2 + y_a^2 + (z_a - t_a)^2}, \tag{9.25}$$

and,

$$\tau \equiv z_a - t - t_b. \tag{9.26}$$

We replace the parameters $f$ and $\tau$ in equation (9.24) and we square it,

$$(f - n\vartheta)^2 = (x_a + \vartheta\wp_x)^2 + (y_a + \vartheta\wp_y)^2 + (\tau + \vartheta\wp_z)^2, \tag{9.27}$$

expanding, and tacking $\wp_x^2 + \wp_y^2 + \wp_z^2 = 1$,

$$\begin{aligned} f^2 - 2fn\vartheta + n^2\vartheta^2 = x_a^2 + 2x_a\wp_x\vartheta + \wp_x^2\vartheta^2 + y_a^2 + 2y_a\wp_y\vartheta + \wp_y^2\vartheta^2 \\ + \tau^2 + 2\tau\wp_z\vartheta + \wp_z^2\vartheta^2, \end{aligned} \tag{9.28}$$

collecting for terms with $\vartheta$,

$$(1 - n^2)\vartheta^2 + [2(fn + x_a\wp_x + y_a\wp_y + \tau\wp_z)]\vartheta + (x_a^2 + y_a^2 + \tau^2 - f^2) = 0. \tag{9.29}$$

Multiplying all the elements by $4(1 - n^2)$,

$$\begin{aligned} &4(1 - n^2)^2\vartheta^2 + 4(1 - n^2)[2(fn + x_a\wp_x + y_a\wp_y + \tau\wp_z)]\vartheta \\ &+ 4(1 - n^2)(x_a^2 + y_a^2 + \tau^2 - f^2) = 0. \end{aligned} \tag{9.30}$$

The next step is to sum $[2(fn + x_a\wp_x + y_a\wp_y + \tau\wp_z)]^2$ in both sides,

$$\begin{aligned} &4(1 - n^2)^2\vartheta^2 + 4(1 - n^2)[2(fn + x_a\wp_x + y_a\wp_y + \tau\wp_z)] \\ &\quad \vartheta + [2(fn + x_a\wp_x + y_a\wp_y + \tau\wp_z)]^2 \\ &= [2(fn + x_a\wp_x + y_a\wp_y + \tau\wp_z)]^2 - 4(1 - n^2)(x_a^2 + y_a^2 + \tau^2 - f^2), \end{aligned} \tag{9.31}$$

completing the square binomial,

$$\begin{aligned} &\left[2(1 - n^2)\vartheta + 2(fn + x_a\wp_x + y_a\wp_y + \tau\wp_z)\right]^2 \\ &= [2(fn + x_a\wp_x + y_a\wp_y + \tau\wp_z)]^2 - 4(1 - n^2)(x_a^2 + y_a^2 + \tau^2 - f^2), \end{aligned} \tag{9.32}$$

applying the square root in both sides,

$$\begin{aligned} &2(1 - n^2)\vartheta + 2(fn + x_a\wp_x + y_a\wp_y + \tau\wp_z) \\ &= \sqrt{[2(fn + x_a\wp_x + y_a\wp_y + \tau\wp_z)]^2 - 4(1 - n^2)(x_a^2 + y_a^2 + \tau^2 - f^2)}. \end{aligned} \tag{9.33}$$

Solving for $\vartheta$,

$$\begin{aligned} \vartheta = &\frac{-(fn + x_a\wp_x + y_a\wp_y + \tau\wp_z)}{(1 - n^2)} \\ &\pm \frac{\sqrt{(fn + x_a\wp_x + y_a\wp_y + \tau\wp_z)^2 - (1 - n^2)(x_a^2 + y_a^2 + \tau^2 - f^2)}}{(1 - n^2)}. \end{aligned} \tag{9.34}$$

Once, we know $\vartheta$, we get the solution which is given by the following expressions,

$$\boxed{\begin{cases} x_b = x_a + \vartheta\wp_x, \\ y_b = y_a + \vartheta\wp_y, \\ z_b = z_a + \vartheta\wp_z. \end{cases}} \tag{9.35}$$

The most important equation in this chapter is equation (9.35). It describes the second surface of a singlet such that it is on-axis stigmatic. We want to remark that equation (9.35) works if and only if the rays inside the lens do not cross each other. Just like the on-axis stigmatic asphere of chapter 4. The lenses designed with equation (9.35) are homeomorphic to the lenses of chapter 4, which means that they are topologically equivalent. They hold the same topological properties.

Another critical remark of this equation is that it only works if at the origin the normal vectors of the first and second surface are parallel to the optical axis.

### 9.2.4 Illustrative examples

The following gallery presents several finite–finite stigmatic freeform singlets. The present examples are of cases when the point object/image can be real or virtual. The information needed to reproduce each example is presented in the respective figure. The top image of each figure is the $x/z$-plane and the bottom image is the $y/z$-plane. The figures of the gallery are 9.2, 9.3, 9.4, 9.5 and 9.6.

## 9.3 The freeform collector lens

The next objective is the freeform collector lens, that is, when $t_a \to -\infty$. To achieve this we need to evaluate some limits on the cosine directors $\wp_x$, $\wp_y$ and $\wp_z$ and the parameter $f$.

Let's recall these parameters by calling equations (9.17), (9.18), (9.19), and (9.23)

$$\wp_x \equiv \frac{x_a\left(z_{a_y}^2 + 1\right) - z_{a_x}\left(y_a z_{a_y} + t_a - z_a\right)}{-\mathrm{sgn}(t_a)n\left(z_{a_x}^2 + z_{a_y}^2 + 1\right)\sqrt{(t_a - z_a)^2 + x_a^2 + y_a^2}} - \frac{z_{a_x}\sqrt{1 - \dfrac{(y_a z_{a_x} - x_a z_{a_y})^2 + \left[z_{a_x}(z_a - t_a) + x_a\right]^2 + \left[z_{a_y}(z_a - t_a) + y_a\right]^2}{n^2\left(z_{a_x}^2 + z_{a_y}^2 + 1\right)\left[(t_a - z_a)^2 + x_a^2 + y_a^2\right]}}}{\sqrt{z_{a_x}^2 + z_{a_y}^2 + 1}}, \tag{9.17}$$

$$\wp_y \equiv \frac{y_a\left(z_{a_x}^2 + 1\right) - z_{a_y}(x_a z_{a_x} + t_a - z_a)}{-\mathrm{sgn}(t_a)n\left(z_{a_x}^2 + z_{a_y}^2 + 1\right)\sqrt{(t_a - z_a)^2 + x_a^2 + y_a^2}} - \frac{z_{a_y}\sqrt{1 - \dfrac{(y_a z_{a_x} - x_a z_{a_y})^2 + \left[z_{a_x}(z_a - t_a) + x_a\right]^2 + \left[z_{a_y}(z_a - t_a) + y_a\right]^2}{n^2\left(z_{a_x}^2 + z_{a_y}^2 + 1\right)\left[(t_a - z_a)^2 + x_a^2 + y_a^2\right]}}}{\sqrt{z_{a_x}^2 + z_{a_y}^2 + 1}}, \tag{9.18}$$

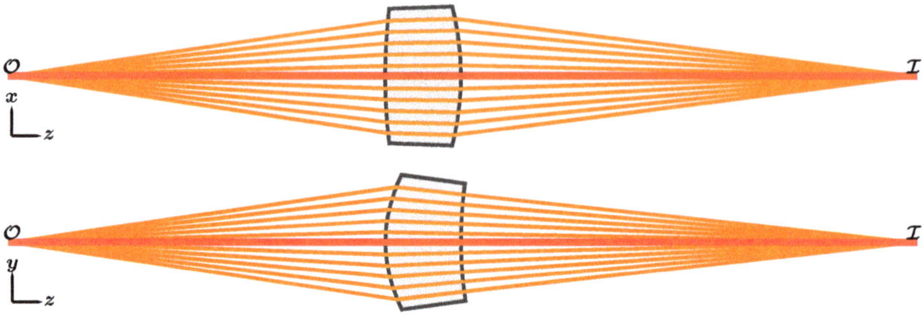


**Figure 9.2.** Design specifications: $n = 1.9$, $t_a = -50$ mm, $t = 10$ mm, $t_b = 60$ mm, $z_a = (x_a^2 + 4y_a^2)/150$ and $z_b =$ equation (9.35).

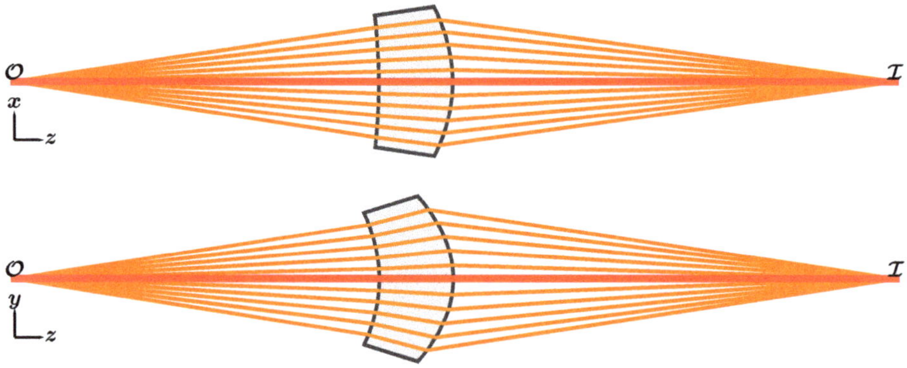


**Figure 9.3.** Design specifications: $n = 1.9$, $t_a = -50$ mm, $t = 10$ mm, $t_b = 60$ mm, $z_a = -(x_a^2 + 4y_a^2)/155$ and $z_b =$ equation (9.35).

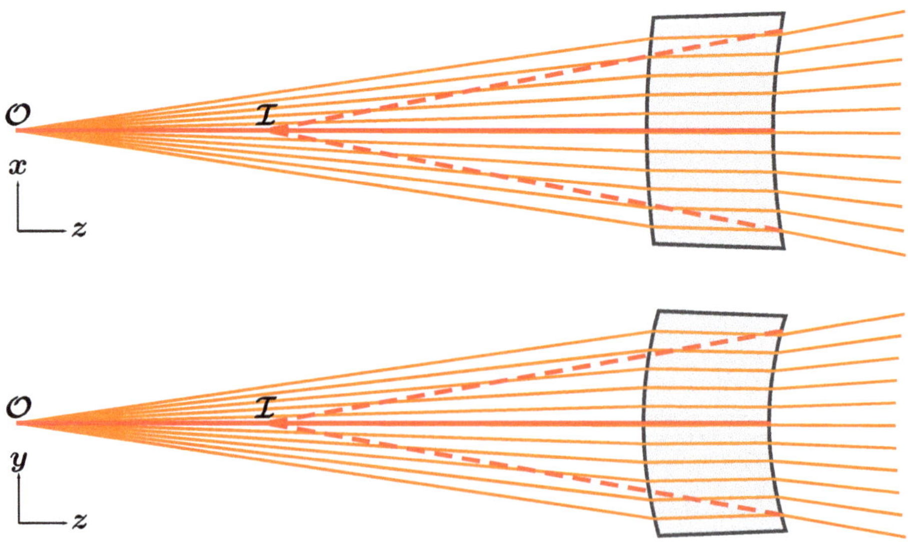


**Figure 9.4.** Design specifications: $n = 1.9$, $t_a = -50$ mm, $t = 10$ mm, $t_b = -40$ mm, $z_a = (x_a^2 + 2.25y_a^2)/150$ and $z_b =$ equation (9.35).

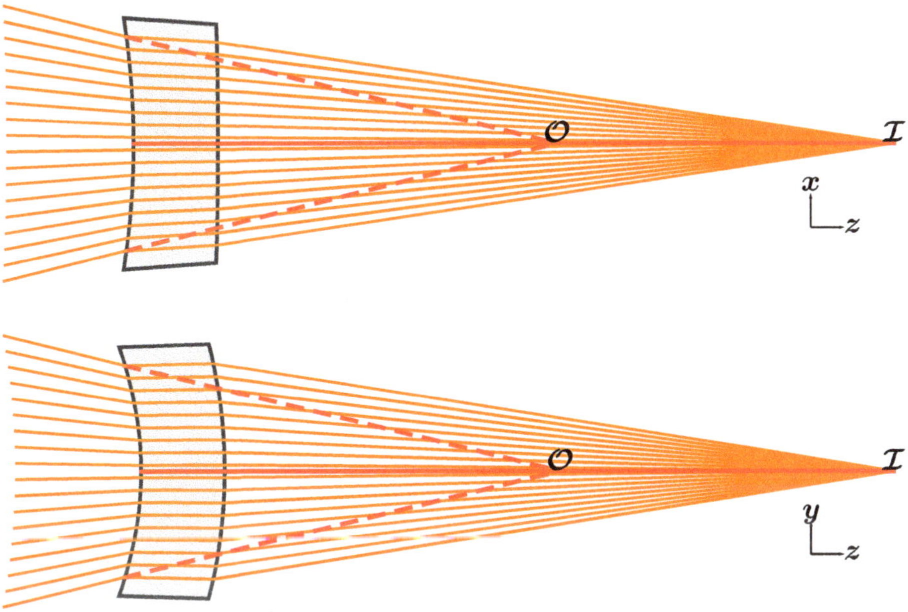


**Figure 9.5.** Design specifications: $n = 1.9$, $t_a = 50$ mm, $t = 10$ mm, $t_b = 80$ mm, $z_a = -(x_a^2 + 2.25y_a^2)/190$ and $z_b =$ equation (9.35).

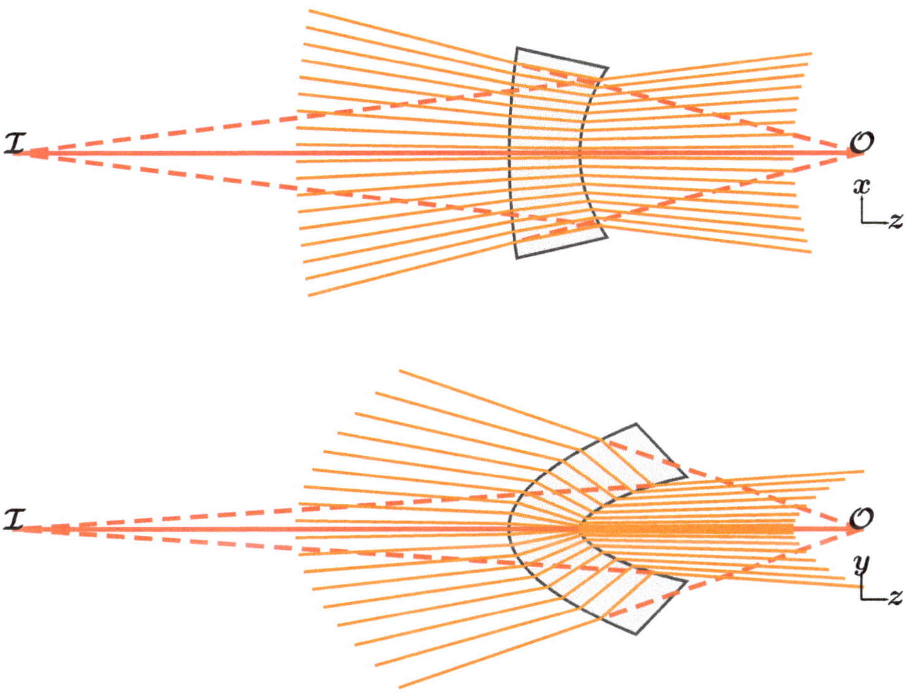


**Figure 9.6.** Design specifications: $n = 1.9$, $t_a = 50$ mm, $t = 10$ mm, $t_b = -80$mm, $z_a = (x_a^2 + 4y_a^2)/200$ and $z_b =$ equation (9.35).

$$\wp_z \equiv \frac{-(t_a - z_a)\left(z_{a_x}^2 + z_{a_y}^2\right) + x_a z_{a_x} + y_a z_{a_y}}{-\text{sgn}(t_a)n\left(z_{a_x}^2 + z_{a_y}^2 + 1\right)\sqrt{(t_a - z_a)^2 + x_a^2 + y_a^2}} + \frac{\sqrt{1 - \dfrac{(y_a z_{a_x} - x_a z_{a_y})^2 + \left[z_{a_x}(z_a - t_a) + x_a\right]^2 + \left[z_{a_y}(z_a - t_a) + y_a\right]^2}{n^2\left(z_{a_x}^2 + z_{a_y}^2 + 1\right)\left[(t_a - z_a)^2 + x_a^2 + y_a^2\right]}}}{\sqrt{z_{a_x}^2 + z_{a_y}^2 + 1}}, \tag{9.19}$$

and,

$$f \equiv -t_a + nt + t_b + \text{sgn}(t_a)\sqrt{x_a^2 + y_a^2 + (z_a - t_a)^2}. \tag{9.23}$$

The next step is to apply the limit when $t_a \rightarrow -\infty$ over them,

$$\lim_{t_a \rightarrow -\infty} (\wp_x) = \frac{z_{a_x}\left(\sqrt{(n^2 - 1)(z_{a_x}^2 + z_{a_y}^2) + n^2} - 1\right)}{n\left(z_{a_x}^2 + z_{a_y}^2 + 1\right)}, \tag{9.36}$$

$$\lim_{t_a \rightarrow -\infty} (\wp_y) = \frac{z_{a_y}\left(\sqrt{(n^2 - 1)(z_{a_x}^2 + z_{a_y}^2) + n^2} - 1\right)}{n\left(z_{a_x}^2 + z_{a_y}^2 + 1\right)}, \tag{9.37}$$

$$\lim_{t_a \rightarrow -\infty} (\wp_z) = \frac{z_{a_x}^2 + z_{a_y}^2 + \sqrt{(n^2 - 1)(z_{a_x}^2 + z_{a_y}^2) + n^2}}{n\left(z_{a_x}^2 + z_{a_y}^2 + 1\right)}, \tag{9.38}$$

and,

$$\lim_{t_a \rightarrow -\infty} (f) = nt + t_b - z_a. \tag{9.39}$$

Therefore, the distance, $\vartheta$, that each ray travels inside the lens when the object is at minus infinity is,

$$\lim_{t_a \rightarrow -\infty} (\vartheta) = \frac{-\beta \pm \sqrt{\beta^2 - (1 - n^2)\left\{x_a^2 + y_a^2 + \tau^2 - \left[\lim\limits_{t_a \rightarrow -\infty} (f)\right]^2\right\}}}{(1 - n^2)}, \tag{9.40}$$

where,

$$\beta \equiv \left\{\left[\lim_{t_a \rightarrow -\infty} (f)\right]n + x_a\left[\lim_{t_a \rightarrow -\infty} (\wp_x)\right] + y_a\left[\lim_{t_a \rightarrow -\infty} (\wp_y)\right] + \tau\left[\lim_{t_a \rightarrow -\infty} (\wp_z)\right]\right\}. \tag{9.41}$$

Then we can describe the second surface of the collector lens as,

$$\boxed{\begin{cases} \lim\limits_{t_a \to -\infty} (x_b) = x_a + \left[\lim\limits_{t_a \to -\infty} (\vartheta)\right]\left[\lim\limits_{t_a \to -\infty} (\wp_x)\right], \\ \lim\limits_{t_a \to -\infty} (y_b) = y_a + \left[\lim\limits_{t_a \to -\infty} (\vartheta)\right]\left[\lim\limits_{t_a \to -\infty} (\wp_y)\right], \\ \lim\limits_{t_a \to -\infty} (z_b) = z_a + \left[\lim\limits_{t_a \to -\infty} (\vartheta)\right]\left[\lim\limits_{t_a \to -\infty} (\wp_z)\right]. \end{cases}} \tag{9.42}$$

Please notice that we applied the limit over $\vartheta$ and the cosine directors, $\wp_x$, $\wp_y$ and $\wp_z$. This is because the first surface only depends on $x_a$ and $y_a$ not on $t_a$.

### 9.3.1 Examples

The next example is an on-axis stigmatic collector lens for real point image. The design parameters are presented in figure 9.7.

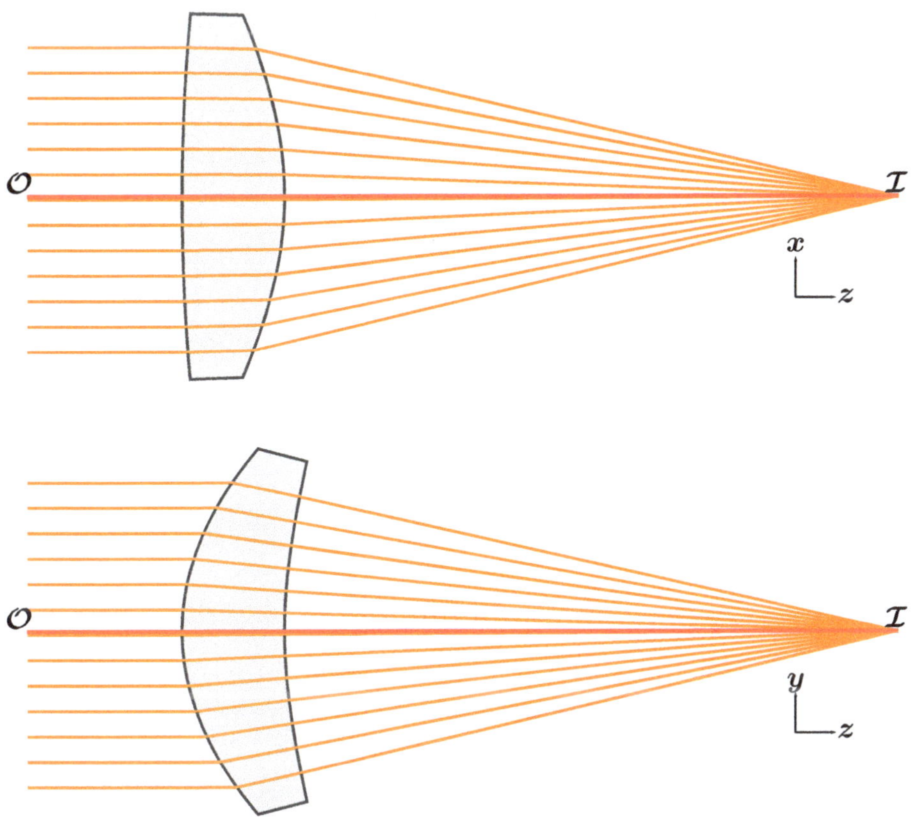


**Figure 9.7.** Design specifications: $n = 1.5$, $t = 10$ mm, $t_b = 60$ mm, $z_a = 200 - \sqrt{200^2 - x_a^2 - 9y_a^2}$ and $z_b =$ equation (9.42).

## 9.4 The freeform collimator lens

Our next goal is called freeform collimator lens; it is the case of the on-axis lens with an image at plus infinity. Since the cosine directors implemented in the finite–finite case are not affected with this limit, we do not need to apply a limit over them. Thus we use the following equations (9.17), (9.18), (9.19) to express $\wp_x$, $\wp_y$ and $\wp_z$, respectively. But we need to modify equation (9.2). The OPL of the reference beam remains the same, but the limit modifies the OPL of any other beam. So, let's remember the equation (9.2),

$$\begin{aligned} -t_a + nt + t_b = &-\operatorname{sgn}(t_a)\sqrt{x_a^2 + y_a^2 + (z_a - t_a)^2} \\ &+ n\sqrt{(x_b - x_a)^2 + (y_b - y_a)^2 + (z_b - z_b)^2} \\ &+ \operatorname{sgn}(t_b)\sqrt{x_b^2 + y_b^2 + (z_b - t - t_b)^2}. \end{aligned} \tag{9.2}$$

Let's insert the distance that a ray travels inside the lens, $\vartheta \equiv \sqrt{(x_b - x_a)^2 + (y_b - y_a)^2 + (z_b - z_a)^2}$ in equation (9.2),

$$\begin{aligned} &-t_a + nt + t_b - \operatorname{sgn}(t_b)\sqrt{x_b^2 + y_b^2 + (z_b - t - t_b)^2} - n\vartheta \\ &= -\operatorname{sgn}(t_a)\sqrt{x_a^2 + y_a^2 + (z_a - t_a)^2}. \end{aligned} \tag{9.43}$$

If we apply the limit when $t_b \to \infty$ on the left side,

$$\begin{aligned} &\lim_{t_b \to \infty}\left[-t_a + nt + t_b - \operatorname{sgn}(t_b)\sqrt{x_b^2 + y_b^2 + (z_b - t - t_b)^2} - n\vartheta\right] \\ &= (-1 + n)t - t_a + \left[\lim_{t_b \to \infty}(z_b)\right] - n\left[\lim_{t_b \to \infty}(\vartheta)\right]. \end{aligned} \tag{9.44}$$

After computing the limit when $t_b \to \infty$ and sending $n\left[\lim_{t_b \to \infty}(\vartheta)\right]$ to the right side,

$$\begin{aligned} &(-1 + n)t - t_a + \left[\lim_{t_b \to \infty}(z_b)\right] \\ &= -\operatorname{sgn}(t_a)\sqrt{x_a^2 + y_a^2 + (z_a - t_a)^2} + n\left[\lim_{t_b \to \infty}(\vartheta)\right]. \end{aligned} \tag{9.45}$$

At this step we are in a very good position to replace $z_b = z_a + \vartheta\wp_z$ in the above equation.

$$\begin{aligned} &(-1 + n)t - t_a + \left[\lim_{t_b \to \infty}(z_a + \vartheta\wp_z)\right] \\ &= -\operatorname{sgn}(t_a)\sqrt{x_a^2 + y_a^2 + (z_a - t_a)^2} + n\left[\lim_{t_b \to \infty}(\vartheta)\right]. \end{aligned} \tag{9.46}$$

The limit when $t_b \to \infty$ does not affect $z_a$ and $\wp_z$. It only affects $\vartheta$, thus,

$$\begin{aligned} &(-1 + n)t - t_a + z_a + \left[\lim_{t_b \to \infty}(\vartheta)\right]\wp_z \\ &= -\operatorname{sgn}(t_a)\sqrt{x_a^2 + y_a^2 + (z_a - t_a)^2} + n\left[\lim_{t_b \to \infty}(\vartheta)\right]. \end{aligned} \tag{9.47}$$

So, from the last equation we can solve for $\left[\lim_{t_b\to\infty}(\vartheta)\right]$,

$$\left[\lim_{t_b\to\infty}(\vartheta)\right] = \frac{(-1+n)t - t_a + z_a + \text{sgn}(t_a)\sqrt{x_a^2 + y_a^2 + (z_a - t_a)^2}}{n - \wp_z}. \tag{9.48}$$

Finally, we know how $\left[\lim_{t_b\to\infty}(\vartheta)\right]$ is, we can replace it in the general solution and we get,

$$\boxed{\begin{cases} \lim_{t_b\to\infty}(x_b) = x_a + \left[\lim_{t_b\to\infty}(\vartheta)\right]\wp_x, \\ \lim_{t_b\to\infty}(y_b) = y_a + \left[\lim_{t_b\to\infty}(\vartheta)\right]\wp_y, \\ \lim_{t_b\to\infty}(z_b) = z_a + \left[\lim_{t_b\to\infty}(\vartheta)\right]\wp_z. \end{cases}} \tag{9.49}$$

The last three equations describe point by point the second surface of an on-axis collimator lens.

### 9.4.1 Illustrative examples

In this section, we put an example collimator lens only computing equation (9.49). The example is in figure 9.8.

## 9.5 The beam-shaper

Our last case to cover in this chapter is the optical beam-shaper. The beam-shaper is a singlet freeform lens where at at the input and output the beam is collimated along the optical axis.

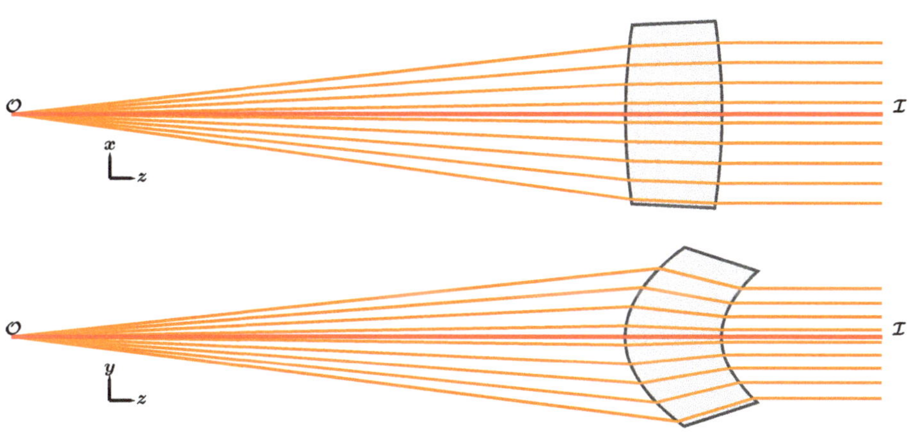


**Figure 9.8.** Design specifications: $n = 1.5$, $t_a = -95$ mm, $t = 15$ mm, $z_a = 100 - \sqrt{100^2 - x_a^2 - 9y_a^2}$ and $z_b$ = equation (9.49).

To obtain the general formula of the beam-shaper, we need to compute the limit when $t_a \to -\infty$ in equation (9.48),

$$\lim_{t_a \to -\infty}\left[\lim_{t_b \to \infty}(\vartheta)\right] = \left[\lim_{t_a \to -\infty}\frac{(-1+n)t - t_a + z_a + \mathrm{sgn}(t_a)\sqrt{x_a^2 + y_a^2 + (z_a - t_a)^2}}{n - \wp_z}\right]. \tag{9.50}$$

After computing the limit, $\lim_{t_a \to -\infty}\left[\lim_{t_b \to \infty}(\vartheta)\right]$ is given by,

$$\lim_{t_a \to -\infty}\left[\lim_{t_b \to \infty}(\vartheta)\right] = \frac{(n-1)t}{n - \left[\lim_{t_a \to -\infty}(\wp_z)\right]}. \tag{9.51}$$

Finally, the second surface of the beam-shaper is expressed with the following formulas,

$$\begin{cases} \lim_{t_a \to -\infty}\left[\lim_{t_b \to \infty}(x_b)\right] = x_a + \left\{\lim_{t_a \to -\infty}\left[\lim_{t_b \to \infty}(\vartheta)\right]\right\}\left[\lim_{t_a \to -\infty}(\wp_x)\right], \\ \lim_{t_a \to -\infty}\left[\lim_{t_b \to \infty}(y_b)\right] = y_a + \left\{\lim_{t_a \to -\infty}\left[\lim_{t_b \to \infty}(\vartheta)\right]\right\}\left[\lim_{t_a \to -\infty}(\wp_y)\right], \\ \lim_{t_a \to -\infty}\left[\lim_{t_b \to \infty}(z_b)\right] = z_a + \left\{\lim_{t_a \to -\infty}\left[\lim_{t_b \to \infty}(\vartheta)\right]\right\}\left[\lim_{t_a \to -\infty}(\wp_z)\right]. \end{cases} \tag{9.52}$$

To properly use equation (9.52) we need to use equation (9.51) for $\lim_{t_a \to -\infty}\left[\lim_{t_b \to \infty}(\vartheta)\right]$ and equations (9.36), (9.37) and (9.38), for $\left[\lim_{t_a \to -\infty}(\wp_z)\right]$, $\left[\lim_{t_a \to -\infty}(\wp_y)\right]$, $\left[\lim_{t_a \to -\infty}(\wp_z)\right]$.

### 9.5.1 Illustrative example

An example of a beam-shaper is presented using equation (9.52) in figure 9.9.

## 9.6 Conclusions

In this chapter, we generalize the on-axis stigmatic aspheric lens to an on-axis stigmatic freeform lens. We derived it step by step and presented it illustratively.

In all the cases, the equation provided correctly designs the lenses, in such a way that the second surface ensures that the singlet is free from spherical aberration.

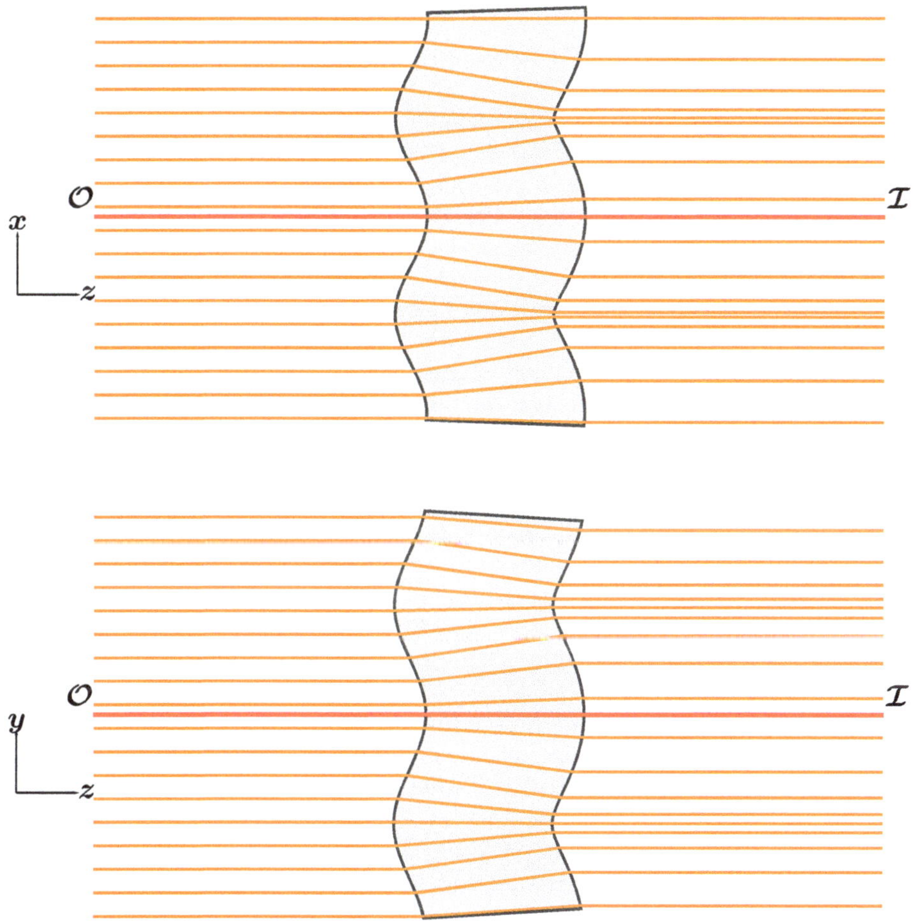


**Figure 9.9.** Specifications of the design: $n = 1.5$, $t = 10$ mm, $z_a = \cos(0.5x_a + 0.4y_a)$ and $z_b$ = equation (9.52).

## Further reading

Acuña R G G and Gutiérrez-Vega J C 2019 General formula of the refractive telescope design free spherical aberration *Novel Optical Systems, Methods, and Applications XXII* 11105 (Bellingham, WA: International Society for Optics and Photonics)

Bauer A and Rolland J P 2015 Design of a freeform electronic viewfinder coupled to aberration fields of freeform optics *Opt. Express* **23** 28141–53

Born M and Wolf E 2013 Principles of Optics: Electromagnetic Theory of Propagation *Interference and Diffraction of Light* (Amsterdam: Elsevier)

Cakmakci O, Moore B, Foroosh H and Rolland J P 2008 Optimal local shape description for rotationally non-symmetric optical surface design and analysis *Opt. Express* **16** 1583–9

Chaves J 2016 *Introduction to Nonimaging Optics* 2nd edn (Boca Raton, FL: CRC Press)

Chen J-J, Wang T-Y, Huang K-L, Liu T-S, Tsai M-D and Lin C-T 2012 Freeform lens design for LED collimating illumination *Opt. Express* **20** 10984–95

Descartes R 1637a *De la nature des lignes courbes*

Descartes R 1637b *La Géométrie*

Ding Y, Liu X, Zheng Z-r and Gu P-f 2008 Freeform LED lens for uniform illumination *Opt. Express* **16** 12958–66

Fang F Z, Zhang X D, Weckenmann A, Zhang G X and Evans C 2013 Manufacturing and measurement of freeform optics *CIRP Ann.* **62** 823–46

Feng Z, Froese B D, Liang R, Cheng D and Wang Y 2017 Simplified freeform optics design for complicated laser beam shaping *Appl. Opt.* **56** 9308–14

Forbes G W 2007 Shape specification for axially symmetric optical surfaces *Opt. Express* **15** 5218–26

Forbes G W 2010 Robust, efficient computational methods for axially symmetric optical aspheres *Opt. Express* **18** 19700–12

Forbes G W 2012 Characterizing the shape of freeform optics *Opt. Express* **20** 2483–99

Forbes G W 2013 Fitting freeform shapes with orthogonal bases *Opt. Express* **21** 19061–81

Fuerschbach K, Rolland J P and Thompson K P 2012 Extending Nodal aberration theory to include mount-induced aberrations with application to freeform surfaces *Opt. Express* **20** 20139–55

Fuerschbach K, Rolland J P and Thompson K P 2014 Theory of aberration fields for general optical systems with freeform surfaces *Opt. Express* **22** 26585–606

González-Acuña R G and Chaparro-Romo H A 2018 General formula for bi-aspheric singlet lens design free of spherical aberration *Appl. Opt.* **57** 9341–5

González-Acuña R G and Guitiérrez-Vega J C 2018 Generalization of the axicon shape: the gaxicon *J. Opt. Soc. Am.* A **35** 1915–8

González-Acuña R G and Gutiérrez-Vega J C 2019a Analytic formulation of a refractive-reflective telescope free of spherical aberration *Opt. Eng.* **58** 085105

González-Acuña R G and Gutiérrez-Vega J C 2019 General formula for aspheric collimator lens design free of spherical aberration *Current Developments in Lens Design and Optical Engineering XX* 11104 (Bellingham, WA: International Society for Optics and Photonics)

González-Acuña R G, Avendaño-Alejo M and Gutiérrez-Vega J C 2019a Singlet lens for generating aberration-free patterns on deformed surfaces *J. Opt. Soc. Am.* A **36** 925–9

González-Acuña R G, Chaparro-Romo H A and Gutiérrez-Vega J C 2019b General formula to design freeform singlet free of spherical aberration and astigmatism *Appl. Opt.* **58** 1010–5

Gonzalez-Utrera D, Su T, Su P and Schwiegerling J 2015 Ophthalmic applications of freeform optics *FFreeform Optics* (Washington, DC: Optical Society of America), p FW4B–1

Huygens C 1690 *Traité de la Lumière*

Kaya I, Thompson K P and Rolland J P 2012 Comparative assessment of freeform polynomials as optical surface descriptions *Opt. Express* **20** 22683–91

Kingslake R and Johnson R B 2009 *Lens Design Fundamentals* (New York: Academic)

Lohn J D, Hornby G S and Linden D S 2005 *An evolved antenna for deployment on NASA's space technology 5 mission Genetic Programming Theory and Practice II* (Berlin: Springer), pp 301–15

Luneburg R K and Herzberger M 1964 *Mathematical Theory of Optics* (Berkeley, CA: University of California Press)

Michaelis D, Schreiber P and Bräuer A 2011 Cartesian oval representation of freeform optics in illumination systems *Opt. Lett.* **36** 918–20

Miñano J C, Benítez P, Zamora P, Buljan M, Mohedano R and Santamaría A 2013 Free-form optics for Fresnel-lens-based photovoltaic concentrators *Opt. Express* **21** A494–502

Moiseev M A, Doskolovich L L and Kazanskiy N L 2011 Design of high-efficient freeform LED lens for illumination of elongated rectangular regions *Opt. Express* **19** A225–33

Muslimov E, Hugot E, Jahn W, Vives S, Ferrari M, Chambion B, Henry D and Gaschet C 2017 Combining freeform optics and curved detectors for wide field imaging: a polynomial approach over squared aperture *Opt. Express* **25** 14598–610

Newton I 1704 *Opticks, or, a Treatise of the Reflections, Refractions, Inflections & Colours of Light* (New York: Dover)

Ochse D 2018 Aberration fields of anamorphic systems *Optical Design and Engineering VII* 10690 (Bellingham, WA: International Society for Optics and Photonics)

Stavroudis O N 2006 *The Mathematics of Geometrical and Physical Optics: The k-Function and Its Ramifications* (New York: Wiley)

Sun H 2016 *Lens Design: A Practical Guide* (Boca Raton, FL: CRC Press)

Zhenrong Z, Xiang H and Xu L 2009 Freeform surface lens for LED uniform illumination *Appl. Opt.* **48** 6627–34

Zhong Y and Gross H 2017 Initial system design method for non-rotationally symmetric systems based on Gaussian brackets and Nodal aberration theory *Opt. Express* **25** 10016–30

Zhong Y and Gross H 2018 Vectorial aberrations of biconic surfaces *J. Opt. Soc. Am.* A **35** 1385–92

**Rafael G González-Acuña, Héctor A Chaparro-Romo and Julio C Gutiérrez-Vega**

# Chapter 10

# On-axis astigmatic freeform lens

We present a lens model such that given the first surface of the singlet, the model calculates a second surface that generates a user-defined astigmatism pattern. The first surface must be continuous and the rays inside the lens must not cross each other.

## 10.1 Introduction

We use lenses to read, to drive or to see better. We often refer to them as a *pair of glasses*, but the real name is spectacle lenses. A spectacle lens consists of a single refractive element which is a constant refractive index material bounded by two polished surfaces. Spectacle lens designers take into account various considerations such as shape, refractive index, design ergonomics.

The spectacles lenses, in the most common cases, are designated with induced astigmatism such that it compensates for the astigmatism of the human eye. Commonly, spectacle lenses are modified toric surfaces with the help of optimization processes.

In this chapter, we present how the second surface of a spectacle lens must be in order that the output wavefront has a predefined astigmatism pattern. The input values of the model are the distance of the object to the first surface, the first surface, the central thickness of the lens, the refraction index and the astigmatism pattern at the output.

## 10.2 Mathematical model

We assume that the spectacle lens is a homogeneous optical element with relative refractive index $n$ and axial thickness $t$, as it can be seen figures 10.1 and 10.2. Its input surface is known, and it is described by the freeform function $z_a(x_a, y_a)$, where the subscript $a$ refers to the coordinates on the input surface. The shape of the output surface is unknown, and it is described by the function $z_b(x_b, y_b)$ to be determined, where the subscript $b$ refers to the coordinates on the output surface. The distance

doi:10.1088/978-0-7503-5774-6ch10  

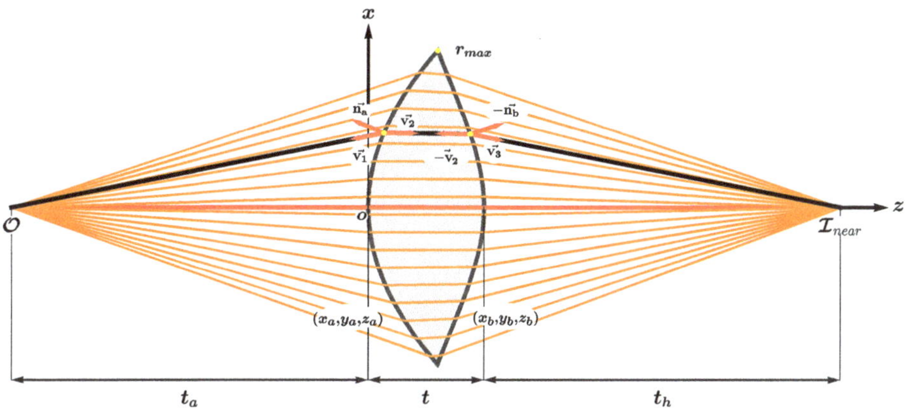


**Figure 10.1.** $x$-plane on-axis freeform astigmatic singlet. The first surface is given by $(x_a, y_a, z_a)$, and the second surface is given by $(x_b, y_b, z_b)$. The gap between the first surface and the object is $t_a$, the thickness at the center of the lens is $t$, and the gap between the second surface and the image near image is $t_h$. $\vec{\mathbf{v}}_1$ is unit vector of the incident ray, $\vec{\mathbf{v}}_2$ is the unit vector of the refracted ray inside the lens, $\vec{\mathbf{v}}_3$ is the output $\vec{\mathbf{n}}_a$ is the normal vector of the first surface and $\vec{\mathbf{n}}_b$ is the normal vector of the second surface.

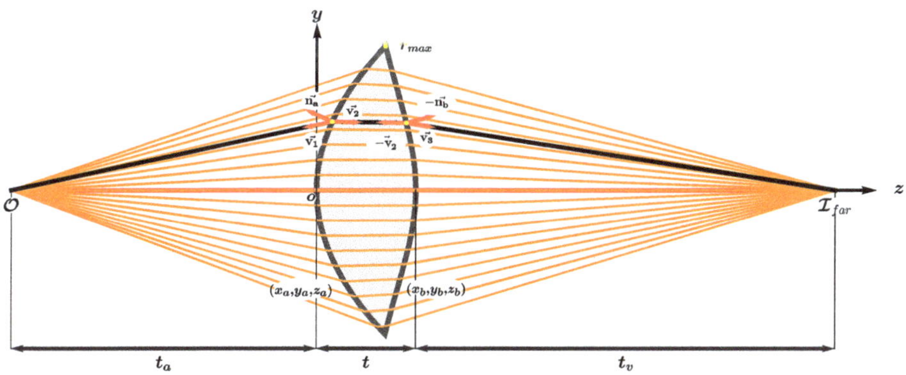


**Figure 10.2.** $y$-plane on-axis freeform astigmatic singlet. The gap between the second surface and the far image is $tv$.

from the object to the first surface is $t_a$. The gap between the second surface and the image is $t_b$. $t_b$ depends on the angle of incidence of the input ray concerning the $x$–$y$-plane. Therefore, $t_b$ is the astigmatic pattern at the output. We will further assume that the normal vector of the input surface at the optical axis points out in direction $z$, i.e. the normal is perpendicular to the tangent plane of the input surface at the origin.

The goal is to determine the output function $z_b(x_b, y_b)$ given the input function $z_a(x_a, y_a)$ in order to give the astigmatic pattern $t_b$ at the output. Therefore, we can use the equation of the on-axis stigmatic singlet and generalize it such that it becomes an on-axis stigmatic singlet. So, let's call equations (9.35), (9.34), (9.25), (9.26), (9.17), (9.18) and (9.18).

$$\begin{cases} x_b = x_a + \vartheta\wp_x, \\ y_b = y_a + \vartheta\wp_y, \\ z_b = z_a + \vartheta\wp_z, \end{cases} \tag{9.35}$$

where, $\vartheta$ is the distance that the rays travel inside the lens.

$$\vartheta \equiv \frac{-(fn + x_a\wp_x + y_a\wp_y + \tau\wp_z)}{(1 - n^2)} \pm \frac{\sqrt{(fn + x_a\wp_x + y_a\wp_y + \tau\wp_z)^2 - (1 - n^2)(x_a^2 + y_a^2 + \tau^2 - f^2)}}{(1 - n^2)}. \tag{9.34}$$

Inside $\vartheta$ are the parameters $f$ and $\tau$ which are defined as,

$$f \equiv -t_a + nt + t_b + \operatorname{sgn}(t_a)\sqrt{x_a^2 + y_a^2 + (z_a - t_a)^2}, \tag{9.25}$$

and,

$$\tau \equiv z_a - t - t_b. \tag{9.26}$$

Finally, the cosine directors of the refracted ray inside the lens, $\wp_x$ is the cosine director in $x$ direction.

$$\wp_x \equiv \frac{x_a\left(z_{a_y}^2 + 1\right) - z_{a_x}\left(y_a z_{a_y} + t_a - z_a\right)}{-\operatorname{sgn}(t_a)n\left(z_{a_x}^2 + z_{a_y}^2 + 1\right)\sqrt{(t_a - z_a)^2 + x_a^2 + y_a^2}} - \frac{z_{a_x}\sqrt{1 - \frac{(y_a z_{a_x} - x_a z_{a_y})^2 + \left[z_{a_x}(z_a - t_a) + x_a\right]^2 + \left[z_{a_y}(z_a - t_a) + y_a\right]^2}{n^2\left(z_{a_x}^2 + z_{a_y}^2 + 1\right)\left[(t_a - z_a)^2 + x_a^2 + y_a^2\right]}}}{\sqrt{z_{a_x}^2 + z_{a_y}^2 + 1}}, \tag{9.17}$$

$\wp_y$ is the cosine director in $y$ direction.

$$\wp_y \equiv \frac{y_a\left(z_{a_x}^2 + 1\right) - z_{a_y}(x_a z_{a_x} + t_a - z_a)}{-\operatorname{sgn}(t_a)n\left(z_{a_x}^2 + z_{a_y}^2 + 1\right)\sqrt{(t_a - z_a)^2 + x_a^2 + y_a^2}} - \frac{z_{a_y}\sqrt{1 - \frac{(y_a z_{a_x} - x_a z_{a_y})^2 + \left[z_{a_x}(z_a - t_a) + x_a\right]^2 + \left[z_{a_y}(z_a - t_a) + y_a\right]^2}{n^2\left(z_{a_x}^2 + z_{a_y}^2 + 1\right)\left[(t_a - z_a)^2 + x_a^2 + y_a^2\right]}}}{\sqrt{z_{a_x}^2 + z_{a_y}^2 + 1}}, \tag{9.18}$$

and, $\wp_z$ in $z$ direction,

$$\wp_z \equiv \frac{-(t_a - z_a)\left(z_{a_x}^2 + z_{a_y}^2\right) + x_a z_{a_x} + y_a z_{a_y}}{-\operatorname{sgn}(t_a) n\left(z_{a_x}^2 + z_{a_y}^2 + 1\right)\sqrt{(t_a - z_a)^2 + x_a^2 + y_a^2}}$$
$$+ \frac{\sqrt{1 - \dfrac{(y_a z_{a_x} - x_a z_{a_y})^2 + [z_{a_x}(z_a - t_a) + x_a]^2 + \left[z_{a_y}(z_a - t_a) + y_a\right]^2}{n^2\left(z_{a_x}^2 + z_{a_y}^2 + 1\right)\left[(t_a - z_a)^2 + x_a^2 + y_a^2\right]}}}{\sqrt{z_{a_x}^2 + z_{a_y}^2 + 1}}. \quad (9.19)$$

We can use the above equations to design on-axis stigmatic freeform lens if and only if $t_b$ is a point image. But what happens if we defined $t_b$ like,

$$\boxed{t_b = \text{a function defined by the user that depends on the angle between } x_a \text{ and } y_a.} \quad (10.1)$$

Equation (10.1) is the most important equation in the chapter, this confirmation may sound odd, but $t_b$ is the predefined astigmatic pattern.

Countless functions accomplish the condition of equation (10.1), for example,

$$t_b = t_h + (t_v - t_h)\theta, \qquad \theta \equiv \frac{2}{\pi}\sin^{-1}\left(\left|\frac{y_a}{\sqrt{x_a^2 + y_a^2}}\right|\right). \quad (10.2)$$

$\theta$ is the angle between $x_a$ and $y_a$ is the $x$–$y$-plane. Remember that the first surface is given by $z_a(x_a, y_a)$ where $x_a$ and $y_a$ are the independent variables. $\frac{2}{\pi}$ is just a constant to normalize $\theta$. $t_h$ is the point image for the horizontal axis of the lens, for $x_a$ equal any value inside the region of the lens and $y_a = 0$. $t_v$ is the point image for the vertical axis of the lens, then $x_a = 0$ and $y_a$ takes any value inside the region of the lens. For this example the image point for $x_a = y_a$, or $x_a = -y_a$ is the mean of $t_h$ and $t_v$, $t_h/2 + t_v/2$.

Since $t_b$ is chosen by the user, $t_h$ and $t_v$ are also chosen by the designer. All the rays of the horizontal axis are in phase with each other. All the rays of the vertical axis are in phase with their group. Please observe, that the rays of the horizontal axis of the singlet, the ones that end in $t_h$ are not in phase with the rays of the vertical axis, the ones that meet in $t_v$. They are in phase if and only if $t_h = t_v$. But the idea is that $t_h \neq t_v$, so they travel different distances with different OPLs. This happens because equation (9.35) only works if at the origin the normal vectors of the first and second surface are parallel to the optical axis.

## 10.3 Gallery of examples

The generality of equation (9.35) allows us to show a large variety of interesting geometries of the singlet lens. In all examples, the input surface is freeform, and the user defines it. The information to reproduce the examples is in the figures. The figures are 10.3, 10.4 and 10.5.

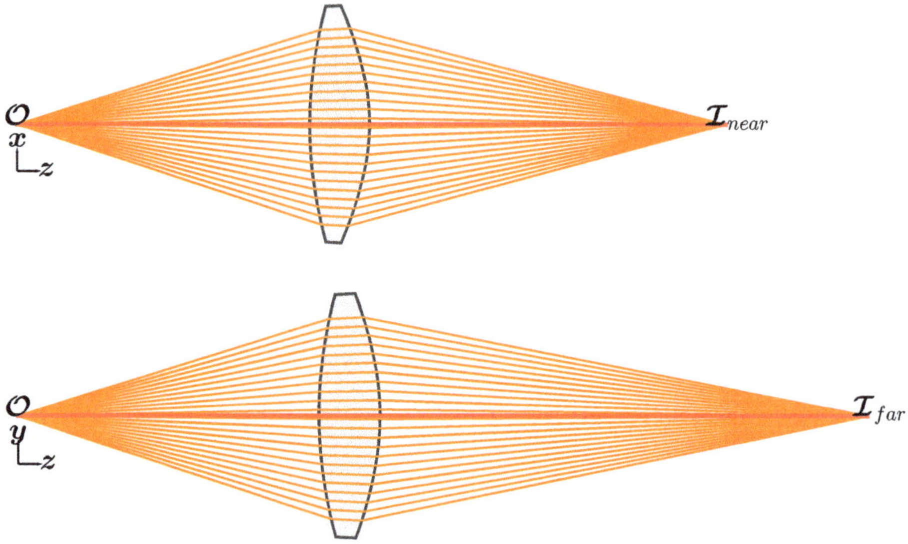


**Figure 10.3.** Design specifications: $n = 1.9$, $t_a = -50$ mm, $t = 10$ mm, $t_h = 80$ mm, $t_v = 80$ mm, $z_a = (x_a^2 + y_a^2)/150$ and $z_b$= equation (9.35).

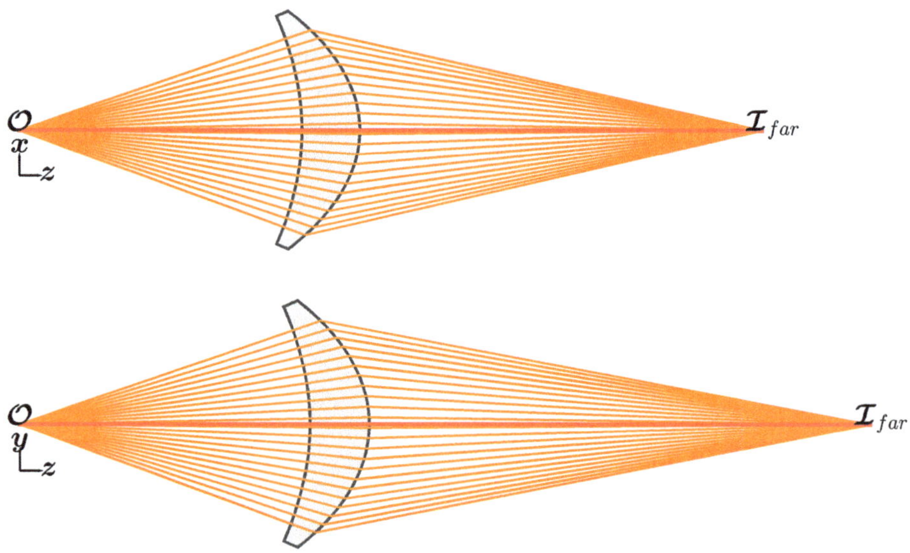


**Figure 10.4.** Design specifications: $n = 1.9$, $t_a = -50$ mm, $t = 10$ mm, $t_h = 70$ mm, $t_v = 85$ mm, $z_a = -(x_a^2 + y_a^2)/90$ and $z_b$= equation (9.35).

## 10.4 Conclusions

In this chapter, we introduced a general formula (10.1) to design a spectacle lens that generates and induced astigmatism pattern. The model works as follows: for a given input surface $(x_a, y_a, z_a)$ and astigmatic pattern $t_b$, the formula yields a second

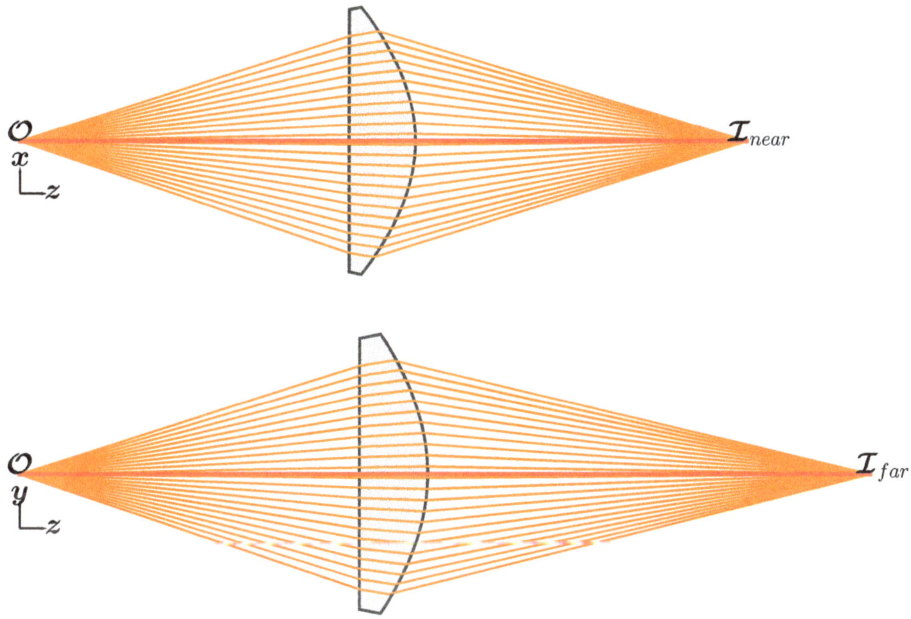


**Figure 10.5.** Design specifications: $n = 1.9$, $t_a = -50$ mm, $t = 10$ mm, $t_h = 50$ mm, $t_v = 65$ mm, $z_a = 0$ and $z_b$= equation (9.35).

surface $(x_b, y_b, z_b)$ that at the output generates an astigmatic pattern defined by $t_b$. We have tested a variety of input surfaces and in all the scenarios equation (9.35) gives the correct behavior.

## Further reading

Acuña R G G and Gutiérrez-Vega J C 2019 General formula of the refractive telescope design free spherical aberration *Novel Optical Systems, Methods, and Applications XXII* 11 105 (Bellingham, WA: International Society for Optics and Photonics)

Born M and Wolf E 2013 Principles of Optics: Electromagnetic Theory of Propagation *Interference and Diffraction of Light* (Amsterdam: Elsevier)

Chaves J 2016 *Introduction to Nonimaging Optics* 2nd edn (Boca Raton, FL: CRC Press)

Davis J K 1974 Geometric optics in ophthalmic lens design *Applications of Geometrical Optics II* vol 39 (Bellingham, WA: International Society for Optics and Photonics) pp 65–100

Fuerter G and Lahres H 1986 Spectacle lens having astigmatic power *US Patent* 4613217

González-Acuña R G and Chaparro-Romo H A 2018 General formula for bi-aspheric singlet lens design free of spherical aberration *Appl. Opt.* **57** 9341–5

González-Acuña R G and Guitiérrez-Vega J C 2018 Generalization of the axicon shape: the gaxicon *J. Opt. Soc. Am.* A **35** 1915–8

González-Acuña R G and Gutiérrez-Vega J C 2019a Analytic formulation of a refractive-reflective telescope free of spherical aberration *Opt. Eng.* **58** 085105

González-Acuña R G and Gutiérrez-Vega J C 2019b General formula for aspheric collimator lens design free of spherical aberration *Current Developments in Lens Design and Optical Engineering XX* vol 11105 (Bellingham, WA: International Society for Optics and Photonics)

González-Acuña R G, Avendaño-Alejo M and Gutiérrez-Vega J C 2019a Singlet lens for generating aberration-free patterns on deformed surfaces *J. Opt. Soc. Am.* A **36** 925–9
González-Acuña R G, Chaparro-Romo H A and Gutiérrez-Vega J C 2019b General formula to design freeform singlet free of spherical aberration and astigmatism *Appl. Opt.* **58** 1010–5
Gross H, Singer W, Totzeck M, Blechinger F and Achtner B 2005 *Handbook of Optical Systems* vol 1 (New York: Wiley Online Library)
Gross H, Zügge H, Peschka M and Blechinger F 2007 Handbook of Optical Systems *Aberration Theory and Correction of Optical Systems* vol 3 (New York: Wiley-VCH)
Gross H, Blechinger F and Achtner B 2008 Handbook of Optical Systems *Survey of Optical Instruments* vol 4 (New York: Wiley-VCH)
Kingslake R 1994 Who? Discovered Coddington's equations? *Opt. Photonics News* **5** 20–3
Kingslake R and Johnson R B 2009 *Lens Design Fundamentals* (New York: Academic)
Landgrave E, Villalobos A and González C 2006 Spectacle lenses incorporating atoric surfaces *US Patent* 7111937
Landgrave J E A and Moya-Cessa J R 1996 Generalized Coddington equations in ophthalmic lens design *J. Opt. Soc. Am.* A **13** 1637–44
Luneburg R K and Herzberger M 1964 *Mathematical Theory of Optics* (Berkeley, CA: University of California Press)
Malacara D 2001 *Handbook of Optical Engineering* (Boca Raton, FL: CRC Press)
Malacara D and Malacara Z 1994 *Handbook of Lens Design* (New York: Dekker)
Malacara-Hernández D and Malacara-Hernández Z 2016 *Handbook of Optical Design* (Boca Raton, FL: CRC Press)
Murray A E 1957 Skew astigmatism at toric surfaces, with special reference to spectacle lenses *J. Opt. Soc. Am.* **47** 599–602
Ogle K N and Ames A 1943 Ophthalmic lens testing instrument *J. Opt. Soc. Am.* **33** 137–42
Pozo A M and Rubiño M 2005 Optical characterization of ophthalmic lenses by means of modulation transfer function determination from a laser speckle pattern *Appl. Opt.* **44** 7744–8
Stavroudis O N 1976 Simpler derivation of the formulas for generalized ray tracing *J. Opt. Soc. Am.* **66** 1330–3
Stavroudis O N 2006 *The Mathematics of Geometrical and Physical Optics: The k-Function and Its Ramifications* (New York: Wiley)
Sun H 2016 *Lens Design: A Practical Guide* (Boca Raton, FL: CRC Press)

# Part IV

## Stigmatic optical systems

**Rafael G González-Acuña, Héctor A Chaparro-Romo and Julio C Gutiérrez-Vega**

# Chapter 11

# On-axis sequential optical systems

We present a comprehensive analytic and closed-form formula to define the form of a surface that adjusts the spherical aberration caused by an arbitrary number of preceding surfaces. This outermost surface makes the optical on-axis system stigmatic.

## 11.1 Introduction

In chapter 4, we introduced the stigmatic lens free of spherical aberration. However, optical systems suffer from other aberrations, and an on-axis stigmatic singlet lens is not sufficient for all the applications. The general strategy in optical design is to set several lenses and start to optimize their specifications to reduce the aberrations. These aberrations typically are the spherical aberration, coma, astigmatism, field of curvature and distortion. This procedure affects the location of the image by way of changing some specification of a lens inside the system to move the resulting image position.

All these aberrations are related, so if someone reduces one aberration, the others are decreased as well. The idea behind this chapter is to eliminate the spherical aberration in an optical system generated by a sequence of refractive surfaces with a singlet last refractive surface that places the image where it needs it. The designer can use the equation proposed in this chapter and focus on the optimization of the other refractive surface to reduce the other aberrations.

## 11.2 Mathematical model

The mathematical model follows the same pattern applied in other chapters. Since now we are working with an arbitrary number of lenses, it is useful to set a notation first.

Please study carefully figure 11.1, where the variables of the system are placed. The optical system is composed of $(N - 1)$ optical elements, where all the surfaces

doi:10.1088/978-0-7503-5774-6ch11

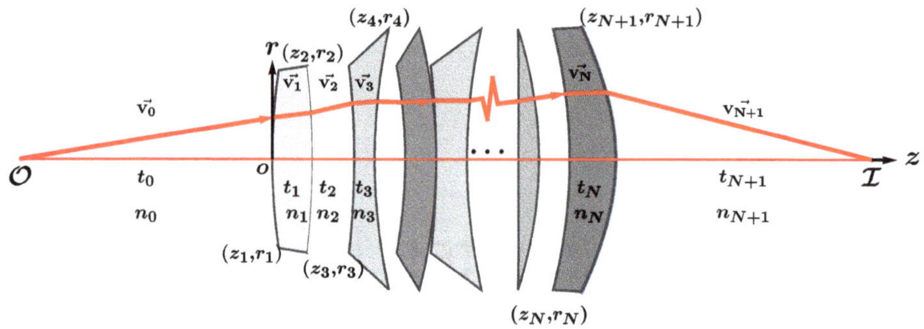


**Figure 11.1.** The original system is composed by $N$ refractive surfaces. An extra corrective surface $N + 1$ can be determined to eliminate the spherical aberration introduced by the original system.

are radially symmetric concerning the optical axis. The first surface of the system is placed on the origin of the Cartesian coordinate system; thus, $z_1(0) = 0$.

We will only work with real point images and real point objects. Therefore, the point object is placed in the second quadrant along the optical axis. The distance between the object and the first surface is $t_0$. The refraction index between the object and the first surface is $n_0$.

Taking this into account, we show optical elements with constant axial thicknesses $t_1, t_2, \ldots, t_{N-1}$ and refractive indices $n_1, n_2, \ldots, n_{N-1}$ in figure 11.1. Therefore, the interfaces between the $N$ thick elements are $N + 1$. These surfaces are defined by the radial functions $z_1(r_1), z_2(r_2), \ldots, z_N(r_N)$, where $r_i$ is the radial coordinate at the $i$th interface.

The course of a light ray moving from the point $(z_i, r_i)$ to the point $(z_{i+1}, r_{i+1})$ inside the $i$th optical section is given by

$$\vec{\mathbf{v}}_i \equiv \wp_{r,i}\vec{\mathbf{e}}_1 + \wp_{z,i}\vec{\mathbf{e}}_2 = \frac{(r_{i+1} - r_i)\vec{\mathbf{e}}_1 + (z_{i+1} - z_i)\vec{\mathbf{e}}_2}{\sqrt{(r_{i+1} - r_i)^2 + (z_{i+1} - z_i)^2}}, \tag{11.1}$$

where $\vec{\mathbf{e}}_1$ is for the $r$ direction and $\vec{\mathbf{e}}_2$ is for the $z$ direction and, $\wp_{r,i}$ and $\wp_{z,i}$ are the Cartesian components of the unit vector $\hat{v}_i$, i.e. its direction cosines, therefore $\wp_{r,\,i}^2 + \wp_{z,\,i}^2 = 1$ for every $i \in \{1, 2, 3, \ldots, N\}$.

Therefore, the distance that a ray travels between the refractive surface $i + 1$ and $i$ is given by

$$\vartheta_i \equiv \sqrt{(r_{i+1} - r_i)^2 + (z_{i+1} - z_i)^2}\,. \tag{11.2}$$

This notation is good for us because the $i$th surface $(z_i, r_i)$ can be related to the previous surface $(z_{i-1}, r_{i-1})$ and so on.

Actually, Snell's law can be seen as a mapping function that maps rays from the surface $(z_{i-1}, r_{i-1})$ to $(z_i, r_i)$. We can apply Snell's law as many times as needed to express $(z_i, r_i)$ in terms of first surface $(z_1, r_1)$. Thus, any point of any refractive surface been expressed in terms of the first surface variables. Our goal is to express the last surface $(z_{N+1}, r_{N+1})$ in terms of the first surface and such that the optical system is free from spherical aberration.

### 11.2.1 Fermat's principle

The refractive surface $(z_{N+1}, r_{N+1})$ that we want to design must be such that the system is free of spherical aberration. Fermat's principle predicts that a system free of spherical aberration has the same OPL for all the rays. Thus, we can obtain an equation from this prediction. We compare the optical path length (OPL) of the reference ray and the OPL of any other ray. Both must be the same. We choose the reference ray as the axial ray because it is the simple ray of the system. The OPL of the reference ray is the following,

$$-t_0 n_0 + \sum_{i=1}^{N+1} n_i t_i = \text{const.} \tag{11.3}$$

The reference ray starts at the image at $-t_0$ because the image is in the second quadrant. Then, it moves in the optical axis until it reaches the image point. This displacement is expressed by $\sum_{i=1}^{N+1} n_i t_i$.

Any other ray has an optical path expressed with the following OPL,

$$\begin{aligned} &n_0\sqrt{r_1^2 + (z_1 - t_0)^2} + \sum_{i=0}^{N-2} n_{i+1}\sqrt{(r_{i+2} - r_{i+1})^2 + (z_{i+2} - z_{i+1})^2} \\ &+ n_N\sqrt{(r_{N+1} - r_N)^2 + (z_{N+1} - z_N)^2} + n_{N+1}\sqrt{r_{N+1}^2 + \left(z_{N+1} - \sum_{i=1}^{N+1} t_i\right)^2} = \text{const.} \end{aligned} \tag{11.4}$$

In the last equation the term $\sqrt{r_1^2 + (z_1 - t_0)^2}$ is the distance between the point object and the first surface. $\sum_{i=0}^{N-2} n_{i+1}\sqrt{(r_{i+2} - r_{i+1})^2 + (z_{i+2} - z_{i+1})^2}$ is the optical path that every ray travels from the first surface to surface $N$. The distance that the light travels from surface $N$ to the unknown surface $N+1$ is $\sqrt{(r_{N+1} - r_N)^2 + (z_{N+1} - z_N)^2}$. Finally, from the last surface to the image the optical path is given by $n_{N+1}\sqrt{r_{N+1}^2 + \left(z_{N+1} - \sum_{i=1}^{N+1} t_i\right)^2}$.

Therefore, we can equal the OPL of the reference ray with the OPL of any other ray,

$$\boxed{\begin{aligned} -t_0 n_0 + \sum_{i=1}^{N+1} n_i t_i = {} & n_0\sqrt{r_1^2 + (z_1 - t_0)^2} \\ &+ \sum_{i=0}^{N-2} n_{i+1}\sqrt{(r_{i+2} - r_{i+1})^2 + (z_{i+2} - z_{i+1})^2} \\ &+ n_N\sqrt{(r_{N+1} - r_N)^2 + (z_{N+1} - z_N)^2} \\ &+ n_{N+1}\sqrt{r_{N+1}^2 + \left(z_{N+1} - \sum_{i=1}^{N+1} t_i\right)^2}. \end{aligned}} \tag{11.5}$$

The above equation is the first fundamental equation in our model. The next step is to study Snell's law.

### 11.2.2 Snell's law

We can express every surface in terms of the first surface. However, we will reveal details of that procedure later. For the moment let's show Snell's law at surface $i$th,

$$\boxed{\vec{\mathbf{v}}_i = \frac{n_{i-1}}{n_i}[\vec{\mathbf{v}}_{i-1} - (\vec{\mathbf{n}}_i \cdot \vec{\mathbf{v}}_{i-1})\vec{\mathbf{n}}_i] - \vec{\mathbf{n}}_i\sqrt{1 + \frac{n_{i-1}^2}{n_i^2}(\vec{\mathbf{n}}_i \wedge \vec{\mathbf{v}}_{i-1})^2} \quad \text{for} \quad \vec{\mathbf{v}}_i, \vec{\mathbf{v}}_{i-1}, \vec{\mathbf{n}}_i \in \mathbb{R}^2,} \tag{11.6}$$

where $\vec{\mathbf{v}}_{i-1}$ is the unit vector of the incident ray, $\vec{\mathbf{v}}_i$ is unit vector of the refracted ray and finally $\vec{\mathbf{n}}_i$ is the normal vector of the $i$th surface,

The unit vectors for the first surface are,

$$\begin{aligned}
\vec{\mathbf{v}}_{i-1} &\equiv \frac{(r_i - r_{i-1})\vec{\mathbf{e}}_1 + (z_i - z_{i-1})\vec{\mathbf{e}}_2}{\sqrt{(r_i - r_{i-1})^2 + (z_i - z_{i-1})^2}},\\
\vec{\mathbf{v}}_i &\equiv \frac{(r_{i+1} - r_i)\vec{\mathbf{e}}_1 + (z_{i+1} - z_i)\vec{\mathbf{e}}_2}{\sqrt{(r_{i+1} - r_i)^2 + (z_{i+1} - z_i)^2}},\\
\vec{\mathbf{n}}_i &\equiv \frac{z_i'\vec{\mathbf{e}}_1 - r_i'\vec{\mathbf{e}}_2}{\sqrt{r_i'^2 + z_i'^2}},
\end{aligned} \tag{11.7}$$

where $z_i' = \mathrm{d}z_i(r_1)/\mathrm{d}r_1$ and $r_i' = \mathrm{d}r_i(r_1)/\mathrm{d}r_1$ are the derivatives of $z_i$ and $r_i$ respect to $r_1$.

We need to replace the unit vectors in Snell's law at the first surface. The procedure is not easy, so let's do it step by step by starting with the most complicated. We start with $(\vec{\mathbf{n}}_i \wedge \vec{\mathbf{v}}_{i-1})$,

$$(\vec{\mathbf{n}}_i \wedge \vec{\mathbf{v}}_{i-1}) = \frac{z_i'(z_i - z_{i-1}) + r_i'(r_i - r_{i-1})}{\sqrt{(r_i - r_{i-1})^2 + (z_i - z_{i-1})^2}\sqrt{r_i'^2 + z_i'^2}}(\vec{\mathbf{e}}_1 \wedge \vec{\mathbf{e}}_2). \tag{11.8}$$

Consider that $(\vec{\mathbf{e}}_1 \wedge \vec{\mathbf{e}}_2)^2 = -1$, therefore $(\vec{\mathbf{n}}_i \wedge \vec{\mathbf{v}}_{i-1})^2$ is given by,

$$(\vec{\mathbf{n}}_i \wedge \vec{\mathbf{v}}_{i-1})^2 = -\frac{[z_i'(z_i - z_{i-1}) + r_i'(r_i - r_{i-1})]^2}{[(r_i - r_{i-1})^2 + (z_i - z_{i-1})^2](r_i'^2 + z_i'^2)}. \tag{11.9}$$

Therefore, the term with the square root, $-\vec{\mathbf{n}}_i\sqrt{1 + \frac{n_{i-1}^2}{n_i^2}(\vec{\mathbf{n}}_i \wedge \vec{\mathbf{v}}_{i-1})^2}$ is expressed as,

$$\begin{aligned}
&-\vec{\mathbf{n}}_i\sqrt{1 + \frac{n_{i-1}^2}{n_i^2}(\vec{\mathbf{n}}_i \wedge \vec{\mathbf{v}}_{i-1})^2}\\
&= -(z_i'\vec{\mathbf{e}}_1 - r_i'\vec{\mathbf{e}}_2)\frac{\sqrt{1 - \frac{n_{i-1}^2[(r_i - r_{i-1})r_i' + (z_i - z_{i-1})z_i']^2}{n_i^2[(r_{i-1} - r_i)^2 + (z_{i-1} - z_i)^2](r_i'^2 + z_i'^2)}}}{\sqrt{r_i'^2 + z_i'^2}}.
\end{aligned} \tag{11.10}$$

Now let's pay attention to $(\vec{\mathbf{n}}_i \cdot \vec{\mathbf{v}}_{i-1})\vec{\mathbf{n}}_i$,

$$(\vec{\mathbf{n}}_i \cdot \vec{\mathbf{v}}_{i-1})\vec{\mathbf{n}}_i = \frac{(z_{i-1} - z_i)r_i' + (r_i - r_{i-1})z_i'}{\sqrt{(r_i - r_{i-1})^2 + (z_i - z_{i-1})^2}\,(r_i'^2 + z_i'^2)}(z_i'\vec{\mathbf{e}}_1 - r_i'\vec{\mathbf{e}}_2). \tag{11.11}$$

The next step is to work with, $(\vec{\mathbf{v}}_{i-1} - (\vec{\mathbf{n}}_i \cdot \vec{\mathbf{v}}_{i-1})\vec{\mathbf{n}}_i)$,

$$\begin{aligned}(\vec{\mathbf{v}}_{i-1} - (\vec{\mathbf{n}}_i \cdot \vec{\mathbf{v}}_{i-1})\vec{\mathbf{n}}_i) \;=\; &+\frac{(r_i - r_{i-1})\vec{\mathbf{e}}_1 + (z_i - z_{i-1})\vec{\mathbf{e}}_2}{\sqrt{(r_i - r_{i-1})^2 + (z_i - z_{i-1})^2}} \\ &- \frac{(z_{i-1} - z_i)r_i' + (r_i - r_{i-1})z_i'}{\sqrt{(r_i - r_{i-1})^2 + (z_i - z_{i-1})^2}\,(r_i'^2 + z_i'^2)}(z_i'\vec{\mathbf{e}}_1 - r_i'\vec{\mathbf{e}}_2).\end{aligned} \tag{11.12}$$

Multiplying by $\frac{n_{i-1}}{n_i}$ and simplifying,

$$\begin{aligned}\frac{n_{i-1}}{n_l}(\vec{\mathbf{v}}_{i-1} - (\vec{\mathbf{n}}_i \cdot \vec{\mathbf{v}}_{i-1})\vec{\mathbf{n}}_i) = &\frac{n_{i-1}r_i'[(r_i - r_{i-1})r_i' + (z_i - z_{i-1})z_i']\vec{\mathbf{e}}_1}{n_i\sqrt{(r_{i-1} - r_i)^2 + (z_{i-1} - z_i)^2}\,(r_i'^2 + z_i'^2)} \\ &+ \frac{n_{i-1}z_i'[(r_i - r_{i-1})r_i' + (z_i - z_{i-1})z_i']\vec{\mathbf{e}}_2}{n_i\sqrt{(r_{i-1} - r_i)^2 + (z_{i-1} - z_1)^2}\,(r_i'^2 + z_i'^2)}.\end{aligned} \tag{11.13}$$

Now, let sum equation (11.10) and (11.13), thus, $\frac{n_{i-1}}{n_i}[\vec{\mathbf{v}}_{i-1} - (\vec{\mathbf{n}}_i \cdot \vec{\mathbf{v}}_{i-1})\vec{\mathbf{n}}_i] -\vec{\mathbf{n}}_i\sqrt{1 + \frac{n_{i-1}^2}{n_i^2}(\vec{\mathbf{n}}_i \wedge \vec{\mathbf{v}}_{i-1})^2}$.

$$\begin{aligned}\vec{\mathbf{v}}_i = &\frac{n_{i-1}r_i'[(r_i - r_{i-1})r_i' + (z_i - z_{i-1})z_i']\vec{\mathbf{e}}_1}{n_i\sqrt{(r_{i-1} - r_i)^2 + (z_{i-1} - z_i)^2}\,(r_i'^2 + z_i'^2)} \\ &+ \frac{n_{i-1}z_i'[(r_i - r_{i-1})r_i' + (z_i - z_{i-1})z_i']\vec{\mathbf{e}}_2}{n_i\sqrt{(r_{i-1} - r_i)^2 + (z_{i-1} - z_1)^2}\,(r_i'^2 + z_i'^2)} \\ &- (z_i'\vec{\mathbf{e}}_1 - r_i'\vec{\mathbf{e}}_2)\frac{\sqrt{1 - \dfrac{n_{i-1}^2[(r_i - r_{i-1})r_i' + (z_i - z_{i-1})z_i']^2}{n_i^2[(r_{i-1} - r_i)^2 + (z_{i-1} - z_i)^2](r_i'^2 + z_i'^2)}}}{\sqrt{r_i'^2 + z_i'^2}}.\end{aligned} \tag{11.14}$$

Since, $\vec{\mathbf{v}}_i \equiv \wp_{r,i}\vec{\mathbf{e}}_1 + \wp_{z,i}\vec{\mathbf{e}}_2$, the cosine directors $\wp_{r,i}$ and $\wp_{z,i}$ are,

$$\begin{aligned}\wp_{r,i} = &\frac{n_{i-1}r_i'[(r_i - r_{i-1})r_i' + (z_i - z_{i-1})z_i']}{n_i\sqrt{(r_{i-1} - r_i)^2 + (z_{i-1} - z_i)^2}\,(r_i'^2 + z_i'^2)} \\ &- \frac{z_i'\sqrt{1 - \dfrac{n_{i-1}^2[(r_i - r_{i-1})r_i' + (z_i - z_{i-1})z_i']^2}{n_i^2[(r_{i-1} - r_i)^2 + (z_{i-1} - z_i)^2](r_i'^2 + z_i'^2)}}}{\sqrt{r_i'^2 + z_i'^2}},\end{aligned} \tag{11.15}$$

and,

$$\wp_{z,i} = \frac{n_{i-1}z_i'[(r_i - r_{i-1})r_i' + (z_i - z_{i-1})z_i']}{n_i\sqrt{(r_{i-1} - r_i)^2 + (z_{i-1} - z_1)^2}(r_i'^2 + z_i'^2)} + \frac{r_i'\sqrt{1 - \frac{n_{i-1}^2[(r_i - r_{i-1})r_i' + (z_i - z_{i-1})z_i']^2}{n_i^2[(r_{i-1} - r_i)^2 + (z_{i-1} - z_i)^2](r_i'^2 + z_i'^2)}}}{\sqrt{r_i'^2 + z_i'^2}}. \tag{11.16}$$

The above equations present how the cosine directors at the surface $i$th must be. This cosine directors start in surface $i$th and point to surface $i + 1$ th.

### 11.2.3 Solution

For the description of the last surface ($z_{N+1}$, $r_{N+1}$) is very convenient to express it in terms of the cosine directions of the $N$ surface. From the procedure of the last section, we can conclude that the cosine directors at $N$ surface are given by,

$$\wp_{r,N} = \frac{n_{N-1}r_N'[(r_N - r_{N-1})r_N' + (z_N - z_{N-1})z_N']}{n_N\sqrt{(r_{N-1} - r_N)^2 + (z_{N-1} - z_N)^2}(r_N'^2 + z_N'^2)} - \frac{z_N'\sqrt{1 - \frac{n_{N-1}^2[(r_N - r_{N-1})r_N' + (z_N - z_{N-1})z_N']^2}{n_N^2[(r_{N-1} - r_N)^2 + (z_{N-1} - z_N)^2](r_N'^2 + z_N'^2)}}}{\sqrt{r_N'^2 + z_N'^2}}, \tag{11.17}$$

and,

$$\wp_{z,N} = \frac{n_{N-1}z_N'((r_N - r_{N-1})r_N' + (z_N - z_{N-1})z_N')}{n_N\sqrt{(r_{N-1} - r_N)^2 + (z_{N-1} - z_1)^2}(r_N'^2 + z_N'^2)} + \frac{r_N'\sqrt{1 - \frac{n_{N-1}^2[(r_N - r_{N-1})r_N' + (z_N - z_{N-1})z_N']^2}{n_N^2[(r_{N-1} - r_N)^2 + (z_{N-1} - z_N)^2](r_N'^2 + z_N'^2)}}}{\sqrt{r_N'^2 + z_N'^2}}. \tag{11.18}$$

Also, we know that the vector $\vec{\mathbf{v}}_N$ is given by,

$$\vec{\mathbf{v}}_N \equiv \wp_{r,N}\vec{\mathbf{e}}_1 + \wp_{z,N}\vec{\mathbf{e}}_2 = \frac{(r_{N+1} - r_N)\vec{\mathbf{e}}_1 + (z_{N+1} - z_i)\vec{\mathbf{e}}_2}{\sqrt{(r_{N+1} - r_i)^2 + (z_{N+1} - z_i)^2}}, \tag{11.19}$$

and the distance $\vartheta_N$ is defined as,

$$\vartheta_N \equiv \sqrt{(r_{N+1} - r_N)^2 + (z_{N+1} - z_N)^2}. \tag{11.20}$$

Therefore, we can express the components of the unit vector $\vec{\mathbf{v}}_N$ as,

$$\frac{z_{N+1} - z_N}{\vartheta_N} = \wp_{z,N}, \tag{11.21}$$

and,

$$\frac{r_{N+1} - r_N}{\vartheta_N} = \wp_{r,N}. \tag{11.22}$$

Solving for $z_N$ and $r_N$,

$$z_{N+1} = z_N + \vartheta \wp_N, \tag{11.23}$$

and,

$$r_{N+1} = r_N + \vartheta_N \wp_N. \tag{11.24}$$

We get the same structure of the solution, we just need to know $\vartheta_N$. So, the next step is to recall equation (11.5),

$$\begin{aligned}
-t_0 n_0 + \sum_{i=1}^{N+1} n_i t_i = {} & n_0\sqrt{r_1^2 + (z_1 - t_0)^2} \\
& + \sum_{i=0}^{N-2} n_{i+1}\sqrt{(r_{i+2} - r_{i+1})^2 + (z_{i+2} - z_{i+1})^2} \\
& + n_N\sqrt{(r_{N+1} - r_N)^2 + (z_{N+1} - z_N)^2} \\
& + n_{N+1}\sqrt{r_{N+1}^2 + \left(z_{N+1} - \sum_{i=1}^{N+1} t_i\right)^2}.
\end{aligned} \tag{11.5}$$

We replace equation (11.20) in equation (11.5), and we replace $z_{N+1} = z_N + \vartheta_N \wp_{z,N}$ and $r_{N+1} = r_N + \vartheta_N \wp_{r,N}$ also in in equation (11.5),

$$\begin{aligned}
\sum_{i=1}^{N+1} n_i t_i = {} & n_0\sqrt{r_1^2 + (z_1 - t_0)^2} + t_0 n_0 \\
& + \sum_{i=0}^{N-2} n_{i+1}\sqrt{(r_{i+2} - r_{i+1})^2 + (z_{i+2} - z_{i+1})^2} \\
& + n_N \vartheta_N \\
& + n_{N+1}\sqrt{(r_N + \vartheta_N \wp_{r,N})^2 + \left(z_N + \vartheta_N \wp_{z,N} - \sum_{i=1}^{N+1} t_i\right)^2}.
\end{aligned} \tag{11.25}$$

To simplify the above expression, we assign parameters to the terms of equation (11.25) without $z_{N+1}$ and $r_{N+1}$,

$$\begin{aligned}
f \equiv {} & -t_0 n_0 + \sum_{i=1}^{N+1} n_i t_i - n_0\sqrt{r_1^2 + (z_1 - t_0)^2} \\
& - \sum_{i=0}^{N-2} n_{i+1}\sqrt{(r_{i+2} - r_{i+1})^2 + (z_{i+2} - z_{i+1})^2},
\end{aligned} \tag{11.26}$$

and,

$$\tau \equiv z_N - \sum_{i=1}^{N+1} t_i, \tag{11.27}$$

replacing equations (11.26) and (11.27) in equation (11.25),

$$(f - n_N \vartheta_N)^2 = n_{N+1}^2 (r_N + \vartheta_N \wp_{r,N})^2 + n_{N+1}^2 (\tau + \vartheta_N \wp_{z,N})^2, \tag{11.28}$$

expanding the binomials,

$$\begin{aligned} f^2 - 2fn_N \vartheta_N + n_N^2 \vartheta_N^2 &= n_{N+1}^2 (r_N^2 + 2r_N \wp_{r,N} \vartheta_N + \wp_{r,\,N}^2 \vartheta_N^2) \\ &\quad + n_{N+1}^2 (\tau^2 + 2\tau \wp_{z,N} \vartheta_N + \wp_{z,\,N}^2 \vartheta_N^2), \end{aligned} \tag{11.29}$$

collecting for $\vartheta_N$,

$$(n_{N+1}^2 - n_N^2)\vartheta_N^2 + 2[fn_N + n_{N+1}^2 (r_N \wp_{r,N} + \tau \wp_{z,N})]\vartheta_N + [n_{N+1}^2 (r_N^2 + \tau^2) - f^2] = 0. \tag{11.30}$$

The next, step is to multiply every element by $4(n_{N+1}^2 - n_N^2)$,

$$\begin{aligned} &4(n_{N+1}^2 - n_N^2)^2 \vartheta_N^2 + 8(n_{N+1}^2 - n_N^2)[fn_N + n_{N+1}^2 (r_N \wp_{r,N} + \tau \wp_{z,N})]\vartheta_N \\ &+ 4(n_{N+1}^2 - n_N^2)[n_{N+1}^2 (r_N^2 + \tau^2) - f^2] = 0. \end{aligned} \tag{11.31}$$

Then we sum $4[fn_N + n_{N+1}^2 (r_N \wp_{r,N} + \tau \wp_{z,N})]^2$ in both sides,

$$\begin{aligned} &4(n_{N+1}^2 - n_N^2)^2 \vartheta_N^2 + 8(n_{N+1}^2 - n_N^2)[fn_N + n_{N+1}^2 (r_N \wp_{r,N} + \tau \wp_{z,N})]\vartheta_N \\ &+ 4[fn_N + n_{N+1}^2 (r_N \wp_{r,N} + \tau \wp_{z,N})]^2 + 4(n_{N+1}^2 - n_N^2)[n_{N+1}^2 (r_N^2 + \tau^2) - f^2] \\ &= 4[fn_N + n_{N+1}^2 (r_N \wp_{r,N} + \tau \wp_{z,N})]^2. \end{aligned} \tag{11.32}$$

From the last equation we can construct square binomial which is given by,

$$\begin{aligned} &\left\{2(n_{N+1}^2 - n_N^2)\vartheta_N + 2[fn_N + n_{N+1}^2 (r_N \wp_{r,N} + \tau \wp_{z,N})]\right\}^2 \\ &= 4[fn_N + n_{N+1}^2 (r_N \wp_{r,N} + \tau \wp_{z,N})]^2 - 4(n_{N+1}^2 - n_N^2)[n_{N+1}^2 (r_N^2 + \tau^2) - f^2]. \end{aligned} \tag{11.33}$$

Applying the square root in both sides,

$$\begin{aligned} &2(n_{N+1}^2 - n_N^2)\vartheta_N + 2[fn_N + n_{N+1}^2 (r_N \wp_{r,N} + \tau \wp_{z,N})] \\ &= \sqrt{4[fn_N + n_{N+1}^2 (r_N \wp_{r,N} + \tau \wp_{z,N})]^2 - 4(n_{N+1}^2 - n_N^2)[n_{N+1}^2 (r_N^2 + \tau^2) - f^2]}. \end{aligned} \tag{11.34}$$

Solving for $\vartheta_N$,

$$\begin{aligned} \vartheta_N = &\frac{-[fn_N + n_{N+1}^2 (r_N \wp_{r,N} + \tau \wp_{z,N})]}{(n_{N+1}^2 - n_N^2)} \\ &\pm \frac{\sqrt{[fn_N + n_{N+1}^2 (r_N \wp_{r,N} + \tau \wp_{z,N})]^2 - (n_{N+1}^2 - n_N^2)[n_{N+1}^2 (r_N^2 + \tau^2) - f^2]}}{(n_{N+1}^2 - n_N^2)}. \end{aligned} \tag{11.35}$$

The last surface of the optical system is given by

$$\boxed{\begin{cases} z_{N+1} = z_N + \vartheta_N \wp_{z,N}, \\ r_{N+1} = r_N + \vartheta_N \wp_{r,N}. \end{cases}} \tag{11.36}$$

Equation (11.36) express the surface $N + 1$. This surface is the unique additional surface that can be placed on the optical systems such that the system is free of spherical aberration.

For the same reason studied in chapter 6, the rays inside the optical system should not cross each other.

In the next section, we finally show the details of how to express all the surface in terms of the first one. The following procedure is needed for correct use of equation (11.36).

### 11.2.4 Surfaces expressed in terms of the refracted rays

The purpose is to express a surface in terms of the rays refracted by the previous surface. For instance, let us do it for a lens formed of two sequential spherical surfaces, whose explicit forms are given by

$$\begin{aligned} z_1 &= R_1 - \sqrt{R_1^2 - r_1^2}, \\ z_2 &= t_1 + R_2 - \sqrt{R_2^2 - r_2^2}, \end{aligned} \tag{11.37}$$

where $R_1$ and $R_2$ are the radii of the first and second surfaces, respectively, and $t_1$ is the central thickness of the lens. We need to write the second surface in terms of the rays refracted by the first surface, for that we use the displacement $\vartheta_1$ of the refracted rays. This scaling is obtained from the algebraic manipulation of the equation (11.38), using the real rays tracing,

$$\begin{aligned} r_2 &= r_1 + \vartheta_1 \wp_{r,1}, \\ z_2 &= z_1 + \vartheta_1 \wp_{z,1}, \end{aligned} \tag{11.38}$$

where, $\wp_{r,1}$ and $\wp_{r,1}$ are direction cosines at the first surface, $\vartheta_1$ is the function that maps the refracted rays by the first surface until they reach the second surface. We now relate both equations (11.37) and (11.38), namely

$$t_1 + R_2 - \sqrt{R_2^2 - (r_1 + \vartheta_1 \wp_{r,1})^2} = z_1 + \vartheta_1 \wp_{z,1}, \tag{11.39}$$

then, finally we solve for $\vartheta_1$,

$$\begin{aligned} \vartheta_1 = - &\sqrt{\begin{aligned} &-2 r_1 \wp_{r,1} \wp_{z,1} (R_2 + t_1 - z_1) + \wp_{z,1}^2 \\ &(R_2 - r_1)(r_1 + R_2) - \wp_{r,1}^2 (t_1 - z_1)(2R_2 + t_1 - z_1) \end{aligned}} \\ &- r_1 \wp_{r,1} + \wp_{z,1}(R_2 + t_1 - z_1). \end{aligned} \tag{11.40}$$

This procedure must be repeated for each surface of the system until the only independent variable is $r_1$. Every surface can be expressed in terms of the rays refracted by the preceding surfaces, so, every surface can be expressed as

$$\begin{aligned} r_{i+1} &= r_i + \vartheta_i \wp_{r,i}, \\ z_{i+1} &= z_i + \vartheta_i \wp_{r,i}, \end{aligned} \tag{11.41}$$

where $\vartheta_i$ is the respective displacement of the refracted ray by the surface $i$ to reach the surface $i + 1$. Once all the surfaces are expressed in the terms of the refracted rays using the form of equation (11.41), we can begin to design the last refractive surface with equation (11.36).

## 11.3 Illustrative examples

It is time to design on-axis stigmatic systems with equation (11.36). In all the following examples, we use the procedure to explain it in subsection 11.2.4. All the examples use equation (11.36), and the general structure of every surface can be seen as,

$$r_{i+1} = r_i + \vartheta_i \wp_{r,i}, \quad z_{i+1} = z_i + \vartheta_i \wp_{r,i}. \tag{11.42}$$

The design specifications of the first example can be seen in table 11.1 and the ray tracing in figure 11.2.

The last surface of the of the first example is given by,

$$z_4 = z_3 + \vartheta_3 \wp_{z,3}, \quad r_4 = r_3 + \vartheta_3 \wp_{r,3}. \tag{11.43}$$

**Table 11.1.** The image is located at $z = 8 + 11 + 9 + 90 = 118$ mm.

| Surface | Profile | Refraction index | Thickness |
|---|---|---|---|
| 0 | Object | $n_0 = 1$ | $t_0 = -50$ mm |
| 1 | $z_1 = r_1^2/f_1$ | $n_1 = 1.5$ | $t_1 = 8$ mm |
| 2 | $z_2 = t_1 + r_2^2/f_2$ | $n_2 = 1$ | $t_2 = 11$ mm |
| 3 | $z_3 = t_1 + t_2 + r_3^2/f_3$ | $n_3 = 1.5$ | $t_3 = 9$ mm |
| 4 | Equation (11.43) | $n_4 = 1$ | $t_4 = 90$ mm |
| 5 | Image | Not apply | Not apply |

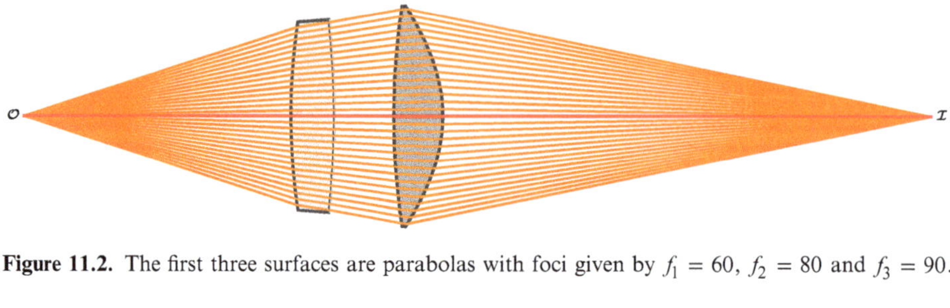


**Figure 11.2.** The first three surfaces are parabolas with foci given by $f_1 = 60$, $f_2 = 80$ and $f_3 = 90$.

where,

$$
\begin{cases}
\vartheta_3 = \dfrac{-[fn_3 + n_4^2(r_3\wp_{r,3} + \tau\wp_{z,3})] \pm \sqrt{[fn_3 + n_4^2(r_3\wp_{r,3} + \tau\wp_{z,3})]^2 - (n_4^2 - n_3^2)[n_4^2(r_3^2 + \tau^2) - f^2]}}{(n_4^2 - n_3^2)} \\
\tau = z_3 - \sum_{i=1}^{4} t_i \\
f = -t_0 n_0 + \sum_{i=1}^{4} n_i t_i - n_0\sqrt{r_1^2 + (z_1 - t_0)^2} \\
\quad - n_1\sqrt{(r_2 - r_1)^2 + (z_2 - z_1)^2} - n_2\sqrt{(r_3 - r_2)^2 + (z_3 - z_2)^2} \\
\vartheta_2 = [2\sqrt{f_3}\sqrt{f_3\wp_{z,2}^2 + r_2\wp_{r,2}\wp_{z,2} + \wp_{r,2}^2(t_1 + t_2 - z_2)} - 2f_3\wp_{z,2} \\
\quad - r_2\wp_{r,2}]/(\wp_{r,2}^2), \\
r_3 = \vartheta_2\wp_{r,2} + r_2, \\
z_3 = \vartheta_2\wp_{z,2} + z_2, \\
\vartheta_1 = [2\sqrt{f_2}\sqrt{f_2\wp_{z,1}^2 + r_1\wp_{r,1}\wp_{z,1} + \wp_{r,1}^2(t_1 - z_1)} - 2f_2\wp_{z,1} - r_1\wp_{r,1}] \\
\quad /(\wp_{r,1}^2), \\
r_2 = \vartheta_1\wp_{r,1} + r_1, \\
z_2 = \vartheta_1\wp_{z,1} + z_1, \\
z_1 = r_1^2/f_1.
\end{cases}
\tag{11.44}
$$

The design specifications of the first example can be seen in table 11.2 and the ray tracing in figure 11.3.

**Table 11.2.** The image is located at $z = 8 + 11 + 9 + 90 = 118$ mm.

| Surface | Profile | Refraction index | Thickness |
|---|---|---|---|
| 0 | Object | $n_0 = 1$ | $t_0 = -40$ mm |
| 1 | $z_1 = r_1^2/f_1$ | $n_1 = 1.5$ | $t_1 = 8$ mm |
| 2 | $z_2 = t_1 + r_2^2/f_2$ | $n_2 = 1$ | $t_2 = 0.1$ mm |
| 3 | $z_3 = t_1 + t_2 + r_3^2/f_3$ | $n_3 = 1.5$ | $t_3 = 9$ mm |
| 4 | Equation (11.45) | $n_4 = 1$ | $t_4 = 90$ mm |
| 5 | Image | Not apply | Not apply |

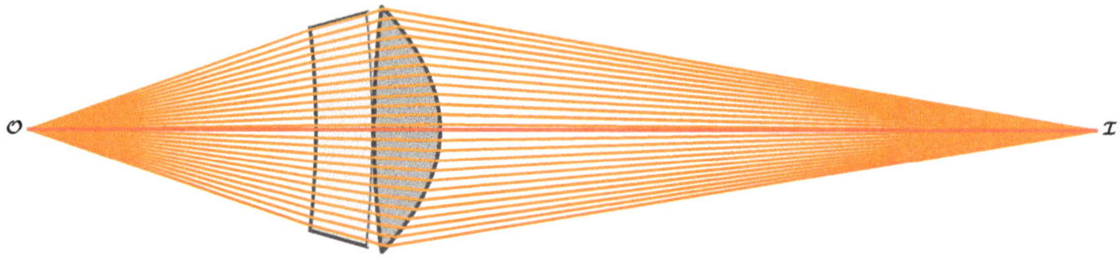


**Figure 11.3.** The first three surfaces are parabolas with focuses given by $f_1 = -60$, $f_2 = 80$ and $f_3 = -70$.

The last surface of the of the first example is given by,

$$z_4 = z_3 + \vartheta_3 \wp_{z,3}, \qquad r_4 = r_3 + \vartheta_3 \wp_{r,3}, \tag{11.45}$$

where,

$$\begin{cases} \vartheta_3 = \dfrac{-[f n_3 + n_4^2(r_3 \wp_{r,3} + \tau \wp_{z,3})] \pm \sqrt{[f n_3 + n_4^2(r_3 \wp_{r,3} + \tau \wp_{z,3})]^2 - (n_4^2 - n_3^2)[n_4^2(r_3^2 + \tau^2) - f^2]}}{(n_4^2 - n_3^2)} \\ \tau = z_3 - \sum_{i=1}^{4} t_i \\ f = -t_0 n_0 + \sum_{i=1}^{4} n_i t_i - n_0\sqrt{r_1^2 + (z_1 - t_0)^2} - n_1\sqrt{(r_2 - r_1)^2 + (z_2 - z_1)^2} \\ \quad - n_2\sqrt{(r_3 - r_2)^2 + (z_3 - z_2)^2} \\ \vartheta_2 = [2\sqrt{f_3}\sqrt{f_3 \wp_{z,2}^2 + r_2 \wp_{r,2} \wp_{z,2} + \wp_{r,2}^2(t_1 + t_2 - z_2)} - 2 f_3 \wp_{z,2} - r_2 \wp_{r,2}] \\ \quad /(\wp_{r,2}^2), \\ r_3 = \vartheta_2 \wp_{r,2} + r_2, \\ z_3 = \vartheta_2 \wp_{z,2} + z_2, \\ \vartheta_1 = [2\sqrt{f_2}\sqrt{f_2 \wp_{z,1}^2 + r_1 \wp_{r,1} \wp_{z,1} + \wp_{r,1}^2(t_1 - z_1)} - 2 f_2 \wp_{z,1} - r_1 \wp_{r,1}] \\ \quad /(\wp_{r,1}^2), \\ r_2 = \vartheta_1 \wp_{r,1} + r_1, \\ z_2 = \vartheta_1 \wp_{z,1} + z_1, \\ z_1 = r_1^2 / f_1. \end{cases} \tag{11.46}$$

The design specifications of the third example can be seen in table 11.3 and the ray tracing in figure 11.4. The last surface of the triplet is given by,

$$z_4 = z_3 + \vartheta_3 \wp_{z,3}, \qquad r_4 = r_3 + \vartheta_3 \wp_{r,3}. \tag{11.47}$$

**Table 11.3.** The image is located at $z = 8 + 4 + 9 + 90 = 111$ mm.

| Surface | Profile | Refraction index | Thickness |
|---|---|---|---|
| 0 | Object | $n_0 = 1$ | $t_0 = -40$ mm |
| 1 | $z_1 = r_1^2/f_1$ | $n_1 = 1.5$ | $t_1 = 8$ mm |
| 2 | $z_2 = t_1 + r_2^2/f_2$ | $n_2 = 1.7$ | $t_2 = 4$ mm |
| 3 | $z_3 = t_1 + t_2 + r_3^2/f_3$ | $n_3 = 1.5$ | $t_3 = 9$ mm |
| 4 | Equation (11.45) | $n_4 = 1$ | $t_4 = 90$ mm |
| 5 | Image | Not apply | Not apply |

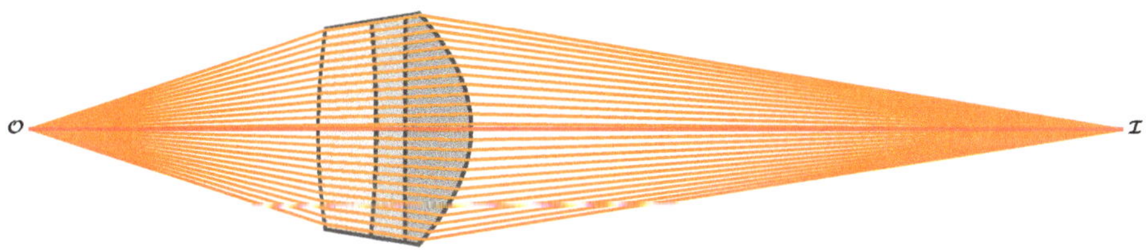


**Figure 11.4.** The first three surfaces are parabolas with foci given by $f_1 = 50$, $f_2 = 80$ and $f_3 = 9000$.

where,

$$\begin{cases}
\vartheta_3 = \dfrac{\begin{aligned}&-[fn_3 + n_4^2(r_3\wp_{r,3} + \tau\wp_{z,3})]\\ &\pm\sqrt{[fn_3 + n_4^2(r_3\wp_{r,3} + \tau\wp_{z,3})]^2 - (n_4^2 - n_3^2)[n_4^2(r_3^2 + \tau^2) - f^2]}\end{aligned}}{(n_4^2 - n_3^2)} \\
\tau = z_3 - \sum_{i=1}^{4} t_i \\
f = -t_0 n_0 + \sum_{i=1}^{4} n_i t_i - n_0\sqrt{r_1^2 + (z_1 - t_0)^2} - n_1\sqrt{(r_2 - r_1)^2 + (z_2 - z_1)^2} \\
\quad - n_2\sqrt{(r_3 - r_2)^2 + (z_3 - z_2)^2} \\
\vartheta_2 = [2\sqrt{f_3}\sqrt{f_3\wp_{z,2}^2 + r_2\wp_{r,2}\wp_{z,2} + \wp_{r,2}^2(t_1 + t_2 - z_2)} - 2f_3\wp_{z,2} \\
\quad - r_2\wp_{r,2}]/(\wp_{r,2}^2), \\
r_3 = \vartheta_2\wp_{r,2} + r_2, \\
z_3 = \vartheta_2\wp_{z,2} + z_2, \\
\vartheta_1 = [2\sqrt{f_2}\sqrt{f_2\wp_{z,1}^2 + r_1\wp_{r,1}\wp_{z,1} + \wp_{r,1}^2(t_1 - z_1)} - 2f_2\wp_{z,1} - r_1\wp_{r,1}] \\
\quad /(\wp_{r,1}^2), \\
r_2 = \vartheta_1\wp_{r,1} + r_1, \\
z_2 = \vartheta_1\wp_{z,1} + z_1, \\
z_1 = r_1^2/f_1.
\end{cases} \tag{11.48}$$

## 11.4 Conclusions

We have generalized the on-axis stigmatic lens into an on-axis stigmatic optical system. The outermost surface of this system is such that the system is free from spherical aberration. The derivation of the outermost surface has been done without any optimization process and paraxial approximation. In the example we use only four refractive surfaces but it could be $N + 1 \in \mathbb{N}$.

## Futher reading

Acuña R G G and Gutiérrez-Vega J C 2019 General formula of the refractive telescope design free spherical aberration *Novel Optical Systems, Methods, and Applications XXII* 11105 (Bellingham, WA: International Society for Optics and Photonics)

Born M and Wolf E 2013 Principles of Optics: Electromagnetic Theory of Propagation *Interference and Diffraction of Light* (Amsterdam: Elsevier)

Chaves J 2016 *Introduction to Nonimaging Optics* 2nd edn (Boca Raton, FL: CRC Press)

González-Acuña R G and Chaparro-Romo H A 2018 General formula for bi-aspheric singlet lens design free of spherical aberration *Appl. Opt.* **57** 9341–5

González-Acuña R G and Guitiérrez-Vega J C 2018 Generalization of the axicon shape: the gaxicon *J. Opt. Soc. Am.* A **35** 1915–8

González-Acuña R G and Gutiérrez-Vega J C 2019a Analytic formulation of a refractive-reflective telescope free of spherical aberration *Opt. Eng.* **58** 085105

González-Acuña R G and Gutiérrez-Vega J C 2019b General formula for aspheric collimator lens design free of spherical aberration *Current Developments in Lens Design and Optical Engineering XX* (Bellingham, WA: International Society for Optics and Photonics) p 111040P

González-Acuña R G, Avendaño-Alejo M and Gutiérrez-Vega J C 2019a Singlet lens for generating aberration-free patterns on deformed surfaces *J. Opt. Soc. Am.* A **36** 925–9

González-Acuña R G, Chaparro-Romo H A and Gutiérrez-Vega J C 2019b General formula to design freeform singlet free of spherical aberration and astigmatism *Appl. Opt.* **58** 1010–5

Gross H, Singer W, Totzeck M, Blechinger F and Achtner B 2005 *Handbook of Optical Systems* vol 1 (New York: Wiley Online Library)

Gross H, Zügge H, Peschka M and Blechinger F 2007 *Handbook of Optical Systems, Aberration Theory and Correction of Optical Systems* vol 3 (New York: Wiley-VCH)

Gross H, Blechinger F and Achtner B 2008 *Handbook of Optical Systems, Survey of Optical Instruments* vol 4 (New York: Wiley-VCH)

Kingslake R and Johnson R B 2009 *Lens Design Fundamentals* (New York: Academic)

Luneburg R K and Herzberger M 1964 *Mathematical Theory of Optics* (Berkeley, CA: University of California Press)

Malacara D 2001 *Handbook of Optical Engineering* (Boca Raton, FL: CRC Press)

Malacara D and Malacara Z 1994 *Handbook of Lens Design* (New York: Dekker)

Malacara-Hernández D and Malacara-Hernández Z 2016 *Handbook of Optical Design* (Boca Raton, FL: CRC Press)

Stavroudis O N 2006 *The Mathematics of Geometrical and Physical Optics: The k-Function and Its Ramifications* (New York: Wiley)

Sun H 2016 *Lens Design: A Practical Guide* (Boca Raton, FL: CRC Press)

Vaskas E M 1957 Note on the Wasserman-Wolf method for designing aspheric surfaces *J. Opt. Soc. Am.* **47** 669–70

Wassermann G D and Wolf E 1949 On the theory of aplanatic aspheric systems *Proc. Phys. Soc. Sect.* B **62** 2

Wolf E 1948 On the designing of aspheric surfaces *Proc. Phys. Soc.* **61** 494

Wolf E and Preddy W S 1947 On the determination of aspheric profiles *Proc. Phys. Soc.* **59** 704

**IOP** Publishing

Analytical Lens Design (Second Edition)

**Rafael G González-Acuña, Héctor A Chaparro-Romo and Julio C Gutiérrez-Vega**

# Chapter 12

# On-axis sequential refractive–reflective telescope

We present a comprehensive analytic and closed-form formula to define the form of a mirror that adjusts the spherical aberration caused by an arbitrary number of preceding surfaces. This outermost mirror makes the refractive–reflective telescope on-axis stigmatic.

## 12.1 Introduction

In this chapter, we are going to get the equations that describe an on-axis stigmatic sequential refractive–reflective telescope. An on-axis stigmatic sequential refractive–reflective telescope is a telescope composed by a sequence of lenses and a reflective mirror such that the system is stigmatic for a point object at minus infinity.

So, the problem in question is, given an arbitrary sequence of lenses, what must the shape of a final aspheric mirror be in order that the optical system is free from spherical aberration. To achieve the mentioned goal, we can use the results obtained in the generalization of the on-axis stigmatic lens to the mathematical model presented in the previous chapter.

### 12.1.1 Mathematical model

As we mention in the introduction, an on-axis sequential refractive–reflective telescope is an extension of the result proposed in the previous chapter. The notation of the problem is almost the same.

The optical system is composed of $(N - 1)$ optical elements, where all the surfaces are radially symmetric concerning the optical axis. The first surface of the system is placed on the origin of the Cartesian coordinate system; thus, $z_1(0) = 0$, see figure 12.1.

The point object is minus infinity since we are designing a telescope. The refraction index within the object and the first surface is $n_0$.

The optical elements of the system have axial thicknesses $t_1, t_2, \ldots, t_{N-1}$ and refractive indices $n_1, n_2, \ldots, n_{N-1}$. This means there are $N + 1$ refractive surfaces and

doi:10.1088/978-0-7503-5774-6ch12  

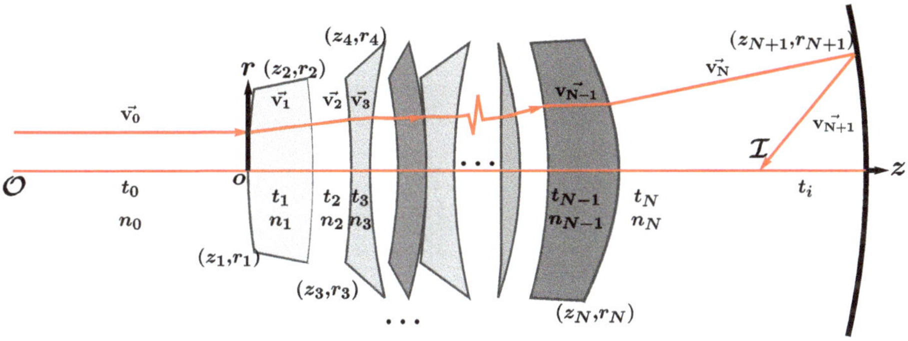


**Figure 12.1.** Telescopic system composed by $N$ lenses with a aspherical corrective mirror $(z_{N+1}, r_{N+1})$.

$N$ thick elements. The surfaces are expressed by the radial functions $z_1(r_1)$, $z_2(r_2)$, … , $z_N(r_N)$, where $r_i$ is the radial coordinate at the $i$th interface.

The path of a ray traveling from the point refractive $(z_i, r_i)$ to the refractive $(z_{i+1}, r_{i+1})$ is,

$$\vec{\mathbf{v}}_i \equiv \wp_{r,i}\vec{\mathbf{e}}_1 + \wp_{z,i}\vec{\mathbf{e}}_2 = \frac{(r_{i+1} - r_i)\vec{\mathbf{e}}_1 + (z_{i+1} - z_i)\vec{\mathbf{e}}_2}{\sqrt{(r_{i+1} - r_i)^2 + (z_{i+1} - z_i)^2}}, \tag{12.1}$$

where $\vec{\mathbf{e}}_1$ is for the $r$ direction and $\vec{\mathbf{e}}_2$ is for the $z$ direction and, $\wp_{r,i}$ and $\wp_{z,i}$ are cosine directors of the unit vector $\hat{v}_i$, thus $\wp_{r,\,i}^2 + \wp_{z,\,i}^2 = 1$ for every $i \in \{1, 2, 3, \dots, N\}$.

$\wp_{r,i}$ and $\wp_{z,i}$ are cosine directors of the previous chapter, so let's recall them,

$$\wp_{r,i} = \frac{n_{i-1}r_i'[(r_i - r_{i-1})r_i' + (z_i - z_{i-1})z_i']}{n_i\sqrt{(r_{i-1} - r_i)^2 + (z_{i-1} - z_i)^2}(r_i'^2 + z_i'^2)} - \frac{z_i'\sqrt{1 - \dfrac{n_{i-1}^2[(r_i - r_{i-1})r_i' + (z_i - z_{i-1})z_i']^2}{n_i^2[(r_{i-1} - r_i)^2 + (z_{i-1} - z_i)^2](r_i'^2 + z_i'^2)}}}{\sqrt{r_i'^2 + z_i'^2}}, \tag{12.2}$$

and,

$$\wp_{z,i} = \frac{n_{i-1}z_i'[(r_i - r_{i-1})r_i' + (z_i - z_{i-1})z_i']}{n_i\sqrt{(r_{i-1} - r_i)^2 + (z_{i-1} - z_1)^2}(r_i'^2 + z_i'^2)} + \frac{r_i'\sqrt{1 - \dfrac{n_{i-1}^2[(r_i - r_{i-1})r_i' + (z_i - z_{i-1})z_i']^2}{n_i^2[(r_{i-1} - r_i)^2 + (z_{i-1} - z_i)^2](r_i'^2 + z_i'^2)}}}{\sqrt{r_i'^2 + z_i'^2}}, \tag{12.3}$$

where $z_i' = \mathrm{d}z_i(r_1)/\mathrm{d}r_1$ and $r_i' = \mathrm{d}r_i(r_1)/\mathrm{d}r_1$ are the derivatives of $z_i$ and $r_i$ with respect to $r_1$.

Like in the last chapter, we define the distance a ray travels between the refractive surface $i + 1$ and $i$ as,

$$\vartheta_i \equiv \sqrt{(r_{i+1} - r_i)^2 + (z_{i+1} - z_i)^2}\,. \tag{12.4}$$

In this chapter, we are going to use the same procedure as the last one, that is, to express any $(z_i, r_i)$ in terms of first surface $(z_1, r_1)$.

Therefore, we recall $\wp_{r,N}$ and $\wp_{z,N}$,

$$\wp_{r,N} = \frac{n_{N-1}r_N'[(r_N - r_{N-1})r_N' + (z_N - z_{N-1})z_N']}{n_N\sqrt{(r_{N-1} - r_N)^2 + (z_{N-1} - z_N)^2}(r_N'^2 + z_N'^2)} - \frac{z_N'\sqrt{1 - \dfrac{n_{N-1}^2[(r_N - r_{N-1})r_N' + (z_N - z_{N-1})z_N']^2}{n_N^2[(r_{N-1} - r_N)^2 + (z_{N-1} - z_N)^2](r_N'^2 + z_N'^2)}}}{\sqrt{r_N'^2 + z_N'^2}}, \tag{12.5}$$

and,

$$\wp_{z,N} = \frac{n_{N-1}z_N'[(r_N - r_{N-1})r_N' + (z_N - z_{N-1})z_N']}{n_N\sqrt{(r_{N-1} - r_N)^2 + (z_{N-1} - z_1)^2}(r_N'^2 + z_N'^2)} + \frac{r_N'\sqrt{1 - \dfrac{n_{N-1}^2[(r_N - r_{N-1})r_N' + (z_N - z_{N-1})z_N']^2}{n_N^2[(r_{N-1} - r_N)^2 + (z_{N-1} - z_N)^2](r_N'^2 + z_N'^2)}}}{\sqrt{r_N'^2 + z_N'^2}}. \tag{12.6}$$

Using the definition of $\vec{\mathbf{v}}_i \equiv \wp_{r,i}\vec{\mathbf{e}}_1 + \wp_{z,i}\vec{\mathbf{e}}_2$, by separating the components of $\vec{\mathbf{v}}_i$ we get,

$$\frac{z_{N+1} - z_N}{\vartheta_N} = \wp_{z,N}, \tag{12.7}$$

and,

$$\frac{r_{N+1} - r_N}{\vartheta_N} = \wp_{r,N}. \tag{12.8}$$

Solving for $z_N$ and $r_N$,

$$z_{N+1} = z_N + \vartheta\wp_N, \tag{12.9}$$

and,

$$r_{N+1} = r_N + \vartheta_N\wp_N. \tag{12.10}$$

We have the same structure as the solution. We need only know the value of $\vartheta_N$. So, the next step, just like in the last chapter, is to use the prediction of Fermat's principle. Fermat's principle predicts the optical path length (OPL) of every ray in an on-axis optical system is constant. This means the OPL of the axial ray is equal to a constant that is equal to the OPL of any other ray.

For derivation proposes, let's assume that the object is located a distance $t_0$ from the first surface. Then, the OPL of the axial ray is the following,

$$\sum_{i=1}^{N} n_i t_i - t_0 n_0 + n_N(-t_I) = \text{const.}, \tag{12.11}$$

where $\sum_{i=1}^{N} n_i t_i$ is the OPL of the axial ray inside the lenses, $-t_I n_N$ is the OPL once it strikes the mirror, and it is reflected to the image point. Finally, $-t_0 n_0$ is the OPL from point object to the first surface.

The OPL of a non-axial ray is,

$$\begin{aligned} &n_N\sqrt{r_{N+1}^2 + \left(-\sum_{i=1}^{N} t_i + z_{N+1} - t_I\right)^2} + \sum_{i=2}^{N} n_{i-1} \\ &\sqrt{(r_i - r_{i-1})^2 + (z_i - z_{i-1})^2} \\ &+ n_N\sqrt{(r_{N+1} - r_N)^2 + (z_{N+1} - z_N)^2} + n_0\sqrt{r_1^2 + (z_1 - t_0)^2} = \text{const.}, \end{aligned} \tag{12.12}$$

where $n_0\sqrt{r_1^2 + (z_1 - t_0)^2}$ is the OPL of the non-axial ray from the object to the first surface. The OPL inside the refractive surface is $\sum_{i=2}^{N} n_{i-1}\sqrt{(r_i - r_{i-1})^2 + (z_i - z_{i-1})^2}$. The OPL from the last refractive surface to the mirror is $n_N\sqrt{(r_{N+1} - r_N)^2 + (z_{N+1} - z_N)^2}$. Finally, the OPL from the mirror to the point image is defined by the following square root, $n_N\sqrt{r_{N+1}^2 + \left(-\sum_{i=1}^{N} t_i + z_{N+1} - t_I\right)^2}$

Since the OPL of the axial ray and the OPL of the non-axial ray are equal to the same constant, we have,

$$\begin{aligned} \sum_{i=1}^{N} n_i t_i - t_0 n_0 + n_N(-t_I) = &\, n_N\sqrt{r_{N+1}^2 + \left(-\sum_{i=1}^{N} t_i + z_{N+1} - t_I\right)^2} \\ &+ \sum_{i=2}^{N} n_{i-1}\sqrt{(r_i - r_{i-1})^2 + (z_i - z_{i-1})^2} \\ &+ n_N\sqrt{(r_{N+1} - r_N)^2 + (z_{N+1} - z_N)^2} \\ &+ n_0\sqrt{r_1^2 + (z_1 - t_0)^2}. \end{aligned} \tag{12.13}$$

Now, to recover the parallelism of the input rays, we let the distance $t_0$ tend to $-\infty$ in equation (12.13). The limit gives,

$$\boxed{\begin{aligned} &-\sum_{i=2}^{N} n_{i-1}\sqrt{(r_i - r_{i-1})^2 + (z_i - z_{i-1})^2} + \sum_{i=1}^{N} n_i t_i + t_I(-n_N) - n_0 z_1 \\ &= n_N\sqrt{(r_{N+1} - r_N)^2 + (z_{N+1} - z_N)^2} \\ &\quad + n_N\sqrt{r_{N+1}^2 + \left(-\sum_{i=2}^{N+1} t_i + z_{N+1} - t_I\right)^2}. \end{aligned}} \tag{12.14}$$

Note that we do not set the limit when $t_0 \to -\infty$ in $z_{N+1}, r_{N+1}$. We avoid it consciously because we are not going to put examples when $t_0$ is finite. In this style, without formulating the limit, the notation appears more manageable. Now, let's replace $r_N + \vartheta_N \wp_{r,N}$ and $z_N + \vartheta_N \wp_{z,N}$ is the above equation,

$$
\begin{aligned}
&-\sum_{i=2}^{N} n_{i-1}\sqrt{(r_i - r_{i-1})^2 + (z_i - z_{i-1})^2} + \sum_{i=1}^{N} n_i t_i + t_I(-n_N) - n_0 z_1 \\
&= n_N \vartheta_N \\
&+ n_N \sqrt{(r_N + \vartheta_N \wp_{r,N})^2 + \left(-\sum_{i=2}^{N+1} t_i + z_N + \vartheta_N \wp_{z,N} - t_I\right)^2},
\end{aligned} \tag{12.15}
$$

where $\vartheta_N = \sqrt{(r_{N+1} - r_N)^2 + (z_{N+1} - z_N)^2}$. Now let $f$ be defined as

$$
f \equiv -\sum_{i=2}^{N} n_{i-1}\sqrt{(r_i - r_{i-1})^2 + (z_i - z_{i-1})^2} + \sum_{i=1}^{N} n_i t_i + t_I(-n_N) - n_0 z_1, \tag{12.16}
$$

and, let $\tau$ be defined as

$$
\tau = z_N - \sum_{i=2}^{N} t_i - t_I, \tag{12.17}
$$

replacing equation (12.16) and (12.17) in equation (12.15),

$$
f - n_N \vartheta_N = n_N \sqrt{(r_N + \vartheta_N \wp_{r,N})^2 + (\vartheta_N \wp_{z,N} + \tau)^2}, \tag{12.18}
$$

squaring,

$$
(f - n_N \vartheta_N)^2 = n_N^2[(r_N + \vartheta_N \wp_{r,N})^2 + (\vartheta_N \wp_{z,N} + \tau)^2], \tag{12.19}
$$

expanding,

$$
f^2 - 2n_N \vartheta_N = n_N^2(r_N^2 + 2r_N \vartheta_N \wp_{r,N} + 2\tau \vartheta_N \wp_{z,N} + \tau^2). \tag{12.20}
$$

Solving for $\vartheta_N$ in the last equation, we get the solution to the problem proposed in the introduction.

$$
\boxed{\begin{cases} z_{N+1} = z_N + \vartheta_N \wp_{z,N}, \\ r_{N+1} = r_N + \vartheta_N \wp_{r,N}, \end{cases}} \tag{12.21}
$$

where,

$$
\vartheta_N = \frac{f^2 - n_N^2(r_N^2 + \tau^2)}{2n_N(n_N(r_N \wp_{r,N} + \tau \wp_{z,N}) + 1)}. \tag{12.22}
$$

Equation (12.21) is the most important equation in the chapter; it describes the shape of the mirror, such that it corrects the spherical aberration generated by the sequence of refractive surfaces. Like in the last chapter, we need to apply the procedure explained in section 11.2.4 to use equation (12.21).

## 12.2 Examples

In this section, we present examples of an on-axis stigmatic telescope, with the particularity that all the refractive surfaces of the telescope are parabolas.

In table 12.1 we include the specifications of the first example depicted in figure 12.2. The mirror of the on-axis stigmatic telescope proposed in this example is given by,

$$z_4 = z_3 + \vartheta_3 \wp_{z,3}, \qquad r_4 = r_3 + \vartheta_3 \wp_{r,3}. \tag{12.23}$$

**Table 12.1.** The image is located at $z = 5 + 5 + 120 - 50 = 80$ mm.

| Surface | Profile | Refraction index | Thickness |
|---|---|---|---|
| 0 | Object | $n_0 = 1$ | $t_0 = -\infty$ mm |
| 1 | $z_1 = r_1^2/f_1$ | $n_1 = 1.5$ | $t_1 = 5$ mm |
| 2 | $z_2 = t_1 + r_2^2/f_2$ | $n_2 = 1.55$ | $t_2 = 5$ mm |
| 3 | $z_3 = \sum_{i-1}^{2} t_i + r_3^2/f_3$ | $n_3 = 1.1$ | $t_3 = 120$ mm |
| 4 | Equation (12.23) | Not applicable | Not applicable |
| 5 | Image | $n_4 = 1$ | $t_I = -50$ mm |

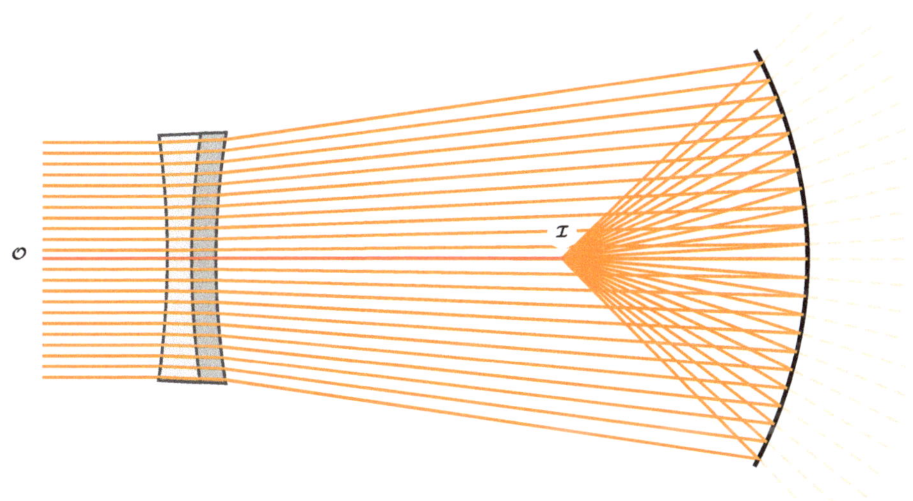


**Figure 12.2.** The first three surfaces are parabolas with foci given by $f_1 = -90$, $f_2 = -90$ and $f_3 = 80$.

where,

$$
\begin{cases}
\vartheta_3 = -(-f^2 + r_3^2 + (t - z_3)^2)/(2(F + r_3\wp_{r,3} + \wp_{z,3}(z_3 - t))) \\
\tau = z_3 - \sum_{i=1}^{3} t_i - t_I \\
f = -\sum_{i=2}^{3} n_{i-1}\sqrt{(r_i - r_{i-1})^2 + (z_i - z_{i-1})^2} + \sum_{i=1}^{3} n_i t_i + t_I(-n_3) - n_0 z_1 \\
\vartheta_2 = \Big[2\sqrt{f_3}\sqrt{f_3\wp_{z,2}^2 + r_2\wp_{r,2}\wp_{z,2} + \wp_{r,2}^2(t_1 + t_2 - z_2)} - 2f_3\wp_{z,2} \\
\qquad - r_2\wp_{r,2}\Big]\Big/\left(\wp_{r,2}^2\right), \\
r_3 = \vartheta_2\wp_{r,2} + r_2, \\
z_3 = \vartheta_2\wp_{z,2} + z_2, \\
\vartheta_1 = \Big[2\sqrt{f_2}\sqrt{f_2\wp_{z,1}^2 + r_1\wp_{r,1}\wp_{z,1} + \wp_{r,1}^2(t_1 - z_1)} - 2f_2\wp_{z,1} - r_1\wp_{r,1}\Big] \\
\qquad \Big/\left(\wp_{r,1}^2\right), \\
r_2 = \vartheta_1\wp_{r,1} + r_1, \\
z_2 = \vartheta_1\wp_{z,1} + z_1, \\
z_1 = r_1^2/f_1.
\end{cases}
\tag{12.24}
$$

In table 12.2 we include the specifications of the second example. The mirror is depicted in figure 12.3 and given by,

$$
z_4 = z_3 + \vartheta_3\wp_{z,3}, \qquad r_4 = r_3 + \vartheta_3\wp_{r,3}, \tag{12.25}
$$

**Table 12.2.** The image is located at $z = 5 + 5 + 120 - 50 = 80$ mm.

| Surface | Profile | Refraction index | Thickness |
|---|---|---|---|
| 0 | Object | $n_0 = 1$ | $t_0 = -\infty$ mm |
| 1 | $z_1 = r_1^2/f_1$ | $n_1 = 1.5$ | $t_1 = 5$ mm |
| 2 | $z_2 = t_1 + r_2^2/f_2$ | $n_2 = 1.75$ | $t_2 = 5$ mm |
| 3 | $z_3 = \sum_{i=1}^{2} t_i + r_2^2/f_3$ | $n_3 = 1.1$ | $t_3 = 120$ mm |
| 4 | Equation (12.25) | Not applicable | Not applicable |
| 5 | Image | $n_4 = 1$ | $t_I = -50$ mm |

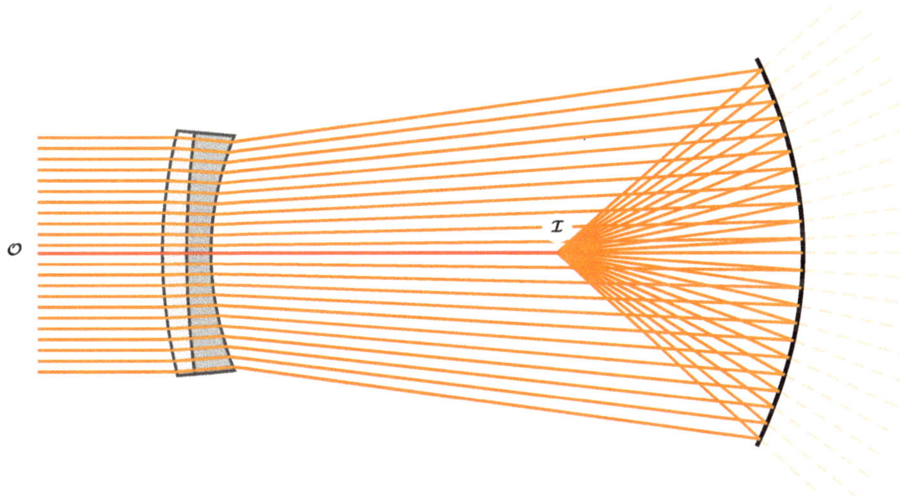


**Figure 12.3.** The first three surfaces are parabolas with focuses given by $f_1 = 50$, $f_2 = -90$ and $f_3 = -30$.

where,

$$\begin{cases}
\vartheta_3 = -\left(-f^2 + r_3^2 + (t - z_3)^2\right) / \left(2\left(F + r_3\wp_{r,3} + \wp_{z,3}\left(z_3 - t\right)\right)\right) \\
\tau = z_3 - \sum_{i=1}^{3} t_i - t_I \\
f = -\sum_{i=2}^{3} n_{i-1}\sqrt{(r_i - r_{i-1})^2 + (z_i - z_{i-1})^2} + \sum_{i=1}^{3} n_i t_i + t_I(-n_3) - n_0 z_1 \\
\vartheta_2 = \left[2\sqrt{f_3}\sqrt{f_3\wp_{z,2}^2 + r_2\wp_{r,2}\wp_{z,2} + \wp_{r,2}^2\left(t_1 + t_2 - z_2\right)} - 2f_3\wp_{z,2} - r_2\wp_{r,2}\right] \\
\qquad \Big/ \left(\wp_{r,2}^2\right), \\
r_3 = \vartheta_2\wp_{r,2} + r_2, \\
z_3 = \vartheta_2\wp_{z,2} + z_2, \\
\vartheta_1 = \left[2\sqrt{f_2}\sqrt{f_2\wp_{z,1}^2 + r_1\wp_{r,1}\wp_{z,1} + \wp_{r,1}^2\left(t_1 - z_1\right)} - 2f_2\wp_{z,1} - r_1\wp_{r,1}\right] \Big/ \left(\wp_{r,1}^2\right), \\
r_2 = \vartheta_1\wp_{r,1} + r_1, \\
z_2 = \vartheta_1\wp_{z,1} + z_1, \\
z_1 = r_1^2 / f_1.
\end{cases} \tag{12.26}$$

**Table 12.3.** The image is located at $z = 5 + 5 + 120 - 50 = 80$ mm.

| Surface | Profile | Refraction index | Thickness |
|---|---|---|---|
| 0 | Object | $n_0 = 1$ | $t_0 = -\infty$ mm |
| 1 | $z_1 = r_1^2/f_1$ | $n_1 = 1.5$ | $t_1 = 5$ mm |
| 2 | $z_2 = t_1 + r_2^2/f_2$ | $n_2 = 1.75$ | $t_2 = 5$ mm |
| 3 | $z_3 = \sum_{i=1}^{2} t_i + r_2^2/f_3$ | $n_3 = 1.1$ | $t_3 = 120$ mm |
| 4 | Equation (12.27) | Not applicable | Not applicable |
| 5 | Image | $n_4 = 1$ | $t_I = -50$ mm |

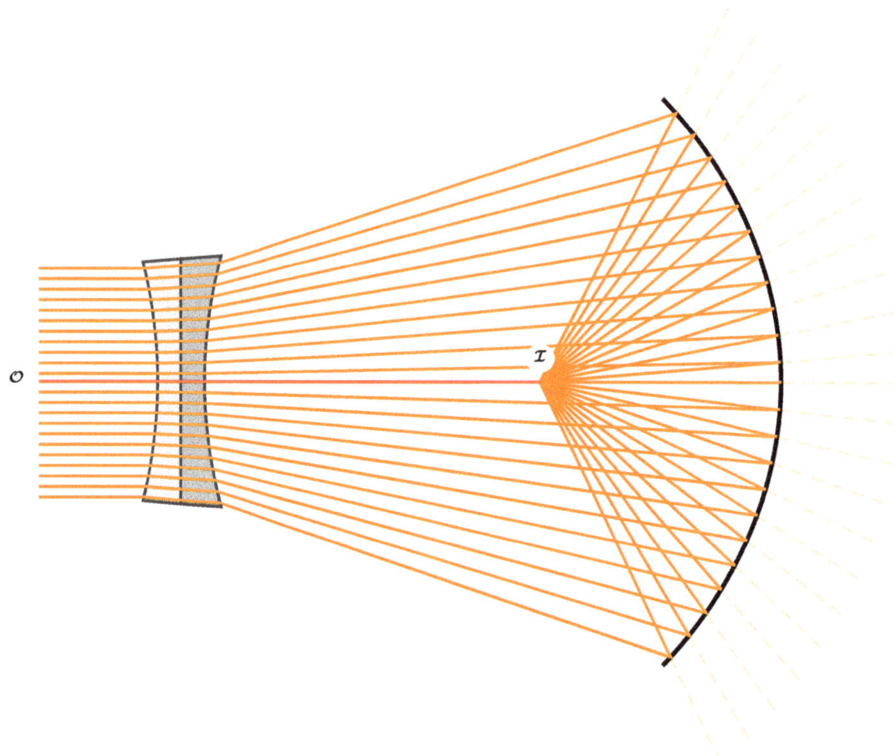


**Figure 12.4.** The first three surfaces are parabolas with focuses given by $f_1 = -50$, $f_2 = -90\,000$ and $f_3 = -50$.

Table 12.3 has the specifications of the last example. The mirror is in figure 12.4 and given by,

$$z_4 = z_3 + \vartheta_3 \wp_{z,3}, \qquad r_4 = r_3 + \vartheta_3 \wp_{r,3}, \tag{12.27}$$

where,

$$\begin{cases} \vartheta_3 = -(-f^2 + r_3^2 + (t - z_3)^2)/(2(F + r_3\wp_{r,3} + \wp_{z,3}(z_3 - t))) \\ \tau = z_3 - \sum_{i=1}^{3} t_i - t_I \\ f = -\sum_{i=2}^{3} n_{i-1}\sqrt{(r_i - r_{i-1})^2 + (z_i - z_{i-1})^2} + \sum_{i=1}^{3} n_i t_i + t_I(-n_3) - n_0 z_1 \\ \vartheta_2 = \left[2\sqrt{f_3}\sqrt{f_3\wp_{z,2}^2 + r_2\wp_{r,2}\wp_{z,2} + \wp_{r,2}^2(t_1 + t_2 - z_2)} - 2f_3\wp_{z,2} \right. \\ \qquad \left. - r_2\wp_{r,2}\right] / \left(\wp_{r,2}^2\right), \\ r_3 = \vartheta_2\wp_{r,2} + r_2, \\ z_3 = \vartheta_2\wp_{z,2} + z_2, \\ \vartheta_1 = \left[2\sqrt{f_2}\sqrt{f_2\wp_{z,1}^2 + r_1\wp_{r,1}\wp_{z,1} + \wp_{r,1}^2(t_1 - z_1)} - 2f_2\wp_{z,1} - r_1\wp_{r,1}\right] \\ \qquad / \left(\wp_{r,1}^2\right), \\ r_2 = \vartheta_1\wp_{r,1} + r_1, \\ z_2 = \vartheta_1\wp_{z,1} + z_1, \\ z_1 = r_1^2/f_1. \end{cases} \tag{12.28}$$

## 12.3 Conclusions

This chapter is a consequence of the results of the previous chapter. Here we focus on the design of an aspheric mirror such that it corrects the spherical aberration generated by an arbitrary number of lenses. Therefore, we can design countless on-axis sequential lenses of refractive, reflective telescopes with the equations of these chapters. Equation (12.21) is robust enough that a designer can quickly achieve aplanatism in his/her optical system if it is used correctly. The reason is that equation (12.21) is such that the mirror that it describes will eliminate the spherical aberration of a sequence of lenses. Therefore, the designer only needs to focus on the optimization of the series of lenses to eliminate coma. Equation (12.21) comprises all the proper knowledge to design an on-axis stigmatic telescope, i.e. the central thicknesses of the lenses, their shapes, refraction indexes, the position of the image and the shape of the aspherical mirror.

## Further reading

Acuña R G G and Gutiérrez-Vega J C 2019 General formula of the refractive telescope design free spherical aberration *Novel Optical Systems, Methods, and Applications XXII* 11 105 (Bellingham, WA: International Society for Optics and Photonics)

Baquero N, Hernández W A and Rincón R A 2006 Diseño y construcción de un telescopio Cassegrain clásico *Orinoquia* **10** 2
Born M and Wolf E 2013 Principles of Optics: Electromagnetic Theory of Propagation *Interference and Diffraction of Light* (Amsterdam: Elsevier)
Chaves J 2016 *Introduction to Nonimaging Optics* 2nd edn (Boca Raton, FL: CRC Press)
González-Acuña R G and Chaparro-Romo H A 2018 General formula for bi-aspheric singlet lens design free of spherical aberration *Appl. Opt.* **57** 9341–5
González-Acuña R G and Guitiérrez-Vega J C 2018 Generalization of the axicon shape: the gaxicon *J. Opt. Soc. Am.* A **35** 1915–8
González-Acuña R G and Gutiérrez-Vega J C 2019a Analytic formulation of a refractive-reflective telescope free of spherical aberration *Opt. Eng.* **58** 085105
González-Acuña R G and Gutiérrez-Vega J C 2019b General formula for aspheric collimator lens design free of spherical aberration *Current Developments in Lens Design and Optical Engineering XX* (Bellingham, WA: International Society for Optics and Photonics)
González-Acuña R G, Avendaño-Alejo M and Gutiérrez-Vega J C 2019a Singlet lens for generating aberration-free patterns on deformed surfaces *J. Opt. Soc. Am.* A **36** 925–9
González-Acuña R G, Chaparro-Romo H A and Gutiérrez-Vega J C 2019b General formula to design freeform singlet free of spherical aberration and astigmatism *Appl. Opt.* **58** 1010–5
Hanany S and Marrone D P 2002 Comparison of designs of off-axis Gregorian telescopes for millimeter-wave large focal-plane arrays *Appl. Opt.* **41** 4666–70
Kingslake R and Johnson R B 2009 *Lens Design Fundamentals* (New York: Academic)
Korsch D 1972 Closed form solution for three-mirror telescopes, corrected for spherical aberration, coma, astigmatism, and field curvature *Appl. Opt.* **11** 2986–7
Lin P D and Tsai C-Y 2012 Determination of unit normal vectors of aspherical surfaces given unit directional vectors of incoming and outgoing rays: reply *J. Opt. Soc. Am.* A **29** 1358
Luneburg R K and Herzberger M 1964 *Mathematical Theory of Optics* (Berkeley, CA: University of California Press)
Pierre Y B 2003 *The Design and Construction of Large Optical Telescopes* (Berlin: Springer)
Romano A and Cavaliere R 2016 *Geometric Optics: Theory and Design of Astronomical Optical Systems Using Mathematica®* (Basel: Birkhäuser)
Saha S K 2007 *Diffraction-Limited Imaging with Large and Moderate Telescopes* (Singapore: World Scientific)
Stavroudis O 2012 *The Optics of Rays, Wavefronts, and Caustics* vol 38 (Amsterdam: Elsevier)
Stavroudis O N 2006 *The Mathematics of Geometrical and Physical Optics: The k-Function and Its Ramifications* (New York: Wiley)
Sun H 2016 *Lens Design: A Practical Guide* (Boca Raton, FL: CRC Press)
Wassermann G D and Wolf E 1949 On the theory of aplanatic aspheric systems *Proc. Phys. Soc. Sect.* B **62** 2
Wolf E 1948 On the designing of aspheric surfaces *Proc. Phys. Soc.* **61** 494
Wolf E and Preddy W S 1947 On the determination of aspheric profiles *Proc. Phys. Soc.* **59** 704

# Part V

## Aplanatic singlets and optical systems

**Rafael G González-Acuña, Héctor A Chaparro-Romo and Julio C Gutiérrez-Vega**

# Chapter 13

## Off-axis stigmatic lens

The very old conjecture of the design of a lens free from coma is revisited. We demonstrate that the analytic closed-form solution of the lens is valid within a mathematical frame.

### 13.1 Introduction

In our previous chapters, we found the analytical closed-form solution of a lens-free of spherical aberration. The input of the general formula presented is the first surface of the lens. The output is the correcting second surface of the singlet; it is such that the singlet is free of spherical aberration. Our main goal is to find the closed-form solution of the aplanatic lens, a lens free from coma and free from spherical aberration. Since in chapter 4 is presented the set of all possible lenses free of spherical aberration, if there exists an aplanatic lens, it must be a subset of it. Something that we need to remark on about the general formula of lenses free from spherical aberration is that if the input is a rationally symmetric, the output as well.

Before we achieve an aplanatic lens, we are going to focus only on the coma. In this chapter, we present the general formula of a singlet free from coma for an off-axis point object. The input of the formula given is the first surface of the lens. The output is the second surface, and it is such that the singlet is free from coma. Then we study the compatibility of the general formulas of coma-free and spherical aberration-free lenses.

### 13.2 Mathematical model

We intend to design a singlet lens free from coma with refraction index $n$. We assume that the lens is surrounded by air, and it is located at the origin $z_a(0) = 0$, as shown in figure 13.1. The input and output surfaces are $z_a(r_a)$ and $z_b(r_b)$, respectively, and $r_a$

doi:10.1088/978-0-7503-5774-6ch13

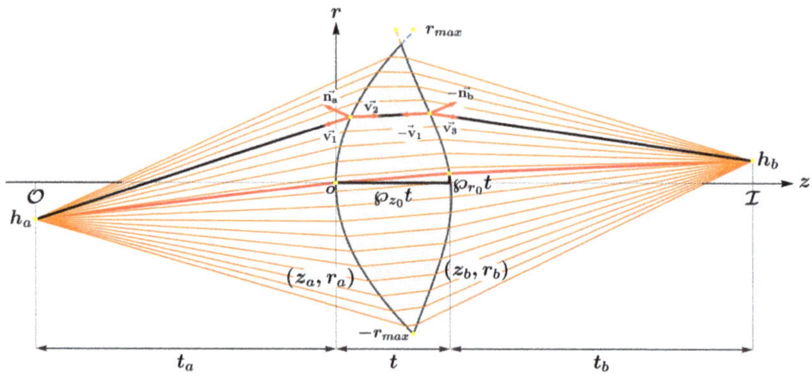


**Figure 13.1.** Diagram of an on-axis stigmatic singlet. The first surface is $(r_a, z_a)$, and the second surface is $(r_b, z_b)$. $\vec{v}_1$ is unit vector of the incident ray, $\vec{v}_2$ is the unit vector of the refracted ray inside the lens, $\vec{v}_3$ is the output $\vec{n}_a$ is the normal vector of the first surface and $\vec{n}_b$ is the normal vector of the second surface.

and $r_b$ are the radii at each surface, respectively. The input surface is known, and the output surface is unknown. We start with Snell's law at the first surface,

$$\boxed{\vec{v}_2 = \frac{1}{n}[\vec{v}_1 - (\vec{n}_a \cdot \vec{v}_1)\vec{n}_a] - \vec{n}_a\sqrt{1 + \frac{1}{n^2}(\vec{n}_a \wedge \vec{v}_1)^2} \quad \text{for} \quad \vec{v}_2, \vec{v}_1, \vec{n}_a \in \mathbb{R}^2,} \tag{13.1}$$

where $\vec{v}_1$ is the unit vector of the incident ray, $\vec{v}_2$ is unit vector of the refracted ray and finally $\vec{n}_a$ is the normal vector of the first surface, see figure 13.1.

The unit vectors related in the refraction suffered by the light in the first surface are,

$$\begin{aligned} \vec{v}_1 &= \frac{(r_a - h_a)\vec{e}_1 + (z_a - t_a)\vec{e}_2}{\sqrt{(r_a - h_a)^2 + (z_a - t_a)^2}}, \quad \vec{v}_2 = \frac{(r_b - r_a)\vec{e}_1 + (z_b - z_a)\vec{e}_2}{\sqrt{(r_b - r_a)^2 + (z_b - z_a)^2}}, \\ \vec{n}_a &= \frac{z_a'\vec{e}_1 - \vec{e}_2}{\sqrt{1 + z_a'^2}}, \end{aligned} \tag{13.2}$$

where $\vec{e}_1$ is for the $r$ direction and $\vec{e}_2$ is for the $z$ direction and $z_a'$ is the derivative respect to $r_a$ of the sagitta of the first surface. We need to replace them in equation (13.1), let's do it step by step. We start with the term, $(\vec{n}_a \wedge \vec{v}_1)^2$,

$$(\vec{n}_a \wedge \vec{v}_1) = \frac{[r_a - h_a + (z_a - t_a)z_a']}{\sqrt{(r_a - h_a)^2 + (z_a - t_a)^2}\sqrt{1 + z_a'^2}}(\vec{e}_1 \wedge \vec{e}_2). \tag{13.3}$$

Then, we square it, remember that $(\vec{e}_1 \wedge \vec{e}_2)^2 = -1$,

$$(\vec{n}_a \wedge \vec{v}_1)^2 = -\frac{[r_a - h_a + (z_a - t_a)z_a']^2}{[(r_a - h_a)^2 + (z_a - t_a)^2](1 + z_a'^2)}, \tag{13.4}$$

therefore, the term squared can be written as,

$$
\begin{aligned}
&-\vec{\mathbf{n}}_a\sqrt{1+\frac{1}{n^2}(\vec{\mathbf{n}}_a\wedge\vec{\mathbf{v}}_1)^2}\\
&\quad=-\frac{(z_a'\vec{\mathbf{e}}_1-\vec{\mathbf{e}}_2)}{\sqrt{1+z_a'^2}}\sqrt{1-\frac{[r_a-h_a+(z_a-t_a)z_a']^2}{n^2[(r_a-h_a)^2+(z_a-t_a)^2](1+z_a'^2)}}.
\end{aligned}
\tag{13.5}
$$

Now let's focus on the term, $\vec{\mathbf{v}}_1-(\vec{\mathbf{n}}_a\cdot\vec{\mathbf{v}}_1)\vec{\mathbf{n}}_a$,

$$
\begin{aligned}
\vec{\mathbf{v}}_1-(\vec{\mathbf{n}}_a\cdot\vec{\mathbf{v}}_1)\vec{\mathbf{n}}_a&=\frac{(r_a-h_a)\vec{\mathbf{e}}_1+(z_a-t_a)\vec{\mathbf{e}}_2}{\sqrt{(r_a-h_a)^2+(z_a-t_a)^2}}\\
&\quad-\left[\frac{(r_a-h_a)z_a'-(z_a-t_a)}{\sqrt{r_a^2+(z_a-t_a)^2}\sqrt{1+z_a'^2}}\right]\frac{(z_a'\vec{\mathbf{e}}_1-\vec{\mathbf{e}}_2)}{\sqrt{1+z_a'^2}}.
\end{aligned}
\tag{13.6}
$$

Simplifying, obtaining the least common divisor and multiplying by $\frac{1}{n}$,

$$
\begin{aligned}
\frac{1}{n}[\vec{\mathbf{v}}_1-(\vec{\mathbf{n}}_a\cdot\vec{\mathbf{v}}_1)\vec{\mathbf{n}}_a]&=\frac{(z_a-t_a)z_a'+r_a-h_a}{n\sqrt{(r_a-h_a)^2+(t_a-z_a)^2}(1+z_a'^2)}\vec{\mathbf{e}}_1\\
&\quad+\frac{[r_a-h_a+(z_a-t_a)z_a']z_a'}{n\sqrt{(r_a-h_a)^2+(t_a-z_a)^2}(1+z_a'^2)}\vec{\mathbf{e}}_2.
\end{aligned}
\tag{13.7}
$$

If we sum equations (13.5) and (13.7) we get $\vec{\mathbf{v}}_2$,

$$
\begin{aligned}
\vec{\mathbf{v}}_2&=\frac{(z_a-t_a)z_a'+r_a-h_a}{n\sqrt{(r_a-h_a)^2+(t_a-z_a)^2}(1+z_a'^2)}\vec{\mathbf{e}}_1\\
&\quad+\frac{[r_a-h_a+(z_a-t_a)z_a']z_a'}{n\sqrt{(r_a-h_a)^2+(t_a-z_a)^2}(1+z_a'^2)}\vec{\mathbf{e}}_2\\
&\quad-\frac{(z_a'\vec{\mathbf{e}}_1-\vec{\mathbf{e}}_2)}{\sqrt{1+z_a'^2}}\sqrt{1-\frac{[r_a-h_a+(z_a-t_a)z_a']^2}{n^2[(r_a-h_a)^2+(z_a-t_a)^2](1+z_a'^2)}}.
\end{aligned}
\tag{13.8}
$$

Separating the coordinates, $\vec{\mathbf{e}}_2$ and $\vec{\mathbf{e}}_2$, we have,

$$
\begin{aligned}
\wp_r\equiv&\frac{-h_a+r_a+(z_a-t_a)z_a'}{n(z_a'^2+1)\sqrt{(h_a-r_a)^2+(t_a-z_a)^2}}\\
&-\frac{z_a'\sqrt{1-\dfrac{[-h_a+r_a+(z_a-t_a)z_a']^2}{n^2(z_a'^2+1)[(h_a-r_a)^2+(t_a-z_a)^2]}}}{\sqrt{z_a'^2+1}},
\end{aligned}
\tag{13.9}
$$

and,

$$
\begin{aligned}
\wp_z \equiv& \frac{z_a'[-h_a + r_a + (z_a - t_a)z_a']}{n(z_a'^2 + 1)\sqrt{(h_a - r_a)^2 + (t_a - z_a)^2}},\\
&+ \frac{\sqrt{1 - \frac{[-h_a + r_a + (z_a - t_a)z_a']^2}{n^2(z_a'^2 + 1)[(h_a - r_a)^2 + (t_a - z_a)^2]}}}{\sqrt{(z_a')^2 + 1}}.
\end{aligned}
\tag{13.10}
$$

Therefore, we can express the components of vector $\vec{\mathbf{v}}_2$ as,

$$
\frac{r_b - r_a}{\sqrt{(r_b - r_a)^2 + (z_b - z_a)^2}} = \wp_r, \tag{13.11}
$$

and,

$$
\frac{z_b - z_a}{\sqrt{(r_b - r_a)^2 + (z_b - z_a)^2}} = \wp_z. \tag{13.12}
$$

Now, let's assign a name to the distance which every ray travels inside the lens, let $\vartheta$ *be*,

$$
\vartheta \equiv \sqrt{(r_b - r_a)^2 + (z_b - z_a)^2}. \tag{13.13}
$$

Then, if we substitute the definition of $\vartheta$ in equations (13.11) and (13.12), we have

$$
\frac{z_b - z_a}{\vartheta} = \wp_z, \tag{13.14}
$$

and,

$$
\frac{r_b - r_a}{\vartheta} = \wp_r, \tag{13.15}
$$

Solving for the unknowns, we get the structure of the solution,

$$
\begin{cases}
z_b = z_a + \vartheta\wp_z,\\
r_b = r_a + \vartheta\wp_r.
\end{cases}
\tag{13.16}
$$

We need to find what $\vartheta$ is in terms of the parameters that we know. For that, we are going to use Fermat's principle as a second equation.

The Fermat principle, for a stigmatic lens, predicts that the optical path of every ray is constant. To use the Fermat principle, we need to assign a reference ray. **The ray that we choose is the ray that strikes at the origin of the lens.** The cosine directors of reference ray traveling inside the lens are,

$$
\wp_{r,0} \equiv \wp_r|_{r_a=0} = -\frac{h_a}{n\sqrt{h_a^2 + t_a^2}}, \tag{13.17}
$$

and,

$$\wp_{z,0} \equiv \wp_z|_{r_a=0} = \sqrt{1 - \frac{(-h_a)^2}{n^2(h_a^2 + t_a^2)}}. \tag{13.18}$$

with $\wp_{z,0}^2 + \wp_{r,0}^2 = 1$.

Please see that in figure 13.1 we are going to equal the path of our reference ray with the optical path of any other ray. First, consider the distance from the point object to the center of the first surface is $(h_a^2 + t_a^2)^{1/2}$. Then, recognize the ray that enters the lens at the origin. This ray is refracted by the input surface and moves inside the lens along a distance $t$. Outside of the lens, the path length from the output point to the image point is $[(\wp_{r,0}t - h_b)^2 + t_b^2]^{1/2}$, where $\wp_{z,0}$ and $\wp_{r,0}$ are the cosine directors of the refracted ray that enters the lens at the origin.

Therefore the optical path length (OPL) of the reference ray is,

$$\sqrt{h_a^2 + t_a^2} + \sqrt{(\wp_{r,0}t - h_b)^2 + t_b^2} + nt = \text{const.} \tag{13.19}$$

The OPL of a non-reference ray is given by,

$$\begin{aligned} &\sqrt{(r_a - h_a)^2 + (z_a - t_a)^2} + \sqrt{(r_b - h_b)^2 + (-t_b + z_b - t\wp_{z,0})^2} \\ &\qquad n\sqrt{(r_b - r_a)^2 + (z_b - z_a)^2} = \text{const.}, \end{aligned} \tag{13.20}$$

where $\sqrt{(r_a - h_a)^2 + (z_a - t_a)^2}$ is the OPL between the point object to the first surface, the OPL inside the lens is $n\sqrt{(r_b - r_a)^2 + (z_b - z_a)^2}$ and finally the following expression $\sqrt{(r_b - h_b)^2 + (-t_b + z_b - t\wp_{z,0})^2}$ is the OPL from the second surface to the point image. Then if we equal the optical path of the reference ray and the optical path of any other ray, we have,

$$\begin{aligned} \sqrt{h_a^2 + t_a^2} + \sqrt{(\wp_{r,0}t - h_b)^2 + t_b^2} + nt &= n\sqrt{(r_b - r_a)^2 + (z_b - z_a)^2} + \\ &\sqrt{(r_a - h_a)^2 + (z_a - t_a)^2} \\ &+ \sqrt{(r_b - h_b)^2 + (-t_b + z_b - t\wp_{z,0})^2}. \end{aligned} \tag{13.21}$$

Replacing $\vartheta$, $z_b$ and, $r_b$ in the above equation, using equation (13.13) and (13.16), respectively,

$$\begin{aligned} &\sqrt{h_a^2 + t_a^2} + \sqrt{(\wp_{r,0}t - h_b)^2 + t_b^2} + nt \\ &\quad = n\vartheta + \sqrt{(r_a - h_a)^2 + (z_a - t_a)^2} \\ &\quad + \sqrt{(r_a + \vartheta\wp_r - h_b)^2 + (-t_b + z_a + \vartheta\wp_z - t\wp_{z,0})^2}. \end{aligned} \tag{13.22}$$

The above equation is still messy, let's define variables that do not have the unknowns,

$$f \equiv \sqrt{h_a^2 + t_a^2} + nt + \sqrt{(\wp_{r,0}t - h_b)^2 + t_b^2} - \sqrt{(r_a - h_a)^2 + (z_a - t_a)^2}, \tag{13.23}$$

and,

$$\tau \equiv z_a - t\wp_{z,0} - t_b. \tag{13.24}$$

Replacing, $f$ and $\tau$ is equation (13.22) and squaring it,

$$(f - n\vartheta)^2 = (r_a - h_b + \vartheta\wp_r)^2 + (\tau + \vartheta\wp_z)^2, \tag{13.25}$$

expanding,

$$f^2 - 2fn\vartheta + n^2\vartheta^2 = (r_a - h_b)^2 + 2(r_a - h_b)\wp_r\vartheta + \wp_r^2\vartheta^2 + \tau^2 + 2\tau\wp_z \vartheta + \wp_z^2\vartheta^2, \tag{13.26}$$

collecting the terms with $\vartheta$,

$$(1 - n^2)\vartheta^2 + 2[fn + (r_a - h_b)\wp_r + \tau\wp_z]\vartheta + [(r_a - h_b)^2 + \tau^2 - f^2] = 0. \tag{13.27}$$

The next step is to multiply every element by $4(1 - n^2)$,

$$4(1 - n^2)^2\vartheta^2 + 8(1 - n^2)[fn + (r_a - h_b)\wp_r + \tau\wp_z]\vartheta + 4(1 - n^2)[(r_a - h_b)^2 + \tau^2 - f^2] = 0. \tag{13.28}$$

Now we sum $4[fn + (r_a - h_b)\wp_r + \tau\wp_z]^2$ in both sides,

$$\begin{aligned}&4(1 - n^2)^2\vartheta^2 + 8(1 - n^2)[fn + (r_a - h_b)\wp_r + \tau\wp_z]\vartheta + 4[fn + (r_a - h_b)\wp_r + \tau\wp_z]^2\\ &= 4[fn + (r_a - h_b)\wp_r + \tau\wp_z]^2 - 4(1 - n^2)[(r_a - h_b)^2 + \tau^2 - f^2],\end{aligned} \tag{13.29}$$

completing the square binomial,

$$\begin{aligned}&\{2(1 - n^2)\vartheta + 2[fn + (r_a - h_b)\wp_r + \tau\wp_z]\}^2\\ &= 4[fn + (r_a - h_b)\wp_r + \tau\wp_z]^2 - 4(1 - n^2)[(r_a - h_b)^2 + \tau^2 - f^2],\end{aligned} \tag{13.30}$$

applying the square root,

$$\begin{aligned}&2(1 - n^2)\vartheta + 2[fn + (r_a - h_b)\wp_r + \tau\wp_z]\\ &= \sqrt{4[fn + (r_a - h_b)\wp_r + \tau\wp_z]^2 - 4(1 - n^2)[(r_a - h_b)^2 + \tau^2 - f^2]}.\end{aligned} \tag{13.31}$$

Finally we solve for $\vartheta$,

$$\vartheta = \frac{-[fn + (r_a - h_b)\wp_r + \tau\wp_z]}{(1 - n^2)} \pm \frac{\sqrt{[fn + (r_a - h_b)\wp_r + \tau\wp_z]^2 - (1 - n^2)[(r_a - h_b)^2 + \tau^2 - f^2]}}{(1 - n^2)}. \tag{13.32}$$

Once, we know $\vartheta$, we get the shape of the second surface,

$$\boxed{\begin{cases} z_b = z_a + \vartheta\wp_z,\\ r_b = r_a + \vartheta\wp_r.\end{cases}} \tag{13.33}$$

Equation (13.33) is the most relevant equation of this chapter. It describes the shape of the second surface of an off-axis stigmatic lens. In the following section, we are going to use it to plot several off-axis stigmatic lenses.

## 13.3 Illustrative examples

In this section, we show some examples of lenses designed with equation (13.33). In the image of the gallery presented in this section are the design parameters of each example. The figures are 13.2–13.7.

In the following tables we show the evolution of the shapes of the lenses. The tables are in figures 13.8–13.18.

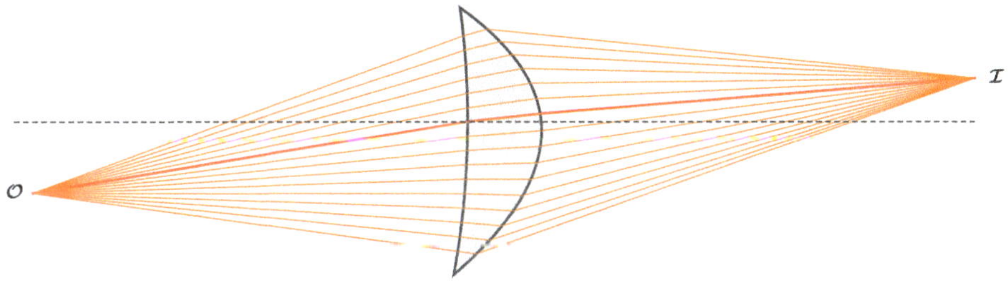


**Figure 13.2.** Design specifications: $n = 1.5$, $t_a = -30$ mm, $t = 5$ mm, $t_b = 30$ mm, $h_a = -5$ mm, $h_b = 3$ mm, $z_a = -r_a^2/120$ and $z_b =$ equation (13.33).

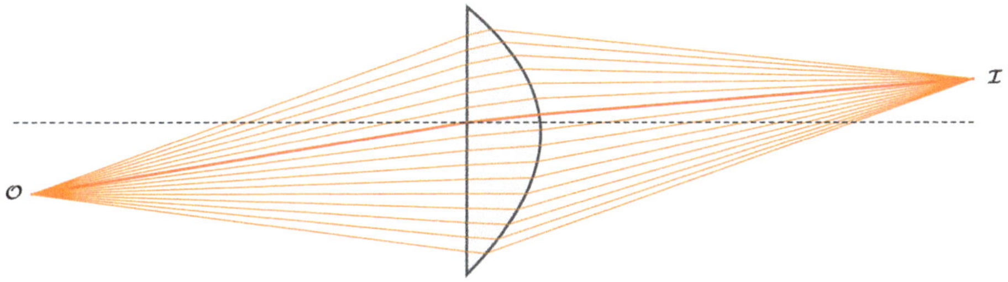


**Figure 13.3.** Design specifications: $n = 1.5$, $t_a = -30$ mm, $t = 5$ mm, $t_b = 30$ mm, $h_a = -5$ mm, $h_b = 3$ mm, $z_a = 0$ and $z_b =$ equation (13.33).

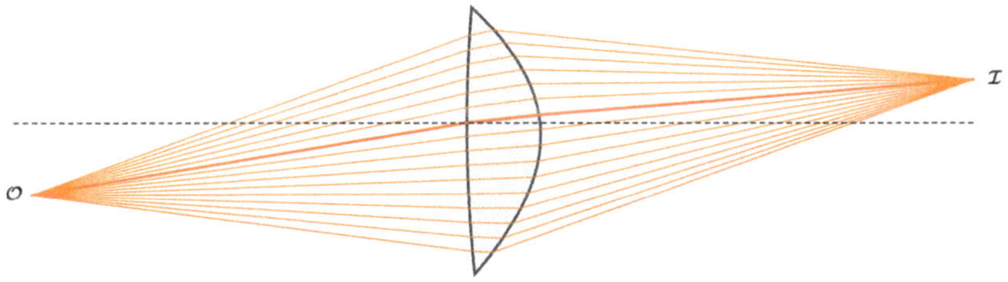


**Figure 13.4.** Design specifications: $n = 1.5$, $t_a = -30$ mm, $t = 5$ mm, $t_b = -60$ mm, $h_a = -5$ mm, $h_b = 3$ mm, $z_a = r_a^2/200$ and $z_b =$ equation (13.33).

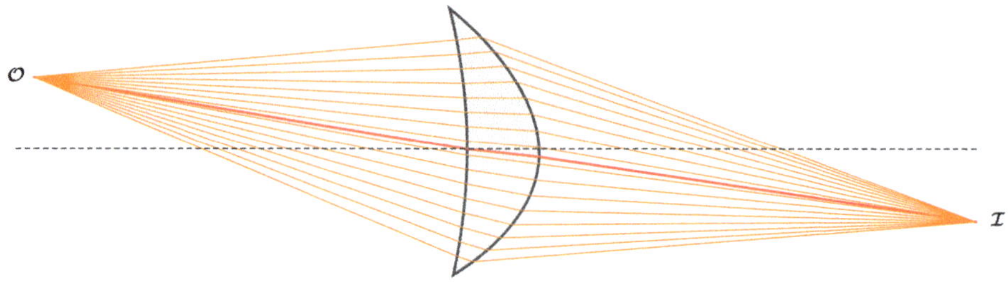


**Figure 13.5.** Design specifications: $n = 1.5$, $t_a = -30$ mm, $t = 5$ mm, $t_b = 30$ mm, $h_a = 5$ mm, $h_b = -5$ mm, $z_a = -r_a^2/80$ and $z_b$ = equation (13.33).

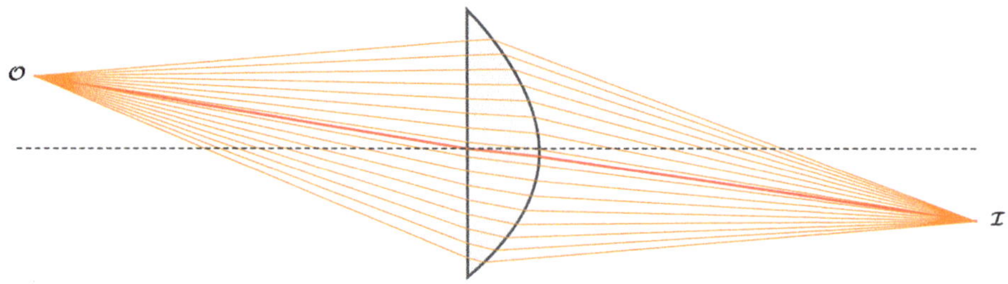


**Figure 13.6.** Design specifications: $n = 1.5$, $t_a = -30$ mm, $t = 5$ mm, $t_b = 30$ mm, $h_a = -5$ mm, $h_b = 5$ mm, $z_a = 0$ and $z_b$ = equation (13.33).

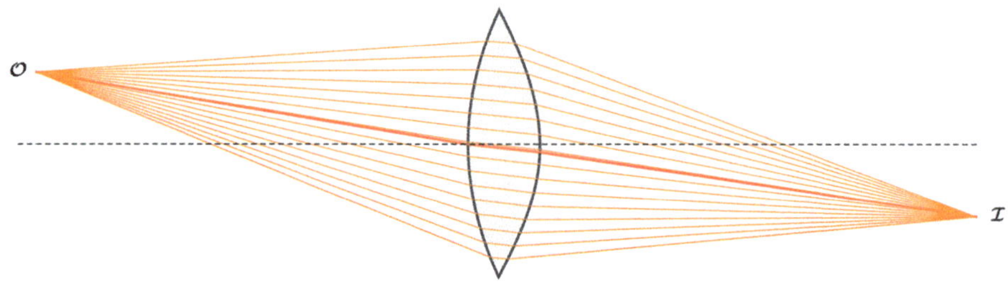


**Figure 13.7.** Design specifications: $n = 1.5$, $t_a = -30$ mm, $t = 5$ mm, $t_b = 30$ mm, $h_a = 5$ mm, $h_b = -5$ mm, $z_a = r_a^2/40$ and $z_b$ = equation (13.33).

From the tables we learn that equation (13.33) is not radially symmetric and it does not matter which input values we choose.

### 13.3.1 A non-symmetric solution

We have plotted and tested equation (13.33). But there is a crucial problem in the solution: it is not radially symmetric. Therefore, the general formula of a lens free from coma does not intersect with the formula of a lens free from spherical aberration.

In this section, we are going to prove that given any radially symmetric input surface, equation (13.33) will not always be radially symmetric.

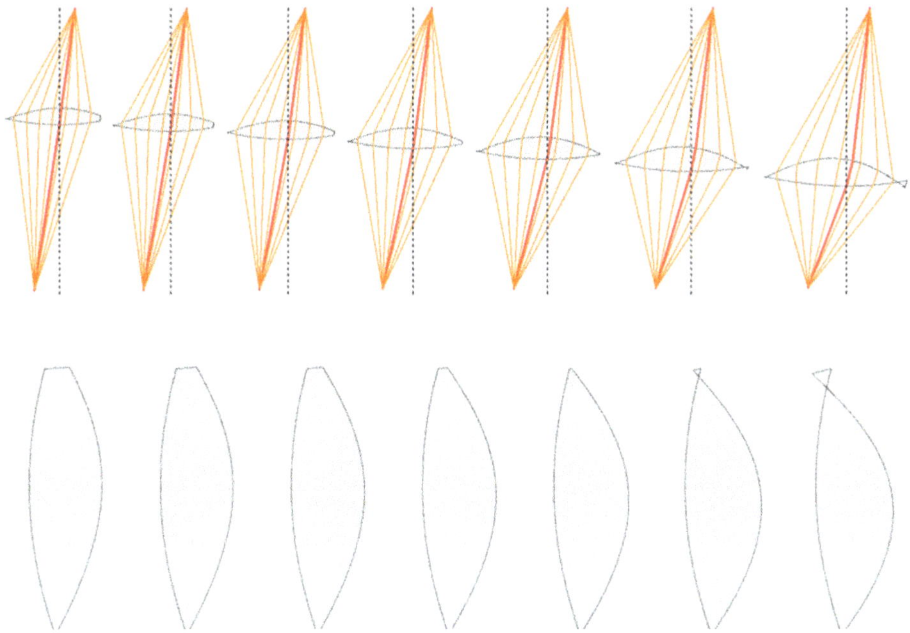

**Figure 13.8.** This table shows the evolution of an aberration-free spherical lens designed with equation (13.33). The input parameters that remain constant are $n = 1.5$, $t = 5$ mm, $t_b = 30$ mm, $h_a = -5$ mm, $h_b = 3$ mm, and $z_a = r_a^2/60$. The variable that changes from left to right is $t_a$, from −50 mm to −20 mm in of steps of 5 mm.

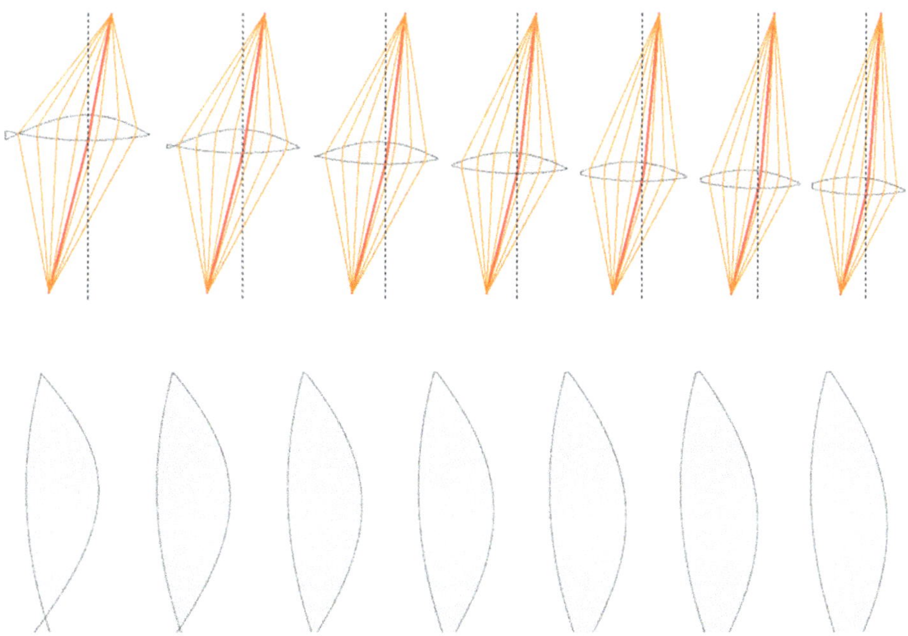

**Figure 13.9.** This table presents the change of an aberration-free coma lens designed with equation (13.33). The input parameters that remain constant are $n = 1.5$, $t = 5$ mm, $t_a = -30$ mm, $h_a = -5$ mm, $h_b = 3$ mm, and $z_a = r_a^2/60$. The input value that varies from left to right is $t_b$, from 20 mm to 50 mm in of steps of 5 mm.

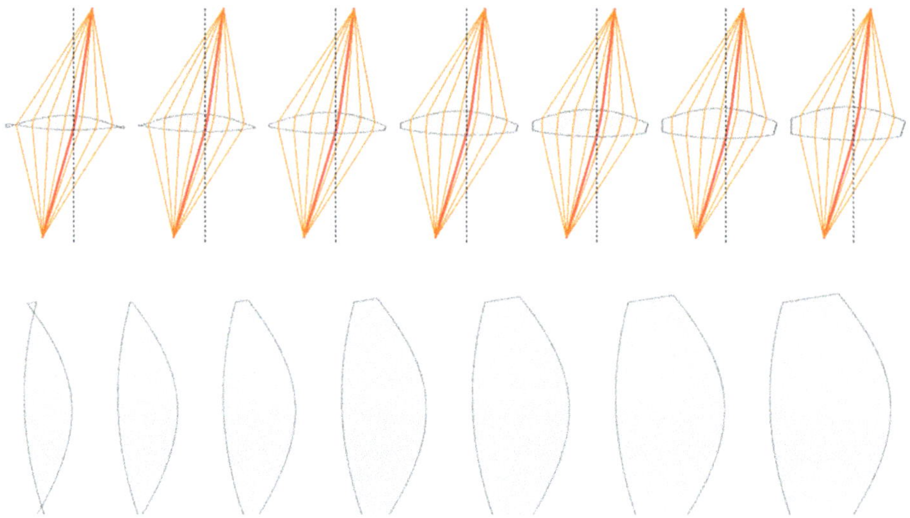

**Figure 13.10.** The table shows the profile of an off-axis stigmatic lens described by equation (13.33). The input parameters that remain constant are $t_a = -30$ mm, $n = 1.5$, $t_b = 30$ mm, $h_a = -5$ mm, $h_b = 3$ mm, and $z_a = r_a^2/60$. The variable that changes from left to right is $t$, from 4 mm to 10 mm in of steps of 1 mm.

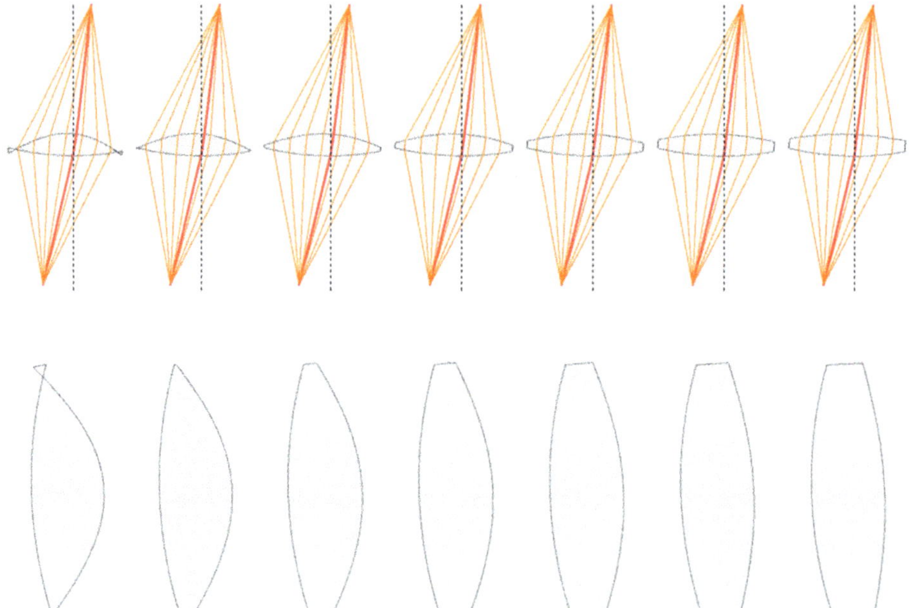

**Figure 13.11.** The table shows the profile of an lens described by equation (13.33). The input parameters that remain constant are $t_a = -30$ mm, $t = 5$, $t_b = 30$ mm, $h_a = -5$ mm, $h_b = 3$ mm, and $z_a = r_a^2/60$. The input value that changes from left to right is $n$, from 1.4 to 2 in of steps of 0.1.

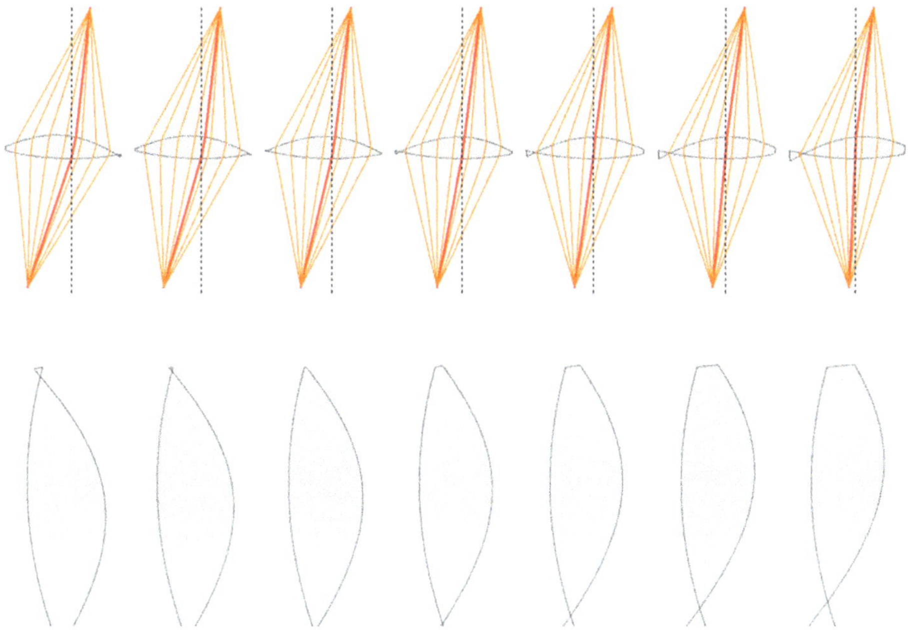

**Figure 13.12.** The table shows lenses of equation (13.33). The input parameters that remain constant are $n = 1.5$, $t_a = -30$ mm, $t = 5$, $t_b = 30$ mm, $h_b = 3$ mm and $z_a = r_a^2/60$. The input value that changes is $h_a$, from $-7$ mm to $-1$ mm in of steps of 1 mm.

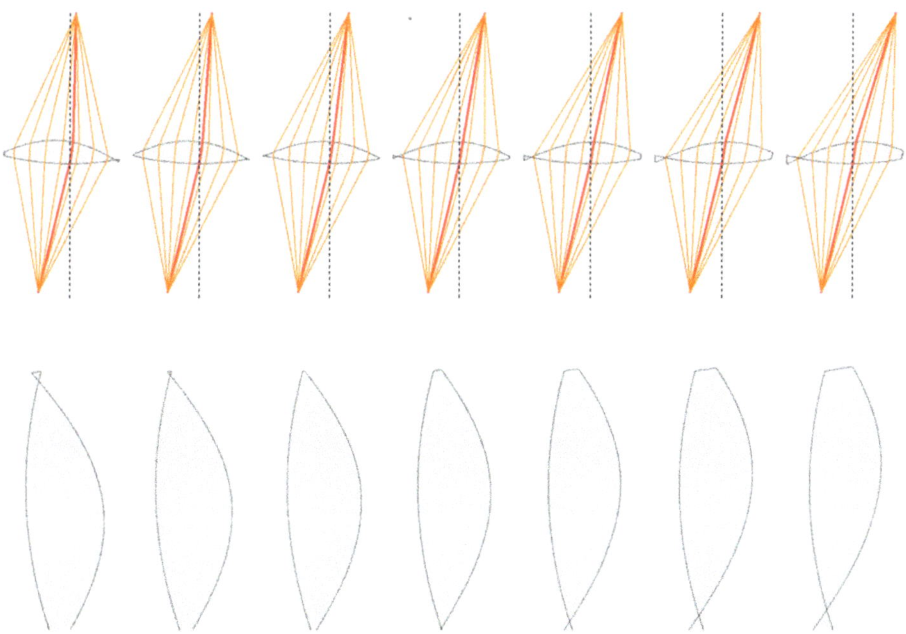

**Figure 13.13.** The table shows lenses of equation (13.33). The input parameters that remain constant are $n = 1.5$, $t_a = -30$ mm, $t = 5$, $t_b = 30$ mm, $h_a = -5$ mm, and $z_a = r_a^2/60$. The input value that changes is $h_b$, from 1 mm to 7 mm in of steps of 1 mm.

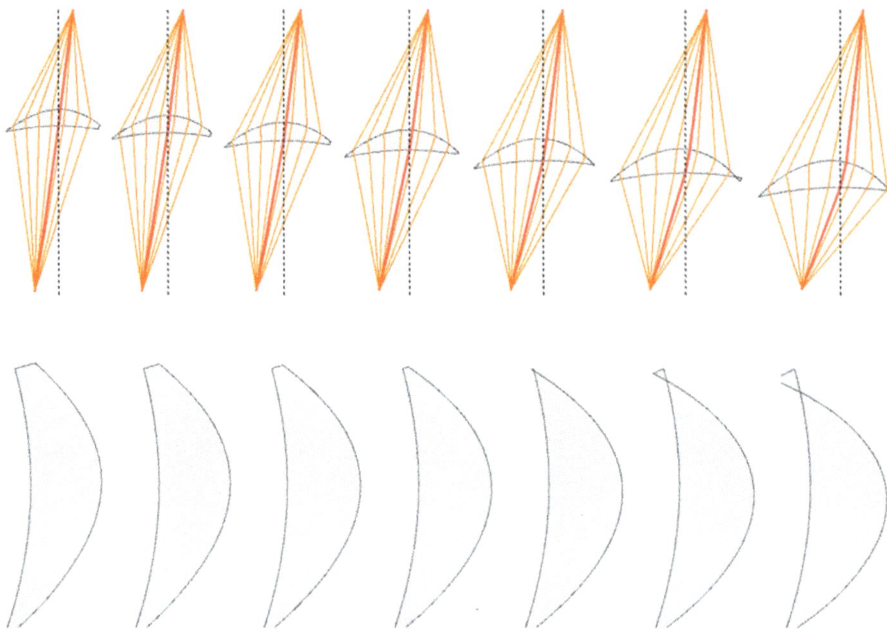

**Figure 13.14.** Lenses designed with equation (13.33). The inputs are $n = 1.5$, $t = 5$ mm, $t_b = 30$ mm, $h_a = -5$ mm, $h_b = 3$ mm, and $z_a = -r_a^2/60$. The variable that changes from left to right is $t_a$, from −50 mm to −20 mm in of steps of 5 mm.

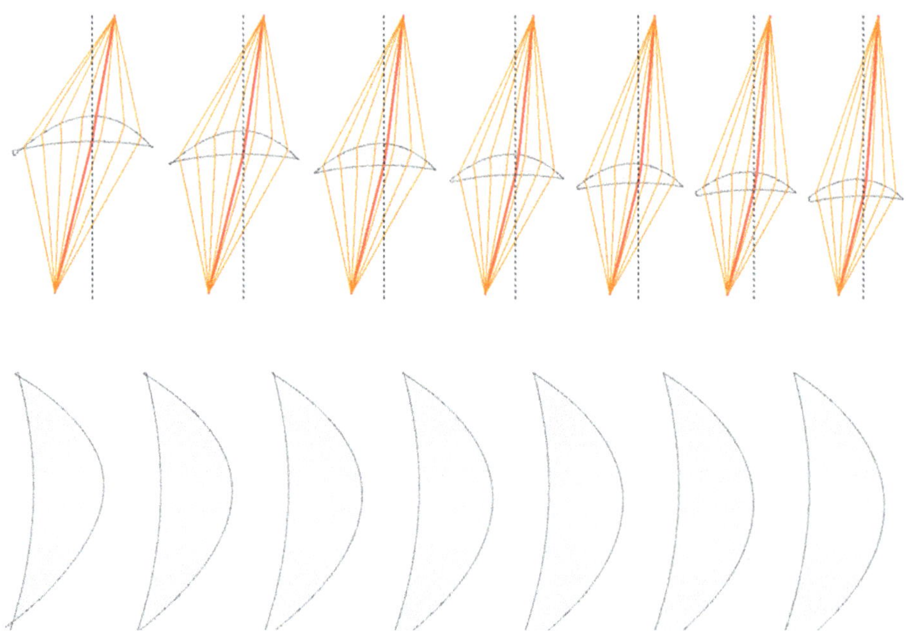

**Figure 13.15.** Lenses designed with equation designed with equation (13.33). The constants are $n = 1.5$, $t = 5$ mm, $t_a = -30$ mm, $h_a = -5$ mm, $h_b = 3$ mm, and $z_a = -r_a^2/60$. The variable is $t_b$, from 20 mm to 50 mm in of steps of 5 mm.

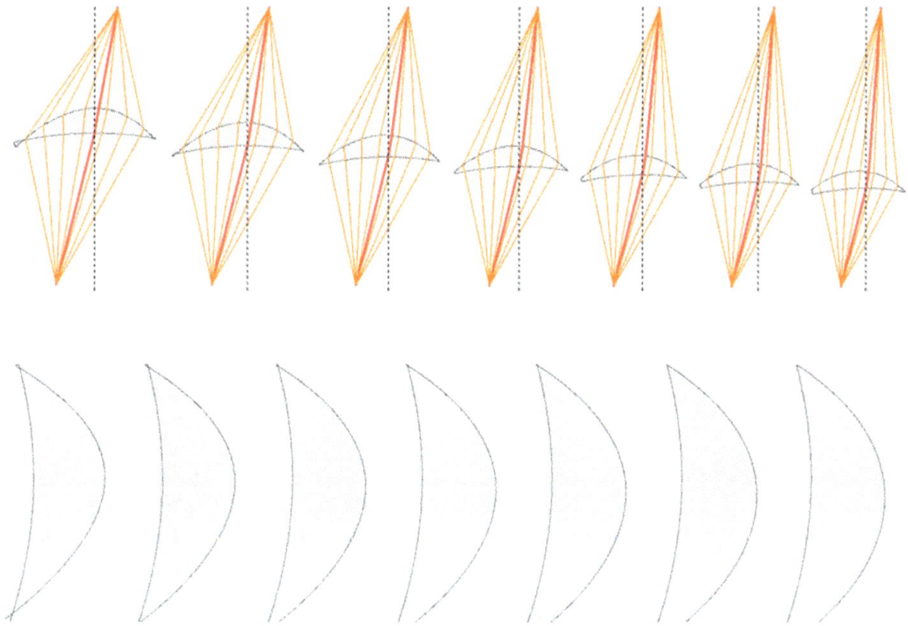

**Figure 13.16.** Profiles described by equation (13.33). The inputs are $n = 1.5$, $t_a = -30$ mm, $t_b = 30$ mm, $h_a = -5$ mm, $h_b = 3$ mm, and $z_a = -r_a^2/60$. The variable that changes from left to right is $t$, from 4 mm to 10 mm in of steps of 1 mm.

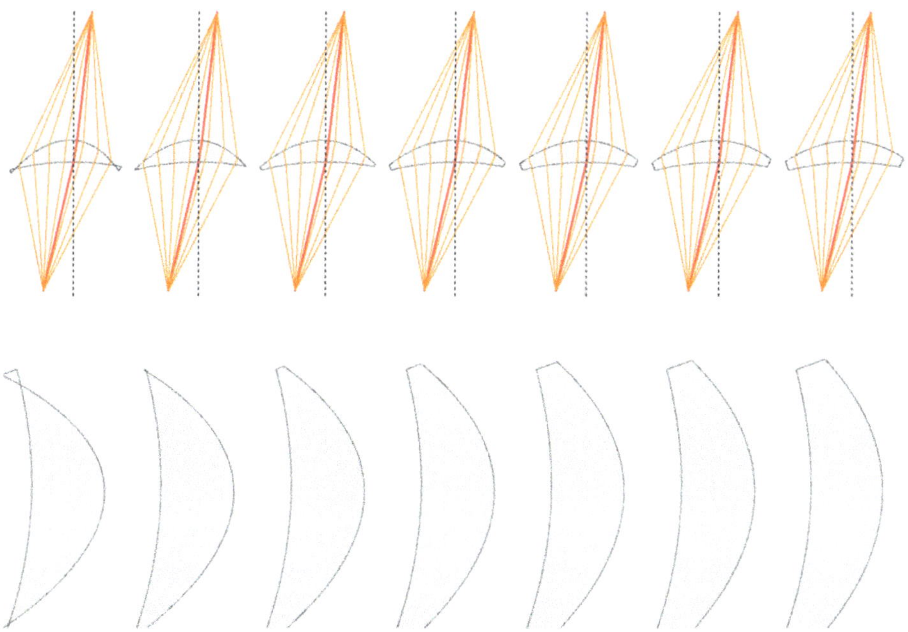

**Figure 13.17.** The table shows lenses described by equation (13.33). The constants are $t_a = -30$ mm, $t = 5$, $t_b = 30$ mm, $h_a = -5$ mm, $h_b = 3$ mm, and $z_a = -r_a^2/60$. The input value that changes is $n$, from 1.4 to 2 in of steps of 0.1.

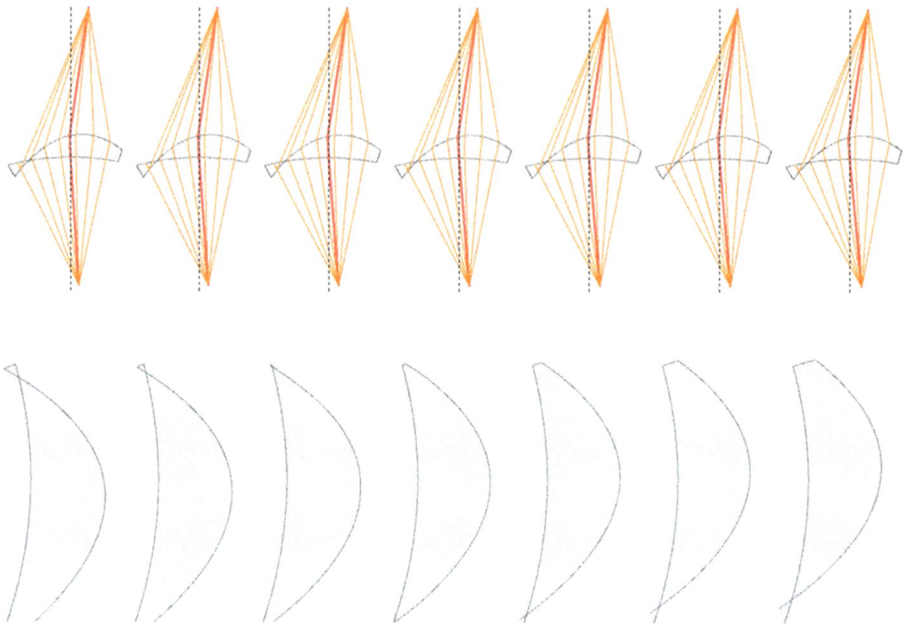

**Figure 13.18.** Lenses of equation (13.33). The input parameters are $n = 1.5$, $t_a = -30$ mm, $t = 5$, $t_b = 30$ mm, $h_b = 3$ mm and $z_a = -r_a^2/60$. The parameter that changes is $h_a$, from $-7$ mm to $-1$ mm in of steps of 1 mm.

Firstly, let us recall the radially symmetric condition for the input surface

$$z_a(-r_a) = z_a(r_a), \tag{13.34}$$

taking $z_a$ as radially symmetric function, so if $z_b$ is radially symmetric it arises that

$$z_b(-r_a) - z_b(r_a) = 0, \tag{13.35}$$

when $z_a$ is radially symmetric.

Nevertheless, this does not happen, since $z_b$ depends on $\wp_r$ and $\wp_z$, and they are odd. they depend on $z_a$, $r_a$, $t_a$ and $h_a$. They are even if and only if $h_a = 0$. Thus they are not radially symmetric. Therefore, we have,

$$z_b(-r_a) - z_b(r_a) \neq 0, \tag{13.36}$$

when $z_a$ is radially symmetric.

## 13.4 Mathematical implications of a non-symmetric solution

Each ray that obeys the Fermat principle hits the lens with a different incident angle that changes the direction of each path. The time that each ray spends inside the lens is different; Therefore, the lens cannot be symmetric. Transit time defines the shape of the lens.

In the case of the singlet lens free of aspherical aberration, it is easy to find that the singlet is symmetric since every path has another path that spends the same time traveling inside the lens. Still, for the coma, there is not that symmetry between the paths.

If we choose a radially symmetric as the input surface, the general formula of the lens free from spherical aberration will give an output surface radially symmetric. Using the same input surface, the output surface provided by the general formula of the lens free from coma will be radially not symmetric. In general, both surfaces do not intersect. Therefore, we analytically cannot design an aplanatic singlet lens without using paraxial or third-order approximations.

Finally, we want to mention that this asymmetry of the off-axis singlet is only a consequence of the fifth Euclidean postulate.

*'Two parallel lines keep a finite distance between them'.*

Since the optical system lies in $\mathbb{R}^2$, it obeys the Euclidean postulates. The Euclidean metric has all this information inside. These lenses follow Fermat's principle that supports the Euclidean metric. Actually, the Fermat principle can be seen as a homotopy.

A path is an application, $\alpha: I \to \mathbb{R}^2$ where $I = [0, 1]$, every path has an initial point and final point, for our case the initial point is $\alpha(0) = (t_a, h_a)$ and the final is $\alpha(1) = (t_b, h_b)$. If two paths have the same initial point and final point is to say that they are homotopic. A homotopy is an application given by $H: I \times I \to \mathbb{R}^2$ such that,

$$\begin{cases} H(t, 0) = \alpha(t), \\ H(t, 1) = \beta(t), \\ H(0, s) = \alpha(0) = \beta(0), \\ H(1, s) = \alpha(1) = \beta(1), \end{cases} \tag{13.37}$$

where $\alpha$ and $\beta$ are homotopic paths. Usually, in topology, we can use any continuous path, but here since we are dealing which a physics problem, the only exciting paths are those that obey Fermat's principle. Fermat's principle is then the homotopy of the problem. The theory of homotopies can describe the lenses presented in this chapter and chapter 4. Lenses with any numerical approximation or optimization process do not have homotopic paths.

## 13.5 Conclusions

In this chapter, we have found the mathematical solution of the conjecture presented in the introduction, equation (13.33). We also found that the output surfaces of the on-axis stigmatic lens and the off-axis stigmatic in general do not overlap since one is radially symmetric, and the other is not.

## References

Acuña R G G and Gutiérrez-Vega J C 2019 General formula of the refractive telescope design free spherical aberration *Novel Optical Systems, Methods, and Applications XXII* vol 11105 (Bellingham, WA: International Society for Optics and Photonics)

Bellucci R, Morselli S and Pucci V 2007 Spherical aberration and coma with an aspherical and a spherical intraocular lens in normal age-matched eyes *J. Cataract Refract. Surg.* **33** 203–9

Born M and Wolf E 2013 *Principles of Optics: Electromagnetic Theory of Propagation Interference and Diffraction of Light* (Amsterdam: Elsevier)

Chaves J 2016 *Introduction to Nonimaging Optics* 2nd edn (Boca Raton, FL: CRC Press)

Duerr F, Benítez P, Minano J C, Meuret Y and Thienpont H 2012 Analytic design method for optimal imaging: coupling three ray sets using two free-form lens profiles *Opt. Express* **20** 5576–85

González-Acuña R G and Chaparro-Romo H A 2018 General formula for bi-aspheric singlet lens design free of spherical aberration *Appl. Opt.* **57** 9341–5

González-Acuña R G and Guitiérrez-Vega J C 2018 Generalization of the axicon shape: the gaxicon *J. Opt. Soc. Am.* A **35** 1915–8

González-Acuña R G and Gutiérrez-Vega J C 2019a Analytic formulation of a refractive-reflective telescope free of spherical aberration *Opt. Eng.* **58** 085105

González-Acuña R G and Gutiérrez-Vega J C 2019b General formula for aspheric collimator lens design free of spherical aberration *Current Developments in Lens Design and Optical Engineering XX* vol 11104 (Bellingham, WA: International Society for Optics and Photonics)

González-Acuña R G, Avendaño-Alejo M and Gutiérrez-Vega J C 2019a Singlet lens for generating aberration-free patterns on deformed surfaces *J. Opt. Soc. Am.* A **36** 925–9

González-Acuña R G, Chaparro-Romo H A and Gutiérrez-Vega J C 2019b General formula to design freeform singlet free of spherical aberration and astigmatism *Appl. Opt.* **58** 1010–5

Kingslake R and Johnson R B 2009 *Lens Design Fundamentals* (New York: Academic)

Luneburg R K and Herzberger M 1964 *Mathematical Theory of Optics* (Berkeley, CA: University of California Press)

Frank E 1935 Lens systems for correcting coma of mirrors *Astrophys. J.* **vol 81** 156

Stavroudis O 2012 The Optics of Rays *Wavefronts, and Caustics* vol 38 (Amsterdam: Elsevier)

Stavroudis O N 2006 *The Mathematics of Geometrical and Physical Optics: The k-Function and Its Ramifications* (New York: Wiley)

Sun H 2016 *Lens Design: A Practical Guide* (Boca Raton, FL: CRC Press)

Tabernero J, Piers P and Artal P 2007 Intraocular lens to correct corneal coma *Opt. Lett.* **32** 406–8

Willy M 1929 Photographic lens, corrected spherically, chromatically, astigmatically, and for coma *US Patent* 1741947

Willy M 1930 Objective corrected spherically, chromatically, astigmatically, and for coma *US Patent* 1786916

Zemlin F, Weiss K, Schiske P, Kunath W and Herrmann K-H 1978 Coma-free alignment of high resolution electron microscopes with the aid of optical diffractograms *Ultramicroscopy* **3** 49–60

# Chapter 14

# Aplanatic singlet lens: general setting part 1

We offer the general configuration and design of an aplanatic optical system consisting of a single aspherical lens and its entrance pupil. The aspheric lens is aplanatic for three-point objects and their respective three-point images when the three-point objects are very far away from the singlet. The entrance pupil is such that the arrangement is aplanatic.

## 14.1 Introduction

In the previous chapters, we have seen the formulas on-axis stigmatic lenses and off-axis stigmatic lenses. We have learned a lot about the stigmatic imaging with the lenses designed with these formulas. We have studied their topological and geometrical properties. We can summarize these properties as:

- The entrance pupil in both cases is the lens itself.
- The general formula on-axis stigmatic lenses is radially symmetric.
- The general formula off-axis stigmatic lenses is not radially symmetric.
- Therefore, an on-axis and off-axis stigmatic lens does not exist if we take the entrance pupil as the lens itself.

However, we can get an aplanatic singlet in certain circumstances. In this chapter, we are going to study the general setting of a stigmatic aplanatic singlet lens for objects far away from the entrance pupil of the singlet. A stigmatic aplanatic singlet is a lens such that it has three-point objects with stigmatic relation with three-point images. The three-point objects are placed in the same plane, and the corresponding three-point images are also aligned in their corresponding plane. See figure 14.1.

We can design stigmatic aplanatic singlet if we merge the formulas of the on-axis and off-axis stigmatic lenses. The first step in the procedure is to get the equation of the off-axis stigmatic lenses for an object far way from the singlet. Then, we get the equation of the on-axis stigmatic lenses of an object very distant from the first

doi:10.1088/978-0-7503-5774-6ch14

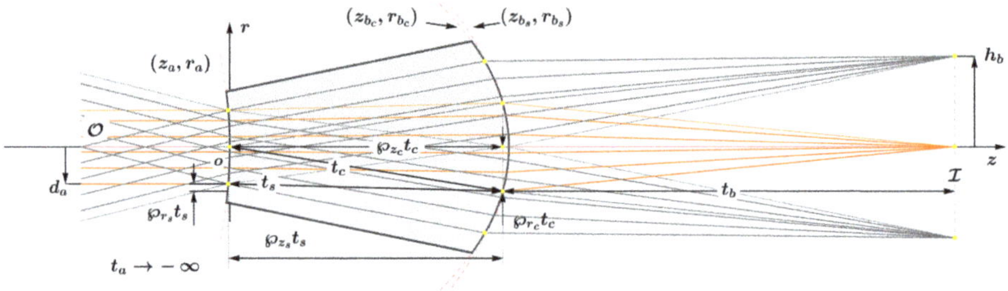

**Figure 14.1.** Aplanatic collector lens. The central section corresponds to equation (14.30) for on-axis objects. The lateral parts correspond to equation (14.9) for off-axis objects. Both equations coexist in the region covered by the entrance pupil. The height of the image is $h_b$ and $d_a$ is the entrance pupil.

surface using an arbitrary ray, instead of choosing the reference ray as the axial ray. Finally, we merge both solutions according to figure 14.1.

In 14.1 is presented a stigmatic aplanatic singlet with first surface $(z_a, r_a)$ and second surface divided into three sections, the central region is described by the general formula on-axis stigmatic lenses and the exterior areas are given by the general formula off-axis stigmatic lenses.

Since we are merging sections, we need to set a precise notation to avoid confusion. We set subscript $c$ as the elements related to the general formula off-axis stigmatic, $c$ is for coma. The subscript $s$ is to the elements related to the general formula on-axis stigmatic, $s$ is for spherical aberration. If a term does not have a subscript, it is because both formulas share it, as can be seen in figure 14.1. For example, the first surface $(z_a, r_a)$.

## 14.2 Off-axis stigmatic collector lens

We begin with the off-axis stigmatic collector lens, which means it receives light from a point source very far away from the first surface, as can be seen in figure 14.1. Therefore, we need to apply two limits. First, the limit when $h_a \to t_a w$ and then, the limit when $t_a \to -\infty$, where $w$ is a real constant related to the angle of incidence of the input rays. If $w = 0$ the point object is at minus infinity and the input rays travel parallel to the optical axis. Therefore, the only thing we need to do is to get the off-axis stigmatic collector lens and apply the mentioned limits on the terms of the formula of off-axis stigmatic lenses. We recall these terms with the subscript $c$. The parameters are taken from equations (13.17), (13.18), (13.9) and (13.10). Therefore, the cosine directors of the reference ray inside the lens,

$$\wp_{r,0_c} \equiv -\frac{h_a}{n\sqrt{h_a^2 + t_a^2}}, \tag{13.17}$$

and,

$$\wp_{z,0_c} \equiv \sqrt{1 - \frac{(-h_a)^2}{n^2(h_a^2 + t_a^2)}}, \tag{13.18}$$

The cosine directors of any other ray inside the lens,

$$\wp_{r_c} \equiv \frac{-h_a + r_a + (z_a - t_a)z_a'}{n(z_a'^2 + 1)\sqrt{(h_a - r_a)^2 + (t_a - z_a)^2}} - \frac{z_a'\sqrt{1 - \frac{[-h_a + r_a + (z_a - t_a)z_a']^2}{n^2[z_a'^2 + 1][(h_a - r_a)^2 + (t_a - z_a)^2]}}}{\sqrt{z_a'^2 + 1}}, \tag{13.9}$$

and,

$$\wp_{z_c} \equiv \frac{z_a'[-h_a + r_a + (z_a - t_a)z_a']}{n(z_a'^2 + 1)\sqrt{(h_a - r_a)^2 + (t_a - z_a)^2}}, + \frac{\sqrt{1 - \frac{[-h_a + r_a + (z_a - t_a)z_a']^2}{n^2(z_a'^2 + 1)[(h_a - r_a)^2 + (t_a - z_a)^2]}}}{\sqrt{z_a'^2 + 1}} \tag{13.10}$$

and, variable $f_c$ of equation (13.23),

$$f_c \equiv \sqrt{h_a^2 + t_a^2} + nt + \sqrt{(\wp_{r,0_c}t - h_b)^2 + t_b^2} - \sqrt{(r_a - h_a)^2 + (z_a - t_a)^2}. \tag{13.23}$$

Therefore, we apply both limits on them,

$$\lim_{t_a \to -\infty}\left[\lim_{h_a \to w t_a}(\wp_{r,0_c})\right] = \frac{w}{n\sqrt{w^2 + 1}}, \tag{14.1}$$

and,

$$\lim_{t_a \to -\infty}\left[\lim_{h_a \to w t_a}(\wp_{z,0_c})\right] = \sqrt{1 - \frac{w^2}{n^2(w^2 + 1)}}. \tag{14.2}$$

Observe that $\lim_{t_a \to -\infty}\left[\lim_{h_a \to w t_a}(\wp_{r,0_c})\right]^2 + \lim_{t_a \to -\infty}\left[\lim_{h_a \to w t_a}(\wp_{z,0_c})\right]^2 = 1$. This is because it is always true that $\wp_{r,0_c^2} + \wp_{r,0_c^2} = 1$. We continue with the cosine directors of the non-reference ray

$$\lim_{t_a \to -\infty}\left[\lim_{h_a \to w t_a}(\wp_{r_c})\right] = \frac{z_a'\left(1 - n\sqrt{w^2 + 1}\sqrt{(z_a')^2 + 1}\sqrt{\frac{z_a'\{[n^2(w^2 + 1) - 1]z_a' - 2w\} + (n^2 - 1)w^2 + n^2}{n^2(w^2 + 1)[(z_a')^2 + 1]}}\right) + w}{n\sqrt{w^2 + 1}[(z_a')^2 + 1]}, \tag{14.3}$$

and,

$$\lim_{t_a\to-\infty}\left[\lim_{h_a\to wt_a}(\wp_{z_c})\right] = \frac{\sqrt{\dfrac{[n^2(w^2+1)-1](z_a')^2+(n^2-1)w^2+n^2-2wz_a'}{n^2(w^2+1)((z_a')^2+1)}}}{\sqrt{(z_a')^2+1}} + \frac{z_a'(z_a'+w)}{n\sqrt{w^2+1}\,((z_a')^2+1)}. \tag{14.4}$$

Again note that $\lim_{t_a\to-\infty}\left[\lim_{h_a\to wt_a}(\wp_{r_c})\right]^2 + \lim_{t_a\to-\infty}\left[\lim_{h_a\to wt_a}(\wp_{z_c})\right]^2 = 1$. Finally, we apply the limits of the parameter $f_c$

$$\lim_{t_a\to-\infty}\left[\lim_{h_a\to wt_a}(f_c)\right] = \sqrt{\left(h_b - t_c\left\{\lim_{t_a\to-\infty}\left[\lim_{h_a\to wt_a}(\wp_{r,0_c})\right]\right\}\right)^2 + t_b^2} + nt_c - \frac{r_a w + z_a}{\sqrt{w^2+1}}. \tag{14.5}$$

Therefore, $\left[\lim_{h_a\to wt_a}(\vartheta_c)\right]$ of equation (13.32), when $h_a \to wt_a$ and then $t_a \to -\infty$ is given by

$$\lim_{t_a\to-\infty}\left[\lim_{h_a\to wt_a}(\vartheta_c)\right] = \frac{-\beta_c \pm \sqrt{\beta_c^2 - (1-n^2)\left\{(r_a-h_a)^2 + \tau_c^2 - \left[\lim_{t_a\to-\infty}(f_c)\right]^2\right\}}}{(1-n^2)}, \tag{14.6}$$

where,

$$\beta_c \equiv n\left\{\lim_{t_a\to-\infty}\left[\lim_{h_a\to wt_a}(f_c)\right]\right\} + (r_a-h_a)\left\{\lim_{t_a\to-\infty}\left[\lim_{h_a\to wt_a}(\wp_z)\right]\right\} + \tau_c\left\{\lim_{t_a\to-\infty}\left[\lim_{h_a\to wt_a}(\wp_{z_c})\right]\right\}, \tag{14.7}$$

and,

$$\tau_c \equiv z_a - t_b - t_c\left\{\lim_{t_a\to-\infty}\left[\lim_{h_a\to wt_a}(\wp_{z,0_c})\right]\right\}. \tag{14.8}$$

Therefore, the shape of the second surface, when the first surface is given, is expressed with the following equations.

$$\begin{cases} \lim_{t_a\to-\infty}\left[\lim_{h_a\to wt_a}(z_{b_c})\right] \\ \quad = z_a + \left\{\lim_{t_a\to-\infty}\left[\lim_{h_a\to wt_a}(\vartheta_c)\right]\right\}\left\{\lim_{t_a\to-\infty}\left[\lim_{h_a\to wt_a}(\wp_{z_c})\right]\right\}, \\ \lim_{t_a\to-\infty}\left[\lim_{h_a\to wt_a}(r_{b_c})\right] = r_a + \left\{\lim_{t_a\to-\infty}\left[\lim_{h_a\to wt_a}(\vartheta_c)\right]\right\}\left\{\lim_{t_a\to-\infty}\left[\lim_{h_a\to wt_a}(\wp_{r_c})\right]\right\} \\ \quad . \end{cases} \tag{14.9}$$

To use the above equations we need to use the following equations (14.1), (14.2), (14.3), (14.4), (14.5), (14.6), (14.7), (14.8) and replace them in (14.9).

Equation (14.9) is the first step of the procedure described in the introduction.

## 14.3 On-axis stigmatic lens for an arbitrary reference path

In this section, we are going to deduce the formula for an on-axis stigmatic lens for an arbitrary reference path. Please see figure 14.1.

In chapter 4, we present a singlet lens such that it collects the rays from an object that is placed at minus infinity. The problem of that expression is that we choose the reference ray to be the axial ray. However, the cosine directors of the non-reference are the same, so we use them, we recall them recalling equations (4.17), and (4.18),

$$\wp_{r_s} \equiv \frac{(z_a - t_a)z_a' + r_a}{-\text{sgn}(t_a)n\sqrt{r_a^2 + (t_a - z_a)^2}(1 + z_a'^2)} - z_a'\frac{\sqrt{1 - \dfrac{[r_a + (z_a - t_a)z_a']^2}{n^2[r_a^2 + (t_a - z_a)^2](1 + z_a'^2)}}}{\sqrt{1 + z_a'^2}}, \tag{4.17}$$

and,

$$\wp_{z_s} \equiv \frac{[r_a + (z_a - t_a)z_a']z_a'}{-\text{sgn}(t_a)n\sqrt{r_a^2 + (t_a - z_a)^2}(1 + z_a'^2)} + \frac{\sqrt{1 - \dfrac{[r_a + (z_a - t_a)z_a']^2}{n^2[r_a^2 + (t_a - z_a)^2](1 + z_a'^2)}}}{\sqrt{1 + z_a'^2}}. \tag{4.18}$$

The next step is to compute the limit when $t_a \to -\infty$ over them,

$$\left[\lim_{t_a\to-\infty}(\wp_{r_s})\right] = -\frac{z_a'\left(n\sqrt{z_a'^2+1}\sqrt{\frac{(n^2-1)z_a^2+n^2}{n^2(z_a'^2+1)}}-1\right)}{n(z_a'^2+1)}, \tag{14.10}$$

and,

$$\left[\lim_{t_a\to-\infty}(\wp_{z_s})\right] = \frac{\sqrt{\frac{(n^2-1)z_a'^2+n^2}{n^2((z_a')^2+1)}}}{\sqrt{(z_a')^2+1}} + \frac{z_a'^2}{nz_a'^2+n}. \tag{14.11}$$

Now, let's get the cosine directors of our reference ray, that is, the ray that will enter with the height of the entrance pupil $d_a$. This means we only need to evaluate $r_a = d_a$ in the above expression,

$$\left[\lim_{t_a\to-\infty}(\wp_{r_s})\Big|_{r_a=d_a}\right] = -\frac{z_a'\Big|_{r_a=d_a}\left(n\sqrt{z_a'^2\big|_{r_a=d_a}+1}\sqrt{\frac{(n^2-1)z_a^2\big|_{r_a=d_a}+n^2}{n^2\left(z_a'^2\big|_{r_a=d_a}+1\right)}}-1\right)}{n\left(z_a'^2\big|_{r_a=d_a}+1\right)}, \tag{14.12}$$

and,

$$\left[\lim_{t_a\to-\infty}(\wp_{z_s})\Big|_{r_a=d_a}\right] = \frac{\sqrt{\frac{(n^2-1)z_a'\Big|_{r_a=d_a}^2+n^2}{n^2\left(z_a'^2\big|_{r_a=d_a}+1\right)}}}{\sqrt{z_a'^2\big|_{r_a=d_a}+1}} + \frac{z_a'^2\big|_{r_a=d_a}}{nz_a'^2\big|_{r_a=d_a}+n}. \tag{14.13}$$

We define the following notation, to remark that they are the cosine directors of the reference ray,

$$\lim_{t_a\to-\infty}(\wp_{z_s})\Big|_{r_a=d_a} \equiv \lim_{t_a\to-\infty}(\wp_{z,0_s}), \tag{14.14}$$

and,

$$\lim_{t_a\to-\infty}(\wp_{r_s})\Big|_{r_a=d_a} \equiv \lim_{t_a\to-\infty}(\wp_{r,0_s}). \tag{14.15}$$

Note the 0 in the subscript; we use the 0 as a symbol of the reference ray in the subscript.

Now for a moment let's suppose that the object is at the finite position. Let's equal the optical path length (OPL) of the reference ray and the OPL of a non-reference ray,

$$\begin{aligned}&\sqrt{d_a^2+(t_a-z_a|_{r_a=d_a})^2}+\sqrt{r_{b_s}^2\Big|_{r_a=d_a}+[t_s(\wp_{z,0_s})-z_{b_s}|_{r_a=d_a}+t_b]^2}+nt_s\\&=\sqrt{[t_s(\wp_{z,0_s})+t_b-z_{b_s}]^2+r_{b_s}^2}+n\vartheta_s+\sqrt{r_a^2+(t_a-z_a)^2}.\end{aligned} \tag{14.16}$$

$\sqrt{d_a^2 + (t_a - z_a|_{r_a=d_a})^2}$ is the distance that the reference ray travels form the entrance pupil to the first surface, $nt_s$ is the OPL of the reference ray inside the lens. Once the reference ray is outside the lens it travels from the second surface to its corresponding image point as distance defined by $\sqrt{r_{b_s}^2|_{r_a=d_a} + [t_s(\wp_{z,0_s}) - z_{b_s}|_{r_a=d_a} + t_b]^2}$.

For the non-reference ray, $\sqrt{r_a^2 + (t_a - z_a)^2}$ is the distance from the point object to the first surface. $\sqrt{[t_s(\wp_{z,0_s}) + t_b - z_{b_s}]^2 + r_{b_s}^2}$ is the distance that a non-reference ray travels from the second surface to the image point. Finally $\vartheta_s$ is the distance that a non-references ray travels inside the lens and it is defined as,

$$\vartheta_s \equiv \sqrt{(r_{b_s} - r_a)^2 + (z_{b_s} - z_a)^2}. \tag{14.17}$$

To recover the parallelism we compute the limit $t_a \to -\infty$ over equation (14.16),

$$\begin{aligned} &\sqrt{r_b^2|_{r_a=d_a} + \left\{t\left[\lim_{t_a \to -\infty}(\wp_{z,0_c})\right] - z_b|_{r_a=d_a} + t_b\right\}^2} + z_a|_{r_a=d_a} + nt - z_a \\ &= \sqrt{\left\{t\left[\lim_{t_a\to-\infty}(\wp_{z,0_s})\right] + t_b - \left[\lim_{t_a\to-\infty}(z_{b_s})\right]\right\}^2 + \left[\lim_{t_a\to-\infty}(r_{b_s})\right]^2} + n\left[\lim_{t_a\to-\infty}(\vartheta_s)\right]. \end{aligned} \tag{14.18}$$

To clean the equation we define the parameter $f_s$ as,

$$f_s \equiv \sqrt{r_b^2|_{r_a=h_a} + \left\{t\left[\lim_{t_a\to-\infty}(\wp_{z_c})|_{r_a=h_a}\right] - z_b|_{r_a=h_a} + t_b\right\}^2} + z_a|_{r_a=d_a} + nt - z_a, \tag{14.19}$$

replacing $f_s$ in equation (14.18),

$$\begin{aligned} f_s = &\sqrt{\left\{t\left[\lim_{t_a\to-\infty}(\wp_{z,0_s})\right] + t_b - \left[\lim_{t_a\to-\infty}(z_{b_s})\right]\right\}^2 + \left[\lim_{t_a\to-\infty}(r_{b_s})\right]^2} \\ &+ n\left[\lim_{t_a\to-\infty}(\vartheta_s)\right]. \end{aligned} \tag{14.20}$$

Manipulating and substituting, $z_{b_s} = z_a + \vartheta_s\wp_z$ and $r_{b_s} = r_a + \vartheta_s\wp_r$ in equation (14.20), we get,

$$\begin{aligned} \left\{f_s - n\left[\lim_{t_a\to-\infty}(\vartheta_s)\right]\right\}^2 = &\left\{t\left[\lim_{t_a\to-\infty}(\wp_{z,0_s})\right] + t_b - \left[\lim_{t_a\to-\infty}(z_a + \vartheta_s\wp_z)\right]\right\}^2 \\ &+ \left[\lim_{t_a\to-\infty}(r_a + \vartheta_s\wp_r)\right]^2. \end{aligned} \tag{14.21}$$

Note that the limit when $t_a \to -\infty$ does not affect on $z_a$ and $r_a$ since they do not depend on $t_a$, thus,

$$\left\{f_s - n\left[\lim_{t_a \to -\infty}(\vartheta_s)\right]\right\}^2 = \left\{t\left[\lim_{t_a \to -\infty}(\wp_{z,0_s})\right] + t_b - z_a - \left[\lim_{t_a \to -\infty}(\vartheta_s)\right]\left[\lim_{t_a \to -\infty}(\wp_{z_s})\right]\right\}^2 + \left\{r_a + \left[\lim_{t_a \to -\infty}(\vartheta_s)\right]\left[\lim_{t_a \to -\infty}(\wp_{r_s})\right]\right\}^2. \tag{14.22}$$

Let's define another parameter to clean the above expression, the new parameter is $\tau_s$,

$$\tau_s \equiv z_a - t\left[\lim_{t_a \to -\infty}(\wp_{z,0_s})\right] - t_b, \tag{14.23}$$

replacing $\tau_s$ in equation (14.21),

$$\left\{f_s - n\left[\lim_{t_a \to -\infty}(\vartheta_s)\right]\right\}^2 = \left\{\tau_s + \left[\lim_{t_a \to -\infty}(\vartheta_s)\right]\left[\lim_{t_a \to -\infty}(\wp_{z_c})\right]\right\}^2 + \left\{r_a + \left[\lim_{t_a \to -\infty}(\vartheta_s)\right]\left[\lim_{t_a \to -\infty}(\wp_{r_c})\right]\right\}^2. \tag{14.24}$$

Squaring the binomials,

$$\begin{aligned} f_s^2 - 2f_s n\left[\lim_{t_a \to -\infty}(\vartheta_s)\right] + n^2\left[\lim_{t_a \to -\infty}(\vartheta_s)\right]^2 \\ = r_a^2 + 2r_a\left[\lim_{t_a \to -\infty}(\wp_{r_c})\right]\left[\lim_{t_a \to -\infty}(\vartheta_s)\right] + \left[\lim_{t_a \to -\infty}(\wp_{r_c})\right]^2\left[\lim_{t_a \to -\infty}(\vartheta_s)\right]^2 \\ + \tau_s^2 + 2\tau_s\left[\lim_{t_a \to -\infty}(\wp_{z_c})\right]\left[\lim_{t_a \to -\infty}(\vartheta_s)\right] + \left[\lim_{t_a \to -\infty}(\wp_{z_c})\right]^2\left[\lim_{t_a \to -\infty}(\vartheta_s)\right]^2. \end{aligned} \tag{14.25}$$

Then, let's collect the terms powered by $\left[\lim_{t_a \to -\infty}(\vartheta_s)\right]$,

$$\begin{aligned} 2\left[\lim_{t_a \to -\infty}(\vartheta_s)\right]\left\{f_s n + r_a\left[\lim_{t_a \to -\infty}(\wp_{r_c})\right] + \tau_s\left[\lim_{t_a \to -\infty}(\wp_{z_c})\right]\right\} \\ \left[\lim_{t_a \to -\infty}(\vartheta_s)\right]^2(1 - n^2) + (r_a^2 + \tau_s^2 - f_s^2) = 0. \end{aligned} \tag{14.26}$$

The last expression, equation (14.26), has the same form of a quadratic formula which is given by,

$$ax^2 + bx + c = 0, \quad x = \frac{-b \pm \sqrt{b^2 - 4ac}}{2a}. \tag{14.27}$$

Using the quadratic formula, we can find the solution for $\left[\lim_{t_a\to-\infty}(\vartheta_s)\right]$ in equation (14.26),

$$\left[\lim_{t_a\to-\infty}(\vartheta_s)\right] = \frac{\beta_s \pm \sqrt{\beta_s^2 - (1-n^2)(r_a^2 + \tau_s^2 - f_s^2)}}{1-n^2}, \tag{14.28}$$

where,

$$\beta_s \equiv \left\{f_s n + r_a\left[\lim_{t_a\to-\infty}(\wp_{r_c})\right] + \tau_s\left[\lim_{t_a\to-\infty}(\wp_{z_c})\right]\right\}. \tag{14.29}$$

Therefore,

$$\boxed{\begin{cases}\lim_{t_a\to-\infty}(z_{b_s}) = z_a + \left[\lim_{t_a\to-\infty}(\vartheta_s)\right]\left[\lim_{t_a\to-\infty}(\wp_{z_c})\right],\\ \lim_{t_a\to-\infty}(r_{b_s}) = r_a + \left[\lim_{t_a\to-\infty}(\vartheta_s)\right]\left[\lim_{t_a\to-\infty}(\wp_{r_c})\right].\end{cases}} \tag{14.30}$$

Equation (14.30) expresses the second surface of the on-axis stigmatic lens for an arbitrary reference path. The path that we take in this case is when $r_a = d_a$. We will see in the next section that we assigned the height of the entrance pupil to be $d_a$.

To properly use equation (14.30), we need to use equations (14.11), (14.12), (14.13), (14.19), (14.23), (14.28) and (14.29).

## 14.4 The merging of two solutions

We mentioned in the introduction that the second surface of the lens is divided into three sections. The central part is described by equation (14.30). The outer region is defined by equation (14.9), and finally a second external region is described by the absolute of equation (14.9). The intersection between the reference rays of the on-axis and off-axis in the second surface is called the meeting point.

The entrance pupil has the height such that both reference rays meet in the meeting point on the second surface.

This condition requires that OPLs of the reference rays meet each other in the second surface. Therefore we have,

$$t_s\wp_{r,0_s} + d_a = t_c\wp_{r,0_c}, \tag{14.31}$$

and,

$$t_s\wp_{z,0_s} + z_a\big|_{r_a=d_a} = t_c\wp_{z,0_c}. \tag{14.32}$$

Solving for the optical path of the on-axis stigmatic lens,

$$t_s = \frac{d_a\wp_{z,0_c} - \wp_{r,0_c}\left(z_a\big|_{r_a=d_a}\right)}{\wp_{z,0_s}\wp_{r,0_c} - \wp_{r,0_s}\wp_{z,0_c}}, \tag{14.33}$$

and solving for the optical path of the of-axis stigmatic lens,

$$t_c = \frac{\wp_{r,0_s}\left(z_a\big|_{r_a=d_a}\right) - d_a\wp_{z,0_s}}{\wp_{r,0_s}\wp_{z,0_c} - \wp_{z,0_s}\wp_{r,0_c}}. \tag{14.34}$$

We get what $t_s$ and $t_c$ must be such that the reference rays meet each other at the meeting point of the second surface.

## 14.5 Examples

In this section, we presented a gallery of stigmatic aplanatic lenses. In the figure of each example is given the design specifications of the corresponding example. Notice that the reference rays meet each other in the meeting point in every case. In the nearby area, $z_{b_s}$ and $z_{b_c}$ overlap each other. The figures are 14.2–14.7.

In all the examples, we choose a $h_b$ such that, we get $z_{b_s}(d_a) = z_{b_c}(0)$. This is to remark the overlapping of equations (14.30) and (14.9) on the region displayed by the entrance pupil.

Recognize that the dashed lines in all the cases are the continuation of the corresponding surfaces. This representation shows how the equations (14.30) and (14.9) are divided among themselves in the areas not displayed by the entrance pupil. Also see that for all the examples, even when the incident angles are huge, the lenses are stigmatic for the predefined point images. These results happen because we didn't accept any paraxial approximation in the derivation.

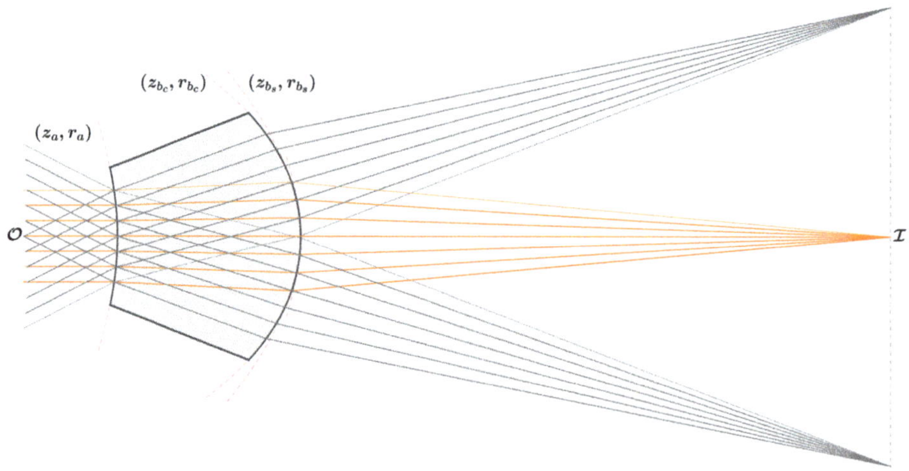


**Figure 14.2.** Specifications of the design: $n = 1.505\,595$, $w = -0.5$, $t_s = 139.17$ mm, $t_c = 142.848$ mm, $d_a = \pm 35.7771$ mm, $t_b = 463.6$ mm, $h_b = \mp 178.885$ mm $z_a = 60 - \sqrt{(r_a/2)^2 + 60^2}$, $z_{b_s}$= equation (14.30) and $z_{b_c}$= equation (14.9).

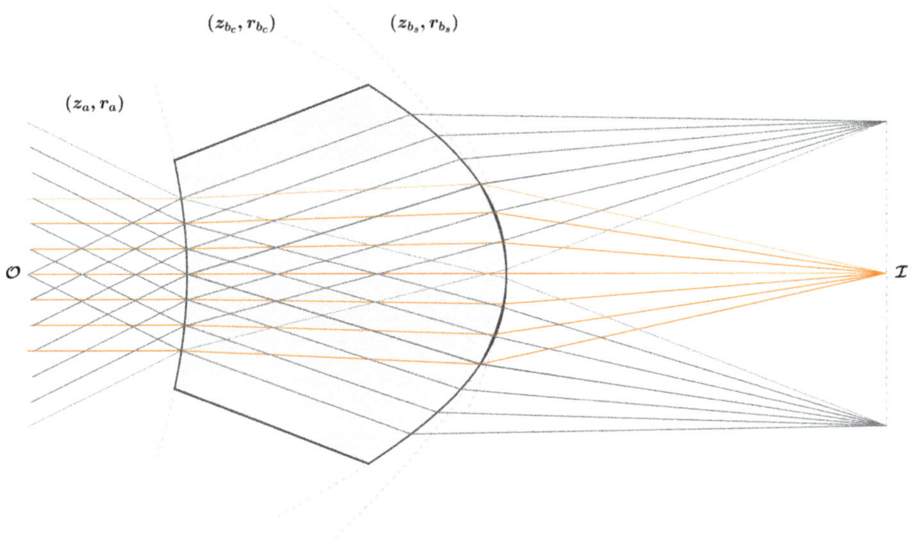


**Figure 14.3.** Design specifications: $n = 1.505\,595$, $w = -0.5$, $t_s = 139.17$ mm, $t_c = 142.848$ mm, $d_a = \pm 35.7771$ mm, $t_b = 189.571$ mm, $h_b = \mp 71.5542$ mm $z_a = 60 - \sqrt{(r_a/2)^2 + 60^2}$, $z_{b_s}$= equation (14.30) and $z_{b_c}$= equation (14.9).

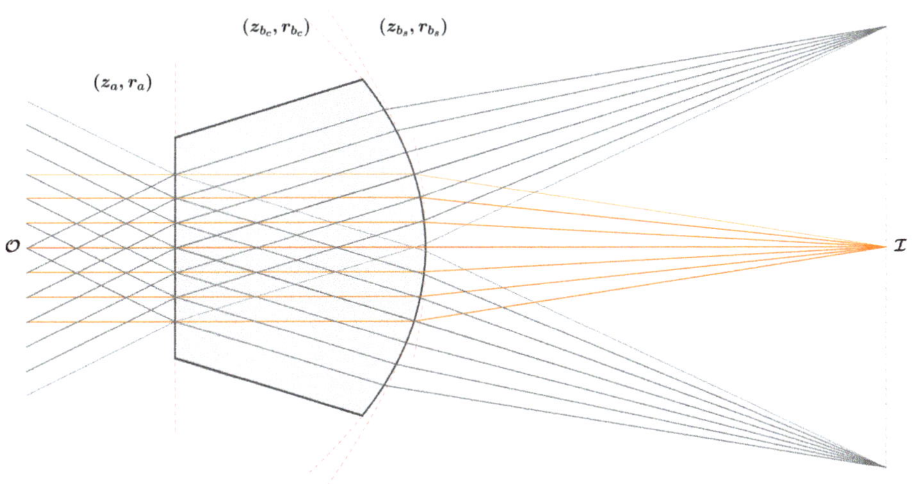


**Figure 14.4.** Design specifications: $n = 1.505\,595$, $w = -0.5$, $t_s = 115.011$ mm, $t_c = 120.448$ mm, $d_a = \pm 35.7771$ mm, $t_b = 228.219$ mm, $h_b = \mp 107.331$ mm $z_a = 0$, $z_{b_s}$= equation (14.30) and $z_{b_c}$= equation (14.9).

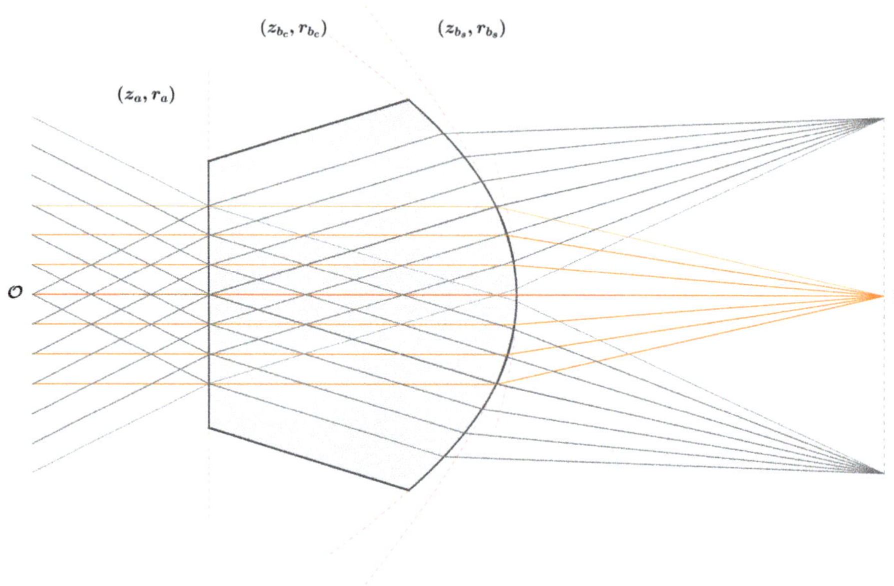


**Figure 14.5.** Design specifications: $n = 1.505\,595$, $w = -0.5$, $t_s = 115.011$ mm, $t_c = 120.448$ mm, $d_a = \pm 35.7771$ mm, $t_b = 154.872$ mm, $h_b = \mp 71.5542$ mm $z_a = 0$, $z_{b_s}$= equation (14.30) and $z_{b_c}$= equation (14.9).

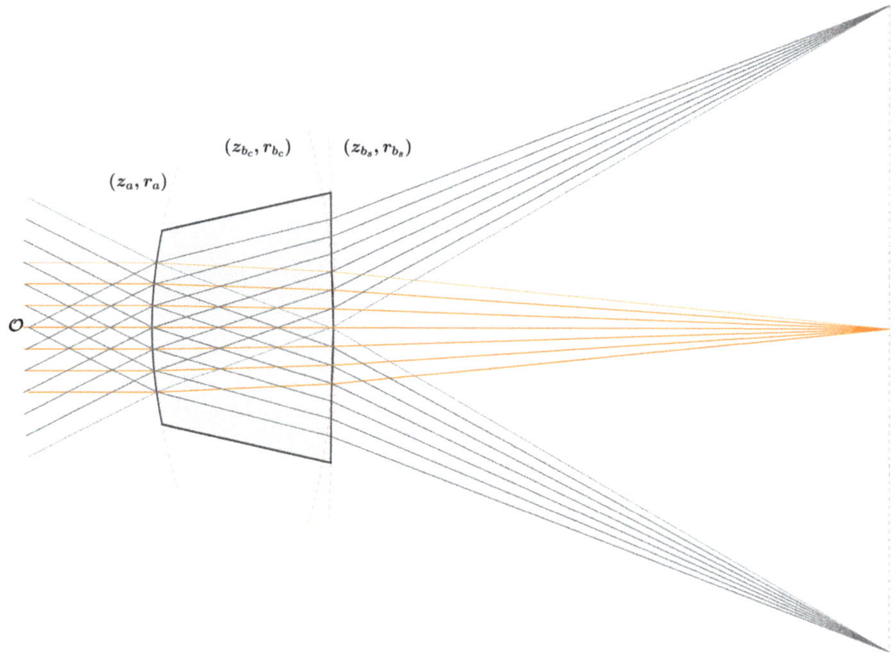


**Figure 14.6.** Design specifications: $n = 1.505\,595$, $w = -0.5$, $t_s = 97.5243$ mm, $t_c = 104.75$ mm, $d_a = \pm 35.7771$ mm, $t_b = 305.788$ mm, $h_b = \mp 178.885$ mm $z_a = -60 + \sqrt{(r_a/2)^2 + 60^2}$, $z_{b_s}$= equation (14.30) and $z_{b_c}$= equation (14.9).

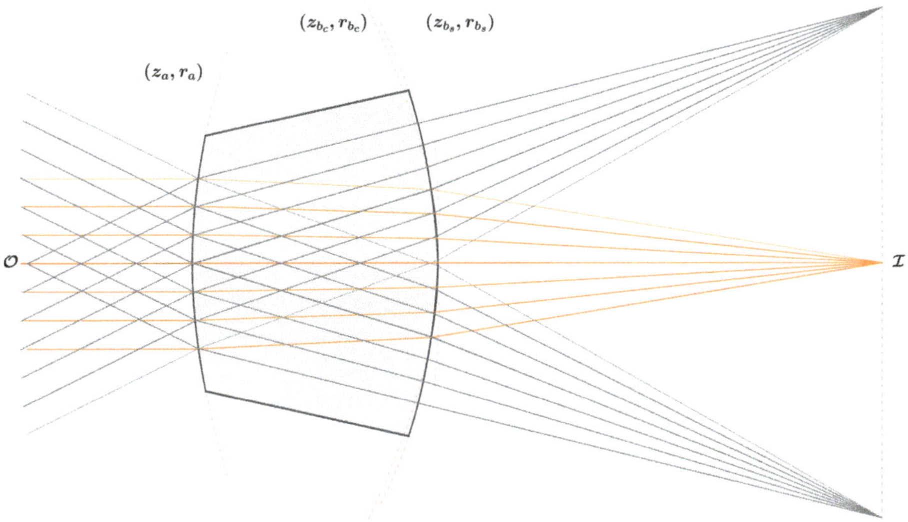


**Figure 14.7.** Design specifications: $n = 1.505\,595$, $w = -0.5$, $t_s = 97.5243$ mm, $t_c = 104.75$ mm, $d_a = \pm 35.7771$ mm, $t_b = 188.903$ mm, $h_b = \mp 107.331$ mm $z_a = -60 + \sqrt{(r_a/2)^2 + 60^2}$, $z_{b_s}$= equation (14.30) and $z_{b_c}$= equation (14.9).

## 14.6 Conclusions

We got a stigmatic aplanatic lens by merging the general equations of the off-axis stigmatic lens and the general equations of the on-axis stigmatic.

The height of the entrance pupil is given by the user. However, it is related to the central thickness of the lens. The height of the entrance pupil is such that the reference rays of both formulas meet in a single point in the second surface.

The gallery presented shows the versatility of the setting of the stigmatic aplanatic lens.

Finally, we want to remark that the procedure and results presented have the same nature of the result along with the treatise; they are analytic and unique.

## References

Acuña R G G and Gutiérrez-Vega J C 2019 General formula of the refractive telescope design free spherical aberration *Novel Optical Systems, Methods, and Applications XXII* vol 11105 (Bellingham, WA: International Society for Optics and Photonics)

Benitez P, Minano J C, Blen J, Mohedano R, Chaves J, Dross O, Hernandez M, Alvarez J L and Falicoff W 2004 SMS design method in 3D geometry: examples and applications *Nonimaging Optics: Maximum Efficiency Light Transfer VII* vol 5185 (Bellingham, WA: International Society for Optics and Photonics) pp 18–29

Benitez P, Miñano J C and Muñoz F 2005 Compact folded-optics illumination lens *US Patent* 6896381

Born M and Wolf E 2013 *Principles of Optics: Electromagnetic Theory of Propagation Interference and Diffraction of Light* (Amsterdam: Elsevier)

Chaves J 2016 *Introduction to Nonimaging Optics* 2nd edn (Boca Raton, FL: CRC Press)

Dross O, Mohedano R, Benitez P, Minano J C, Chaves J, Blen J, Hernandez M and Munoz F 2004 Review of SMS design methods and real-world applications *Nonimaging Optics and Efficient Illumination Systems* vol 5529 (Bellingham, WA: International Society for Optics and Photonics) pp 35–48

Duerr F, Benítez P, Minano J C, Meuret Y and Thienpont H 2012a Analytic design method for optimal imaging: coupling three ray sets using two free-form lens profiles *Opt. Express* **20** 5576–85

Duerr F, Benítez P, Miñano J C, Meuret Y and Thienpont H 2012b Analytic free-form lens design in 3D: coupling three ray sets using two lens surfaces *Opt. Express* **20** 10839–46

González-Acuña R G and Chaparro-Romo H A 2018 General formula for bi-aspheric singlet lens design free of spherical aberration *Appl. Opt.* **57** 9341–5

González-Acuña R G and Guitiérrez-Vega J C 2018 Generalization of the axicon shape: the gaxicon *J. Opt. Soc. Am.* A **35** 1915–8

González-Acuña R G and Gutiérrez-Vega J C 2019a Analytic formulation of a refractive-reflective telescope free of spherical aberration *Opt. Eng.* **58** 085105

González-Acuña R G and Gutiérrez-Vega J C 2019b General formula for aspheric collimator lens design free of spherical aberration *Current Developments in Lens Design and Optical Engineering XX* vol 11104 (Bellingham, WA: International Society for Optics and Photonics)

González-Acuña R G, Avendaño-Alejo M and Gutiérrez-Vega J C 2019a Singlet lens for generating aberration-free patterns on deformed surfaces *J. Opt. Soc. Am.* A **36** 925–9

González-Acuña R G, Chaparro-Romo H A and Gutiérrez-Vega J C 2019b General formula to design freeform singlet free of spherical aberration and astigmatism *Appl. Opt.* **58** 1010–5

Kingslake R and Johnson R B 2009 *Lens Design Fundamentals* (New York: Academic)

Lin W, Benítez P, Miñano J C, Infante J and Biot G 2011 Advances in the SMS design method for imaging optics *Optical Design and Engineering IV* vol 8167 (Bellingham, WA: International Society for Optics and Photonics)

Luneburg R K and Herzberger M 1964 *Mathematical Theory of Optics* (Berkeley, CA: University of California Press)

Miñano J C, Benítez P, Lin W, Infante J, Muñoz F and Santamaría A 2009a An application of the SMS method for imaging designs *Opt. Express* **17** 24036–44

Miñano J C, Benítez P and Santamaría A 2009b Free-form optics for illumination *Opt. Rev.* **16** 99–102

Miñano J C, Benítez P, Grabovickic D, Narasimhan B, Nikolic M and Infante J 2017 Freeform aplanatism *Freeform Optics* (Washington, DC: Optical Society of America) p JTu1C–2

Muñoz F, Benítez P and Miñano J C 2008 High-order aspherics: the SMS nonimaging design method applied to imaging optics *Novel Optical Systems Design and Optimization XI* vol 7061 (Bellingham, WA: International Society for Optics and Photonics)

Miñano J C and Benítez P 2016 Freeform aplanatic systems as a limiting case of SMS *Opt. Express* **24** 13173–8

Stavroudis O 2012 The Optics of Rays *Wavefronts, and Caustics* vol 38 (Amsterdam: Elsevier)

Stavroudis O N 2006 *The Mathematics of Geometrical and Physical Optics: The k-Function and Its Ramifications* (New York: Wiley)

Sun H 2016 *Lens Design: A Practical Guide* (Boca Raton, FL: CRC Press)

**IOP** Publishing

Analytical Lens Design (Second Edition)

**Rafael G González-Acuña, Héctor A Chaparro-Romo and Julio C Gutiérrez-Vega**

# Chapter 15

# Aplanatic singlet lens: general setting part 2

We offer the general setting and design of an aplanatic optical system comprised of a single aspheric lens and its entrance pupil. The aspheric lens is aplanatic for finite three-point objects and their respective finite three-point images. The images fall in the same plane, and the objects are in their respective plane.

## 15.1 Introduction

This is the last chapter of the treatise, here we are going to get the mathematical model of the stigmatic aplanatic lens for finite object and finite image. This model includes the entrance pupil, the position of the three-point objects, the three-point images, the refractive index and the first surface of the singlet. Three-point objects are placed in the same plane, and the corresponding three-point images are also aligned in their corresponding image plane.

This result has been a long road for us, from the first chapter to this one. This result is the crown of our theory. Potentially it has countless applications in science and engineering. This result is the answer to a myth in optical design, asked by many and even by Sir Isaac Newton. Newton placed a sketch of the mentioned lens in figure 3 of the first book of *Opticks* in part 1. See figure 15.1.

The prominent theoretical optician and optical designer Orestes N Stavroudis mentioned one time.

> *It is clear that general solutions to the lens equation, at least in terms of the methods we have been applying, are unobtainable. In our frustration we turn now to a method for obtaining particular solutions, solutions applicable to a particular optical design.*

Stavroudis expressed his frustration about not finding a general formula of a lens such that it is aplanatic. This passage is from his book *The Optics of Rays,*

doi:10.1088/978-0-7503-5774-6ch15  

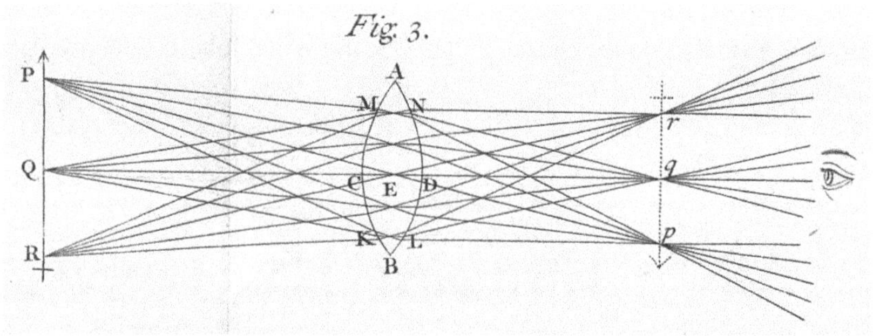


**Figure 15.1.** A stigmatic aplanatic lens for three finite point-objects, from figure 3 of the first book of *Opticks* in part 1, 1704. Source: Smithsonian Libraries https://archive.org/details/optickstreatise00newta/page/n95.

*Wavefronts, and Caustics* chapter 14, named *The lens equation.* Stavroudis also wrote the following paragraph.

> *Like so many other myths in geometrical optics, a prefect lens is so rare as to be almost nonexistent. It should be rotationally symmetric and be able to image certain points in object space perfectly in image space.*

The passage is taken from his book *The Mathematics of Geometrical and Physical Optics*, in chapter 14 called, *The perfect lenses of Gauss and Maxwell.* The chapter is about the attempts of Carl Frederic Gauss and James Clark Maxwell to design an aplanatic lens.

Giants like Gauss, Maxwell and Newton were behind, but they fail in finding it.

Donald Dilworth, a prominent optical designer and entrepreneur, mentioned the following about the lens design problem in his book, *Lens Design: Automatic, and quasi-autonomous computational methods and techniques*:

> *Thus the job is hard for most, and difficult even for the experts, most of the time. The core problem here is the fact that, except in a few simple cases,* ***there is no closed-form solution to the lens design problem.*** *That means there is no formula you can simply plug numbers into and obtain a great design. You have to think, try things, learn from experience, and iterate.*

What we are going to do in this chapter is to get the closest formula such that *you can simply plug numbers into it and obtain a great design.* We are going to design the stigmatic aplanatic singlet. Just like the previous chapter, by merging the general formula off-axis stigmatic lens and the general formula on-axis stigmatic lens in a single expression but now for finite distances.

Like in the previous chapter, we set subscript $c$ as the elements related to the general formula off-axis stigmatic. The subscript $s$ is to the elements related to the general formula on-axis stigmatic. Remember that if a term does not have a subscript, it is because both formulas share it. For example the refraction index $n$.

First, we are going to introduce the general formula off-axis stigmatic. Then, we are going to derive the general formula on-axis stigmatic for an arbitrary reference ray. Finally, we are going to set the entrance pupil and merge both solutions in the stigmatic aplanatic singlet for finite objects and finite images.

## 15.2 Off-axis stigmatic lens

We are going to use the general formula off-axis stigmatic lens presented in chapter 13. We recall it and submit it with the subscript $c$. The second surface of the off-axis stigmatic lens,

$$\begin{cases} z_{b_c} = z_a + \vartheta_c \wp_{z_c}, \\ r_{b_c} = r_a + \vartheta_c \wp_{r_c}, \end{cases} \tag{13.33}$$

where $z_a$ is the first surface of the lens, $\wp_{z_c}$ and $\wp_{r_c}$ are the cosine directors inside the lens,

$$\wp_{r_c} \equiv \frac{-h_a + r_a + (z_a - t_a)z_a'}{n(z_a'^2 + 1)\sqrt{(h_a - r_a)^2 + (t_a - z_a)^2}} - \frac{z_a'\sqrt{1 - \dfrac{[-h_a + r_a + (z_a - t_a)z_a']^2}{n^2(z_a'^2 + 1)[(h_a - r_a)^2 + (t_a - z_a)^2]}}}{\sqrt{z_a'^2 + 1}}, \tag{13.9}$$

and,

$$\wp_{z_c} \equiv \frac{z_a'[-h_a + r_a + (z_a - t_a)z_a']}{n(z_a'^2 + 1)\sqrt{(h_a - r_a)^2 + (t_a - z_a)^2}}, + \frac{\sqrt{1 - \dfrac{[-h_a + r_a + (z_a - t_a)z_a']^2}{n^2(z_a'^2 + 1)[(h_a - r_a)^2 + (t_a - z_a)^2]}}}{\sqrt{z_a'^2 + 1}}, \tag{13.10}$$

and, $\vartheta_c$ is the distance that the ray travels inside the lens,

$$\vartheta_c = \frac{-\left[f_c n + (r_a - h_b)\wp_{r_c} + \tau\wp_{z_c}\right]}{(1 - n^2)} \pm \frac{\sqrt{\left[f_c n + (r_a - h_b)\wp_{r_c} + \tau_c\wp_{z_c}\right]^2 - \left(1 - n^2\right)\left[(r_a - h_b)^2 + \tau_c^2 - f_c^2\right]}}{(1 - n^2)}, \tag{13.32}$$

where, the parameters inside $\vartheta_c$ are,

$$f_c \equiv \sqrt{h_a^2 + t_a^2} + nt_c + \sqrt{(\wp_{r,0_c} t_c - h_b)^2 + t_b^2} - \sqrt{(r_a - h_a)^2 + (z_a - t_a)^2}, \tag{13.23}$$

$$\tau_c \equiv z_a - t_c\wp_{z,0_c} - t_b, \tag{13.24}$$

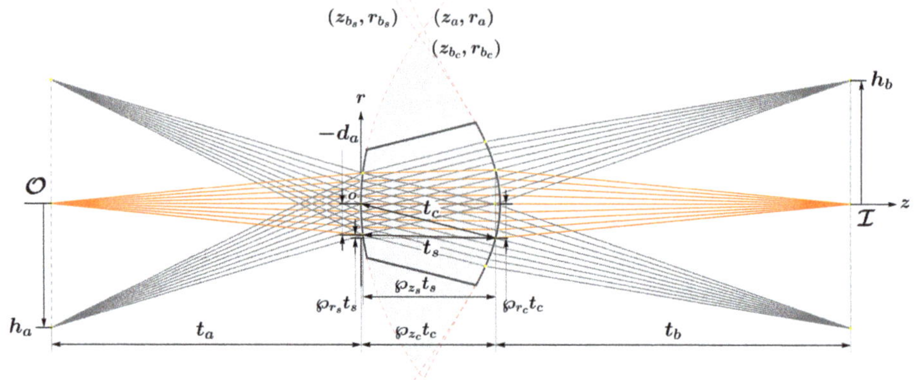


**Figure 15.2.** Aplanatic singlet lens. The central section corresponds to equation (15.15) for on-axis objects. The lateral parts correspond to equation (13.33) for off-axis objects. Both equations coexist in the region covered by the entrance pupil. The transverse displacements of the object and images with respect to the optical axis are $h_a$ and $h_b$, respectively. The coordinate system is placed at the origin, i.e. $z_a(0) = 0$.

$$\wp_{r,0_c} \equiv -\frac{h_a}{n\sqrt{h_a^2 + t_a^2}}, \tag{13.17}$$

and,

$$\wp_{z,0_c} \equiv \sqrt{1 - \frac{(-h_a)^2}{n^2(h_a^2 + t_a^2)}}. \tag{13.18}$$

Please see figures 15.2 and 15.3, the refraction index of the lens is $n$. The horizontal distance between the object and the first surface is $t_a$; the height of the object is $h_a$. The distance the reference ray travels inside the lens is $t_c$. The horizontal distance the reference ray travels from the second surface to the image is $t_b$, and the height of the image is $h_b$. Remember that the reference ray is the one that strikes the first surface at the origin.

## 15.3 On-axis stigmatic lens for an arbitrary reference path

In chapter 4, we worked with the on-axis stigmatic lens when the reference ray is the axial ray. Now, we are going to write the same formula but for an arbitrary reference ray. From the mentioned chapter, we saw that the structure of second surface $(z_{b_s}, r_{b_s})$ is given by,

$$z_{b_s} = z_a + \vartheta_s \wp_{z_s}, \tag{4.22}$$

and,

$$r_{b_s} = r_a + \vartheta_s \wp_{r_s}, \tag{4.23}$$

where $z_a$, where the cosine directors of the refracted ray inside the lens are, $\wp_{z_s}$, and $\wp_{r_s}$,

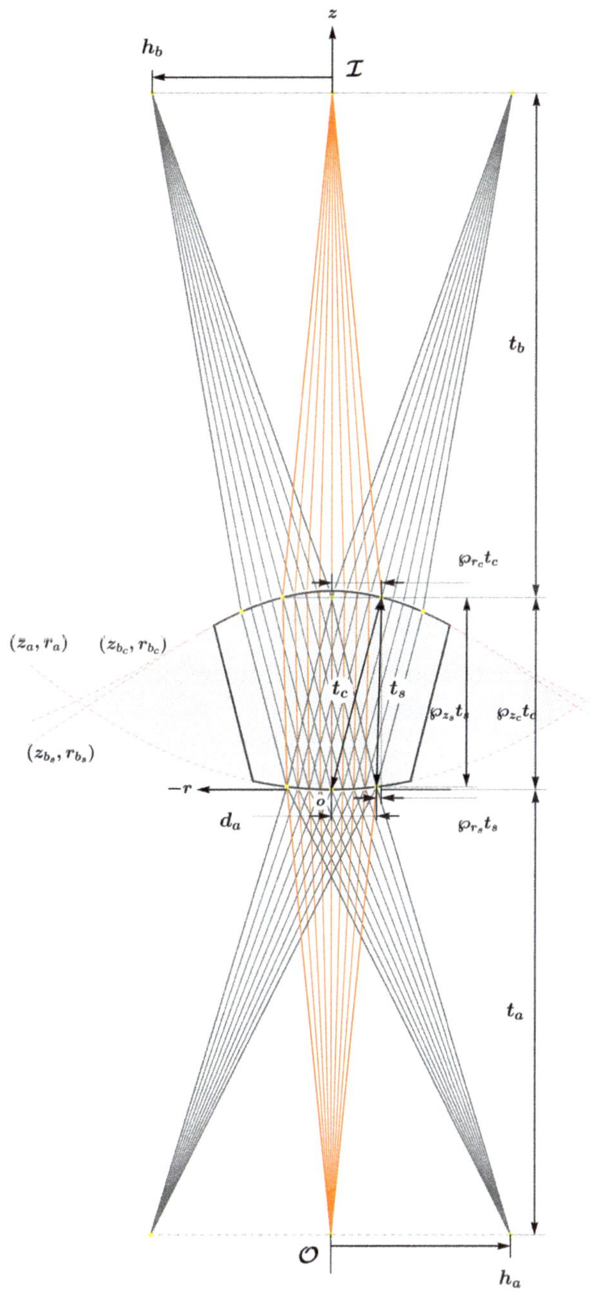


**Figure 15.3.** The diagram of figure 15.2 is rotated and amplified to show all the details.

$$\wp_{r_s} \equiv \frac{(z_a - t_a)z_a' + r_a}{-\operatorname{sgn}(t_a)n\sqrt{r_a^2 + (t_a - z_a)^2}\,(1 + z_a'^2)} - z_a'\frac{\sqrt{1 - \frac{[r_a + (z_a - t_a)z_a']^2}{n^2[r_a^2 + (t_a - z_a)^2](1 + z_a'^2)}}}{\sqrt{1 + z_a'^2}}, \tag{4.17}$$

and,

$$\wp_{z_s} \equiv \frac{[r_a + (z_a - t_a)z_a']z_a'}{-\operatorname{sgn}(t_a)n\sqrt{r_a^2 + (t_a - z_a)^2}\,(1 + z_a'^2)} + \frac{\sqrt{1 - \frac{[r_a + (z_a - t_a)z_a']^2}{n^2[r_a^2 + (t_a - z_a)^2](1 + z_a'^2)}}}{\sqrt{1 + z_a'^2}}, \tag{4.18}$$

where $\vartheta_s$ is the distance that each refracted ray travels inside the lens,

$$\vartheta_s \equiv \sqrt{(z_b - z_a)^2 + (r_b - r_a)^2}. \tag{4.19}$$

The goal in this section is to know what $\vartheta_s$ is in terms of an arbitrary reference ray, for that we will use Fermat's principle.

Please see figures 15.2 and 15.3, we know from chapter 4 that the distance between the object and the first surface is $t_a$. If we choose the reference ray to be the ray that strikes at $z_a|_{r_a=d_a}$, where $d_a$ is an arbitrary value taken by $r_a$, then, the optical path length (OPL) of this ray is,

$$\sqrt{d_a^2 + (t_a - z_a|_{r_a=d_a})^2} + \sqrt{r_{b_s}^2\big|_{r_a=d_a} + (t_s\wp_{z_s}|_{r_a=d_a} - z_{b_s}|_{r_a=d_a} + t_b)^2} + nt_s = \text{const.}, \tag{15.1}$$

where $t_s$ is the distance the reference ray travels inside the lens. $\sqrt{d_a^2 + (t_a - z_a|_{r_a=d_a})^2}$ is the distance from the object to the first surface and, finally, the following expression $\sqrt{r_{b_s}^2|_{r_a=d_a} + (t_s\wp_{z_s}|_{r_a=d_a} - z_{b_s}|_{r_a=d_a} + t_b)^2}$ is the distance from the second surface to the point image. Since, the singlet is stigmatic we can equate the OPL of the reference with the OPL with any other ray,

$$\begin{aligned}&\sqrt{d_a^2 + (t_a - z_a|_{r_a=d_a})^2} + \sqrt{r_{b_s}^2\big|_{r_a=d_a} + (t_s\wp_{z_s}|_{r_a=d_a} - z_{b_s}|_{r_a=d_a} + t_b)^2} + nt_s\\ &= \sqrt{(t_s\wp_{z_s}|_{r_a=d_a} + t_b - z_{b_s})^2 + r_{b_s}^2} + n\sqrt{(r_a - r_{b_s})^2 + (z_a - z_{b_s})^2} + \sqrt{r_a^2 + (t_a - z_a)^2}.\end{aligned} \tag{15.2}$$

Let's define $f_s$ to clear the last equation,

$$\begin{aligned}f_s \equiv &\sqrt{d_a^2 + (t_a - z_a|_{r_a=d_a})^2} + \sqrt{r_b^2\big|_{r_a=d_a} + (t_s\wp_{z_s}|_{r_a=d_a} - z_b|_{r_a=d_a} + t_b)^2}\\ &+ nt_s - \sqrt{r_a^2 + (t_a - z_a)^2}.\end{aligned} \tag{15.3}$$

Replacing equation (15.3) in (15.2),

$$f_s = \sqrt{(t_s \wp_{z_s}|_{r_a=d_a} + t_b - z_{b_s})^2 + r_{b_s}^2} + n\sqrt{(r_a - r_{b_s})^2 + (z_a - z_{b_s})^2}. \tag{15.4}$$

We can evaluate the structure of the second surface when $r_a = d_a$,

$$r_b|_{r_a=d_a} = d_a + t_s \wp_{r_s}|_{r_a=d_a}, \tag{15.5}$$

and,

$$z_b|_{r_a=d_a} = z_a|_{r_a=d_a} + t_s \wp_{z_s}|_{r_a=d_a}. \tag{15.6}$$

Remember that $t_s$ is the distance that the reference ray travels inside the lens, thus

$$\vartheta_s|_{r_a=d_a} \equiv t_s. \tag{15.7}$$

Therefore, we can replace equations (15.5) and (15.6) in (15.3) and we get,

$$\begin{aligned} f_s = \sqrt{d_a^2 + (t_a - z_a|_{r_a=d_a})^2} + n\, t_s - \sqrt{r_a^2 + (t_a - z_a)^2} \\ + \sqrt{(d_a + t_s \wp_{r_s}|_{r_a=d_a})^2 + (t_b - z_a|_{r_a=d_a})^2}, \end{aligned} \tag{15.8}$$

and we can replace equations (15.5), (15.6) and (15.3) in equation (15.4),

$$f_s = \sqrt{(t_s \wp_{z_s}|_{r_a=d_a} - (z_a + \vartheta_s \wp_{z_s}) + t_b)^2 + (\vartheta_s \wp_{r_s} + r_a)^2} + n\vartheta_s. \tag{15.9}$$

We assign another parameter to clear the last expression,

$$\tau_s \equiv -(t_s \wp_{z_s}|_{r_a=d_a} - z_a + t_b), \tag{15.10}$$

replacing $\tau_s$ and squaring equation (15.9),

$$(f_s - n\vartheta_s)^2 = (\tau_s + \vartheta_s \wp_{z_s})^2 + (\vartheta_s \wp_{r_s} + r_a)^2. \tag{15.11}$$

Expanding,

$$f_s^2 - 2f_s n\vartheta_s + n^2\vartheta_s^2 = \tau_s^2 + 2\tau_s\vartheta_s\wp_{z_s} + \vartheta_s^2\wp_{z_s}^2 + \vartheta_s^2\wp_{r_s}^2 + 2\vartheta\wp_{r_s}r_a + r_a^2. \tag{15.12}$$

Collecting the elements multiplied by $\vartheta_s$ and $\vartheta_s^2$,

$$\left(1 - n^2\right)\vartheta_s^2 + 2\left(f_s n + \tau_s\wp_{z_s} + r_a\wp_{r_s}\right)\vartheta_s + \left(r_a^2 + \tau_s^2 - f_s^2\right) = 0. \tag{15.13}$$

The last expression has the form of a quadratic equation; thus its solution is given by,

$$\vartheta_s = \frac{-(f_s n + \tau_s\wp_{z_s} + r_a\wp_{r_s}) \pm \sqrt{(f_s n + \tau_s\wp_{z_s} + r_a\wp_{r_s})^2 - \left(1 - n^2\right)\left(r_a^2 + \tau_s^2 - f_s^2\right)}}{1 - n^2}. \tag{15.14}$$

Therefore, we have the second surface as,

$$\boxed{\begin{cases} z_{b_s} = z_a + \vartheta_s \wp_{z_s}, \\ r_{b_s} = r_a + \vartheta_s \wp_{r_s}. \end{cases}} \tag{15.15}$$

The last equation is fundamental and useful for us because in this form it is easy to merge the on-axis stigmatic singlet with the off-axis stigmatic singlet.

## 15.4 The merging of the two solutions

To merge both solutions, we apply the same strategy implemented in the last chapter. We ensure that the distances $t_c$ and $t_s$ are such that their reference paths touch each other in the second surface, the meeting point. For that, we need that $t_c$ and $t_s$ fulfill the following conditions,

$$t_s \wp_{r,0_s} + d_a = t_c \wp_{r,0_c}, \tag{15.16}$$

and,

$$t_s \wp_{z,0_s} + z_a\big|_{r_a=d_a} = t_c \wp_{z,0_c}. \tag{15.17}$$

Solving $t_s$

$$t_s = \frac{d_a \wp_{z,0_c} - \wp_{r,0_c}\left(z_a\big|_{r_a=d_a}\right)}{\wp_{z,0_s}\wp_{r,0_c} - \wp_{r,0_s}\wp_{z,0_c}}, \tag{15.18}$$

and solving $t_c$

$$t_c = \frac{\wp_{r,0_s}\left(z_a\big|_{r_a=d_a}\right) - d_a \wp_{z,0_s}}{\wp_{r,0_s}\wp_{z,0_c} - \wp_{z,0_s}\wp_{r,0_c}}. \tag{15.19}$$

Thus, we now know what $t_s$ and $t_c$ must be such that the reference rays meet each other in the meeting point of the second surface.

## 15.5 Examples

In this section, we present several examples of lenses that are designed using the equations (13.33) and (15.15) directly, with the conditions proposed in equations (15.16) and (15.17). The figures are 15.4–15.11.

All the lenses are free from spherical aberration and coma for the three-point objects given by the user. The user can choose the entrance pupil, the first surface of the lens and the position of the object.

In all the figures are the design parameters to reproduce them.

We set the image $h_b$ such that, $z_{b_s}(d_a) = z_{b_c}(0)$ to emphasize the overlapping of equations (13.33) and (15.15) on the region exposed by the entrance pupil.

Observe that the dashed lines in all the examples are the extension of the corresponding surfaces. This notation shows how equations (13.33) and (15.15) separate each other in the regions not exposed by the entrance pupil. Also note that

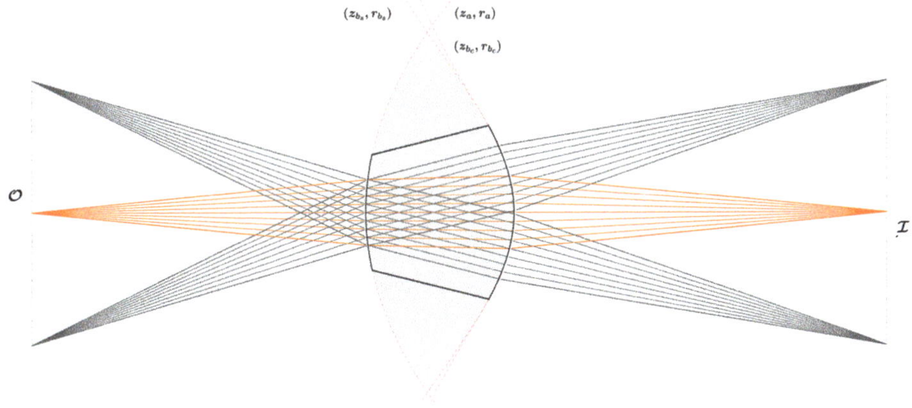


**Figure 15.4.** Design specifications: $n = 1.5$, $t_a = -10$ mm, $t_b = 11.327\,89$ mm, $t_s = 4.249\,02$ mm, $t_c = 4.448\,83$ mm, $d_a = \pm 1$ mm, $h_a = \pm 4$ mm, $h_b = \mp 3.999\,98$ mm $z_a = r_a^2/20$, $z_{b_s}$= equation (15.15) and $z_{b_c}$= equation (13.33).

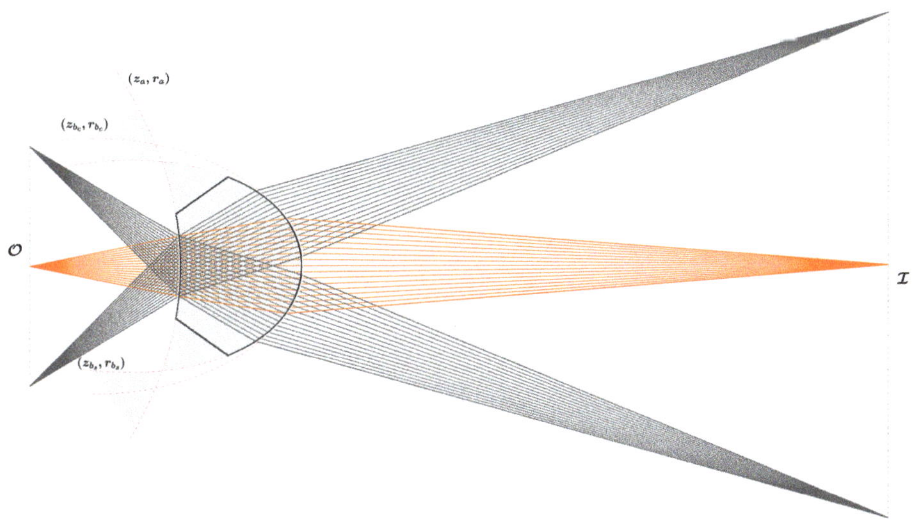


**Figure 15.5.** Design specifications: $n = 1.5$, $t_a = -5$ mm, $t_b = 20$ mm, $t_s = 3.570\,51$ mm, $t_c = 3.818\,28$ mm, $d_a = \pm 1$ mm, $h_a = \pm 4$ mm, $h_b = \mp 8.447\,06$ mm $z_a = -r_a^2/20$, $z_{b_s}$= equation (15.15) and $z_{b_c}$= equation (13.33).

for all the examples, even when the incident angles are enormous, the lenses are stigmatic for the predefined point images, this happens because we didn't use any paraxial approximation.

The method proposed has some limitations that need to be discussed. The first and most important one is that $d_a$ could not be bigger than $h_a$. We have tested in hundreds of examples that the singlet behaves as expected if and only if $d_a < h_a$. The other limitation is that, as mentioned before, we do not choose $t_s$, $t_c$ and $h_b$. They are

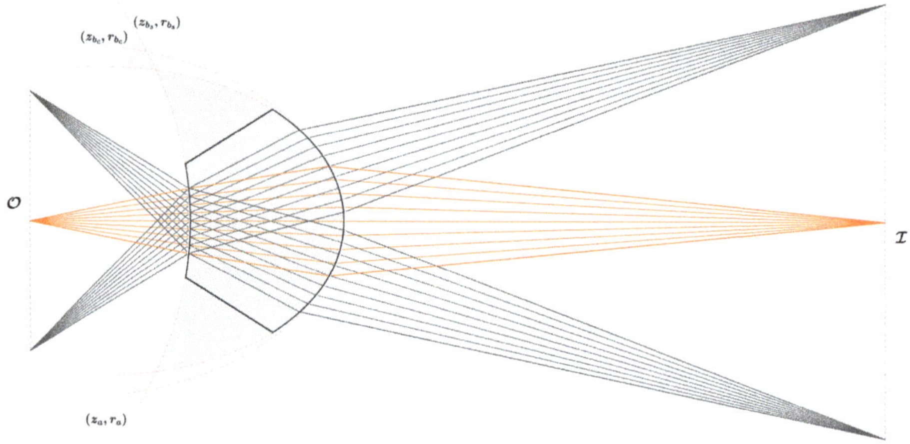


**Figure 15.6.** Design specifications: $n = 1.7$, $t_a = -5$ mm, $t_b = 17$ mm, $t_s = 4.384\,14$ mm, $t_c = 4.6013$ mm, $d_a = \pm 1$ mm, $h_a = \pm 4$ mm, $h_b = \mp 6.705\,22$ mm $z_a = -r_a^2/20$, $z_{b_s}$= equation (15.15) and $z_{b_c}$= equation (13.33).

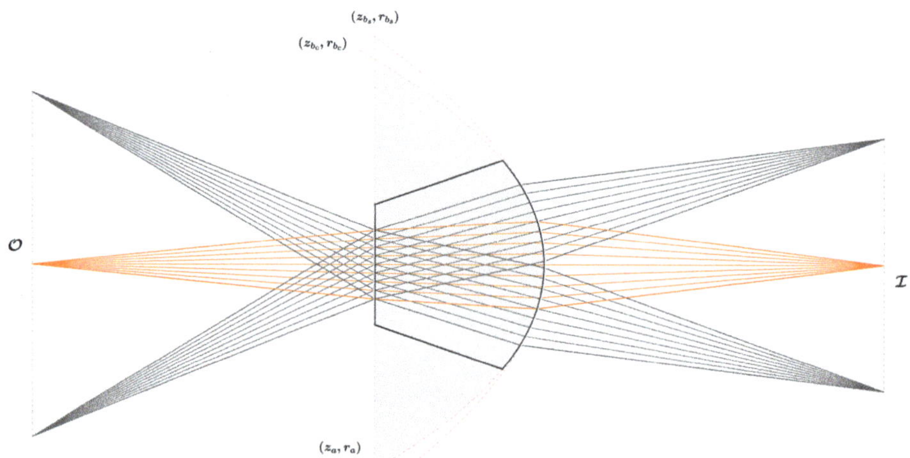


**Figure 15.7.** Design specifications: $n = 1.7$, $t_a = -10$ mm, $t_b = 10$ mm, $t_s = 4.680\,07$ mm, $t_c = 4.842\,62$ mm, $d_a = \pm 1$ mm, $h_a = \pm 5$ mm, $h_b = \mp 3.673\,04$ mm $z_a = 0$, $z_{b_s}$= equation (15.15) and $z_{b_c}$= equation (13.33).

computed in terms of the other variables. Thus, we do not control the thickness of the lens and the height of the image. $z_a$ can be convex or concave.

## 15.6 Conclusions

This was the last chapter of this part and it shows analytical closed-form equations that design a set of aplanatic lenses for three-point finite object images. With this single result, we conclude this chapter of the treatise. In the following chapters analytical closed-form solutions and numerical results will be merged to get semi-analytical optical designs.

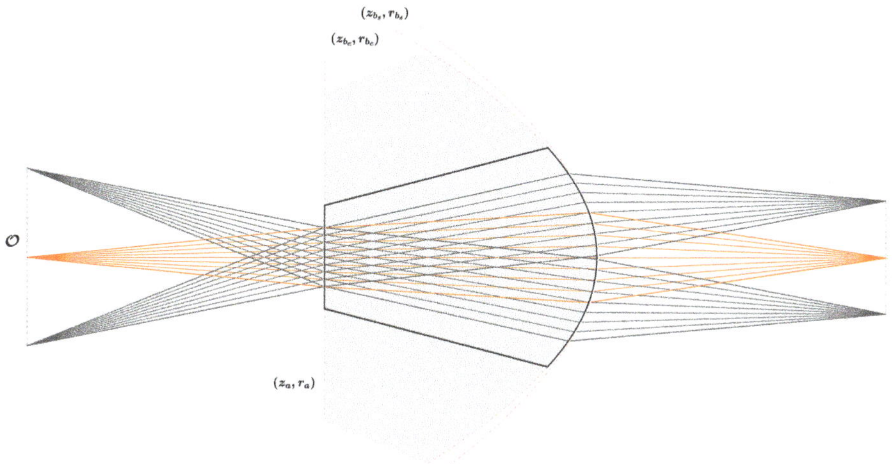


**Figure 15.8.** Design specifications: $n = 1.7$, $t_a = -10$ mm, $t_b = 10$ mm, $t_s = 8.875\,46$ mm, $t_c = 8.9896$ mm, $d_a = \pm 1$ mm, $h_a = \pm 3$ mm, $h_b = \mp 1.914\,53$ mm $z_a = 0$, $z_{b_s}$= equation (15.15) and $z_{b_c}$= equation (13.33).

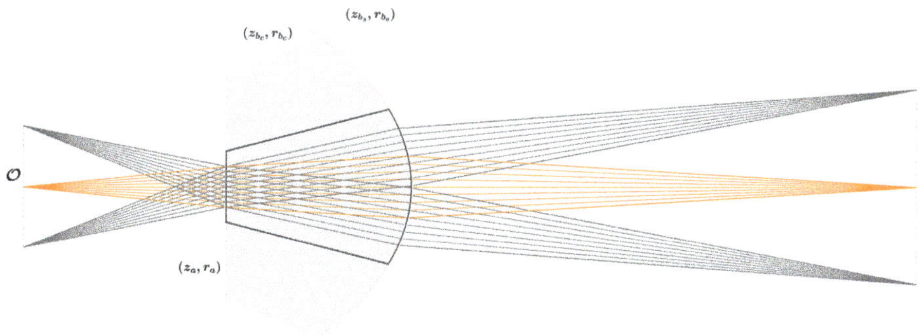


**Figure 15.9.** Design specifications: $n = 1.7$, $t_a = -10$ mm, $t_b = 10$ mm, $t_s = 8.875\,46$ mm, $t_c = 8.9896$ mm, $d_a = \pm 1$ mm, $h_a = \pm 3$ mm, $h_b = \mp 4.815\,45$ mm $z_a = 0$, $z_{b_s}$= equation (15.15) and $z_{b_c}$= equation (13.33).

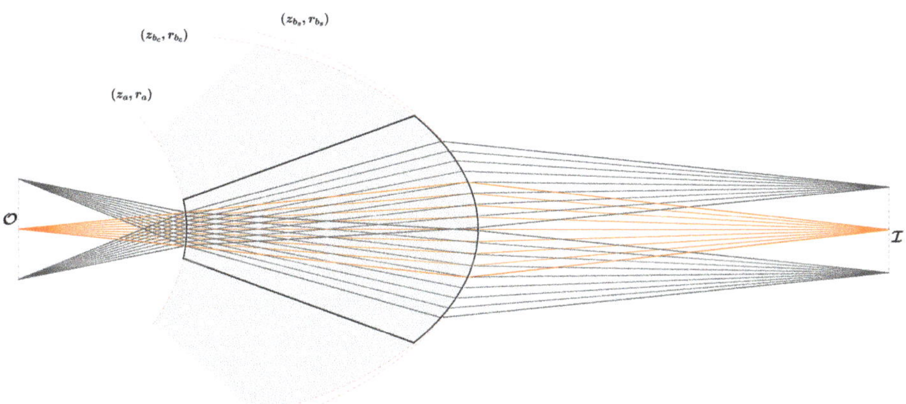


**Figure 15.10.** Design specifications: $n = 1.7$, $t_a = -10$ mm, $t_b = 25$ mm, $t_s = 16.7123$ mm, $t_c = 16.79$ mm, $d_a = \pm 1$ mm, $h_a = \pm 3$ mm, $h_b = \mp 2.557\,62$ mm $z_a = -r_a^2/20$, $z_{b_s}$= equation (15.15) and $z_{b_c}$= equation (13.33).

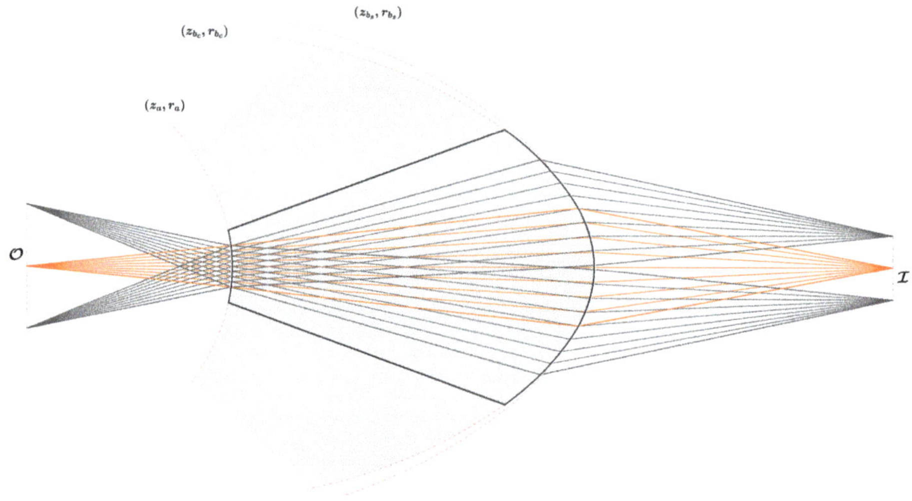


**Figure 15.11.** Design specifications: $n = 1.7$, $t_a = -10$ mm, $t_b = 15$ mm, $t_s = 16.7123$ mm, $t_c = 16.79$ mm, $d_a = \pm 1$ mm, $h_a = \pm 3$ mm, $h_b = \mp 1.5451$ mm $z_a = -r_a^2/20$, $z_{b_s}$= equation (15.15) and $z_{b_c}$= equation (13.33).

# Further reading


Acuña R G G and Gutiérrez-Vega J C 2019 General formula of the refractive telescope design free spherical aberration *Novel Optical Systems, Methods, and Applications XXII* 11105 (Bellingham, WA: International Society for Optics and Photonics)

Benitez P, Minano J C, Blen J, Mohedano R, Chaves J, Dross O, Hernandez M, Alvarez J L and Falicoff W 2004 SMS design method in 3D geometry: examples and applications *Nonimaging Optics: Maximum Efficiency Light Transfer VII* 5185 (Bellingham, WA: International Society for Optics and Photonics) pp 18–29

Benitez P, Miñano J C and Muñoz F 2005 Compact folded-optics illumination lens *US Patent* 6896381

Born M and Wolf E 2013 Principles of Optics: Electromagnetic Theory of Propagation *Interference and Diffraction of Light* (Amsterdam: Elsevier)

Chaves J 2016 *Introduction to Nonimaging Optics* 2nd edn (Boca Raton, FL: CRC Press)

Descartes R 1637a *De la nature des lignes courbes*

Descartes R 1637b *La Géométrie*

Dilworth D 2018 *Lens Design* (Bristol: IOP Publishing)

Dross O, Mohedano R, Benitez P, Minano J C, Chaves J, Blen J, Hernandez M and Munoz F 2004 *Review of SMS design methods and real-world applications Nonimaging Optics and Efficient Illumination Systems* 5529 (Bellingham, WA: International Society for Optics and Photonics) pp 35–48

Duerr F, Benítez P, Minano J C, Meuret Y and Thienpont H 2012a Analytic design method for optimal imaging: coupling three ray sets using two free-form lens profiles *Opt. Express* **20** 5576–85

Duerr F, Benítez P, Miñano J C, Meuret Y and Thienpont H 2012b Analytic free-form lens design in 3D: coupling three ray sets using two lens surfaces *Opt. Express* **20** 10839–46

Euler L 1770 *Dioptricae*

González-Acuña R G and Chaparro-Romo H A 2018 General formula for bi-aspheric singlet lens design free of spherical aberration *Appl. Opt.* **57** 9341–5

González-Acuña R G and Guitiérrez-Vega J C 2018 Generalization of the axicon shape: the gaxicon *J. Opt. Soc. Am.* A **35** 1915–8

González-Acuña R G and Gutiérrez-Vega J C 2019a Analytic formulation of a refractive-reflective telescope free of spherical aberration *Opt. Eng.* **58** 085105

González-Acuña R G and Gutiérrez-Vega J C 2019b General formula for aspheric collimator lens design free of spherical aberration *Current Developments in Lens Design and Optical Engineering XX* (Bellingham, WA: International Society for Optics and Photonics)

González-Acuña R G, Avendaño-Alejo M and Gutiérrez-Vega J C 2019a Singlet lens for generating aberration-free patterns on deformed surfaces *J. Opt. Soc. Am.* A **36** 925–9

González-Acuña R G, Chaparro-Romo H A and Gutiérrez-Vega J C 2019b General formula to design freeform singlet free of spherical aberration and astigmatism *Appl. Opt.* **58** 1010–5

Gross H, Singer W, Totzeck M, Blechinger F and Achtner B 2005 *Handbook of Optical Systems* vol 1 (New York: Wiley)

Gross H, Zügge H, Peschka M and Blechinger F 2007 Handbook of Optical Systems *Aberration Theory and Correction of Optical Systems* vol 3 (New York: Wiley-VCH)

Gross H, Blechinger F and Achtner B 2008 Handbook of Optical Systems *Survey of Optical Instruments* vol 4 (New York: Wiley-VCH)

Huygens C 1690 *Traité de la lumière*

Kingslake R and Johnson R B 2009 *Lens Design Fundamentals* (New York: Academic)

Luneburg R K and Herzberger M 1964 *Mathematical Theory of Optics* (Berkeley, CA: University of California Press)

Malacara D 2001 *Handbook of Optical Engineering* (Boca Raton, FL: CRC Press)

Malacara D and Malacara Z 1994 *Handbook of Lens Design* (New York: Dekker)

Malacara-Hernández D and Malacara-Hernández Z 2016 *Handbook of Optical Design* (Boca Raton, FL: CRC Press)

Miñano J C, Benítez P, Grabovickic D, Narasimhan B, Nikolic M and Infante J 2017 *Freeform Optics* vol 2 (Washington, DC: Optical Society of America) p JTu1C–2

Miñano J C and Benítez P 2016 Freeform aplanatic systems as a limiting case of SMS *Opt. Express* **24** 13173–8

Newton I 1704 *Opticks, or, a Treatise of the Reflections, Refractions, Inflections & Colours of Light* (New York: Dover)

Stavroudis O 2012 The Optics of Rays *Wavefronts, and Caustics* 38 (Amsterdam: Elsevier)

Stavroudis O N 2006 *The Mathematics of Geometrical and Physical Optics: The k-Function and Its Ramifications* (New York: Wiley)

Sun H 2016 *Lens Design: A Practical Guide* (Boca Raton, FL: CRC Press)

# Chapter 16

## Abbe aplanatic singlet

Here we present the exact equations for designing a stigmatic singlet lens that meets Abbe's sine condition for each ray. These lenses are designed from the solution of a system of nonlinear equations that can be solved numerically.

### 16.1 Introduction

In the previous chapters, we focused on merging the on-axis stigmatic lens solution with the off-axis stigmatic lens solution. So far, we have called these lens singlets aplanatic, which is not correct rigorously but gives us an idea of what we can expect from stigmatic designs. In order to have a rigorously aplanatic singlet we need to satisfy the Abbe sine condition.

The Abbe sine condition, sometimes also called the optical sine theorem, states that for a rotationally symmetric optical system to be aplanatic, i.e. free of spherical aberration and coma, the following relation must satisfy the following expression,

$$\frac{\sin\alpha}{\sin\beta} = M = \text{const.}, \tag{16.1}$$

where $\alpha$ is the angle relative to the optical axis of the input ray, the ray that leaves the on-axis object point $\mathcal{O}$. $\beta$ is the angle of the same ray, but at the system's output, as it reaches the on-axis image point $\mathcal{I}$. The magnification $M$ is a constant that describes how much the image of the object is magnified with respect to the object size, see figure 16.1. Please observe that the fulfillment of the condition implies that the ratio is constant for any value of $\alpha$ within the range defined by the largest angle admitted by the system aperture. This constant ratio is the magnification $M$.

In principle, optical systems that satisfy the Abbe sine condition are free of spherical aberration and coma. Therefore, to be free of spherical aberration they should be on-axis stigmatic. The Abbe sine condition was introduced by Ernst Abbe in 1873 and soon later demonstrated independently by Hermann von Helmholtz. Nowadays, it is a well-established result in the theory of aberrations under the

doi:10.1088/978-0-7503-5774-6ch16  

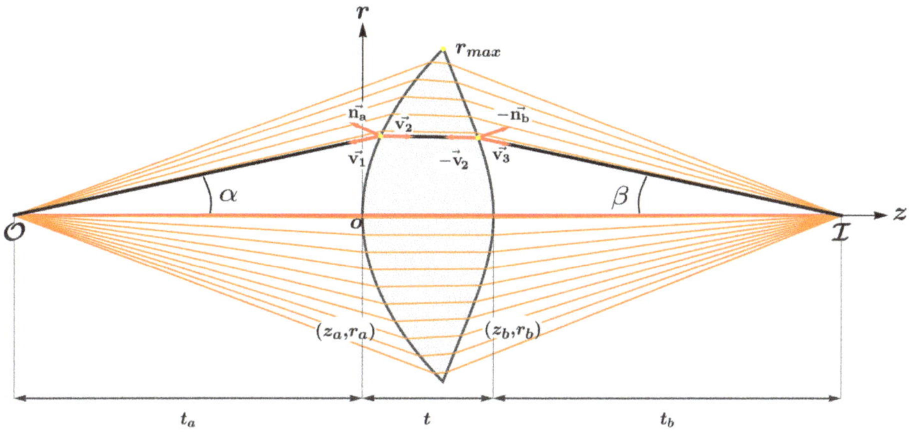


**Figure 16.1.** Diagram of an on-axis stigmatic singlet. The first surface is $(z_a, r_a)$, and the second surface is $(z_b, r_b)$. $\vec{\mathbf{v}}_1$ is the unit vector of the incident ray, $\vec{\mathbf{v}}_2$ is the unit vector of the refracted ray inside the lens, $\vec{\mathbf{v}}_3$ is the output $\vec{\mathbf{n}}_a$ is the normal vector of the first surface and $\vec{\mathbf{n}}_b$ is the normal vector of the second surface. The angle of the input ray is $\alpha$ and the angle of the output ray is $\beta$.

geometrical optics approach. Its derivation is well-known and it will not be shown in this book. The reader is encouraged to read the book *Principles of Optics: Electromagnetic Theory of Propagation, Interference and Diffraction of Light* for detailed proof of the Abbe sine condition. The Abbe sine condition is so often used as a measure of the quality of an optical system that the discrepancy between the ideal magnification and the ratio $\sin\alpha/\sin\beta$ is called offence against the Abbe sine condition.

In this chapter, we present the exact equations to design a stigmatic singlet lens that meets the Abbe sine condition for each ray. These equations come from the simultaneous fulfillment of the stigmatic condition discussed in the Abbe sine condition, therefore we are going to use some of the results that we have previously found in chapter 4. The equation system is nonlinear but can be numerically solved with high precision. Finally, some illustrated examples are presented.

## 16.2 The finite object finite image aplanatic singlet

In this section we first focus on the finite object finite image aplanatic singlet, which is a lens that is free of spherical aberration and meets (16.1). Therefore, we need first to recall the equation of the on-axis stigmatic lens and the express equation (16.1) in terms of that equation.

### 16.2.1 The stigmatic lens for finite object finite image

In order to establish notation and required relations, let us recall the basic equations of the stigmatic singlet from chapter 4. The shape and relevant parameters are described in figure 16.1. In this model, we consider that the lens, surrounded by air, has a uniform refractive index $n$ and axial thickness $t$. The first surface is given by the user and it is denoted by $(z_a, r_a)$, where $r_a$ is the radial coordinate. As can be confirmed in figure 16.1 the origin of the coordinate system is at the vertex of the first surface $(z_a, r_a)$. The distance from the object to the first surface is $t_a$. The second surface $(z_b, r_b)$ is such that all emerging rays from the point object $\mathcal{O}$ meet at the image point $\mathcal{I}$. The distance from the second surface to the image is $t_b$, thus the image $\mathcal{I}$ is located at $z_I = t + t_b$.

The system is stigmatic because it follows that the optical path of each ray is the same, so from chapter 4 we have,

$$\begin{aligned} -t_a + nt + t_b &= \sqrt{r_a^2 + (z_a - t_a)^2} \\ &\quad + n\sqrt{(z_b - z_a)^2 + (r_b - r_a)^2} + \sqrt{r_b^2 + (z_b - t_b - t)^2}. \end{aligned} \tag{4.6}$$

The explicit equation $(z_b, r_b)$ for the second surface that meets (4.6) is given by the following expressions,

$$\begin{cases} z_b = z_a + \vartheta \wp_z, \\ r_b = r_a + \vartheta \wp_r, \end{cases} \tag{4.36}$$

where,

$$\vartheta \equiv \frac{-(fn + r_a\wp_r + \tau\wp_z) \pm \sqrt{(fn + r_a\wp_r + \tau\wp_z)^2 - (1 - n^2)(r_a^2 + \tau^2 - f^2)}}{(1 - n^2)}, \tag{4.35}$$

$$f \equiv -t_a + nt + t_b + \operatorname{sgn}(t_a)\sqrt{r_a^2 + (z_a - t_a)^2}, \tag{4.25}$$

and,

$$\tau \equiv z_a - t - t_b. \tag{4.26}$$

The cosine directors $\wp_r$, $\wp_z$ are given by

$$\begin{aligned} \wp_r \equiv &\frac{(z_a - t_a)(z_a') + r_a}{-\operatorname{sgn}(t_a)n\sqrt{r_a^2 + (t_a - z_a)^2}\,(1 + (z_a')^2)} \\ &- (z_a')\frac{\sqrt{1 - \dfrac{[r_a + (z_a - t_a)(z_a')]^2}{n^2[r_a^2 + (t_a - z_a)^2](1 + (z_a')^2)}}}{\sqrt{1 + (z_a')^2}}, \end{aligned} \tag{4.17}$$

and,

$$\begin{aligned} \wp_z \equiv &\frac{[r_a + (z_a - t_a)(z_a')](z_a')}{-\operatorname{sgn}(t_a)n\sqrt{r_a^2 + (t_a - z_a)^2}\,(1 + (z_a')^2)} \\ &+ \frac{\sqrt{1 - \dfrac{[r_a + (z_a - t_a)(z_a')]^2}{n^2[r_a^2 + (t_a - z_a)^2](1 + (z_a')^2)}}}{\sqrt{1 + (z_a')^2}}. \end{aligned} \tag{4.18}$$

Observe that they follow $\wp_r^2 + \wp_z^2 = 1$.

The derivation of the above equations can be found in chapter 4 of this book. Here we have limited ourselves to addressing only the required relations for the purposes of this chapter. Finally, it is convenient to remember the sign convention. $t_a$ can be positive or negative. If it is negative, then the object is real, otherwise the object is virtual. This is because the origin of the coordinate system is located at the vertex of the first surface. $t$ is positive since it is the thickness of the lens. If $t_b > t$, then the image is real, and if $t_b < t$ the image is virtual. In this case, we are only working with real objects and real images.

### 16.2.2 The aplanatic condition in terms of the on-axis stigmatic singlet

Now, observe in figure 16.1 that we can set the Abbe sine condition since for every meridian ray it strikes the lens at height $r_a$ and it leaves it at height $r_b(r_a)$. Thus using some basic trigonometric relationships we can express equation (16.1) as,

$$\boxed{\frac{r_a}{\sqrt{r_a^2 + (z_a - t_a)^2}} = M\frac{r_b}{\sqrt{r_b^2 + (z_b - t_b - t)^2}}.} \tag{16.2}$$

Please observe that we know what the shape of the second surface $(r_b, z_b)$ must be (see last subsection). $z_b$ and $r_b$ are expressed in terms of $z_a$ and $z_a'$, and $z_a$ is a function given freely by the user. Therefore, the new unknown is $z_a$ in equation (16.2). The problem here is that equation (16.2) becomes a nonlinear differential equation which can be solved only numerically. The actual size of equation (16.2) is enormous and it may take several pages to write it, so hoping to find an analytical solution to it is out of the scope. Here unfortunately we are only going to solve it with numerical methods.

### 16.2.3 Illustrative examples

We now show some examples of lenses designed by numerically solving for $z_a$ in equation (16.2) and finally obtaining the second surface with equation (4.36). The design parameters of each example are presented in the captions of each image. The figures for this section are 16.2 and 16.3.

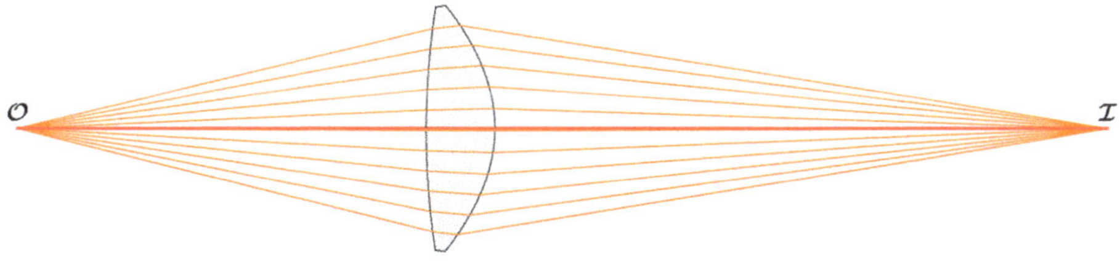


**Figure 16.2.** Design specifications: $n = 1.5$, $t_a = -30$ mm, $t = 5$ mm, $t_b = 30$ mm, $M = 1.2$.

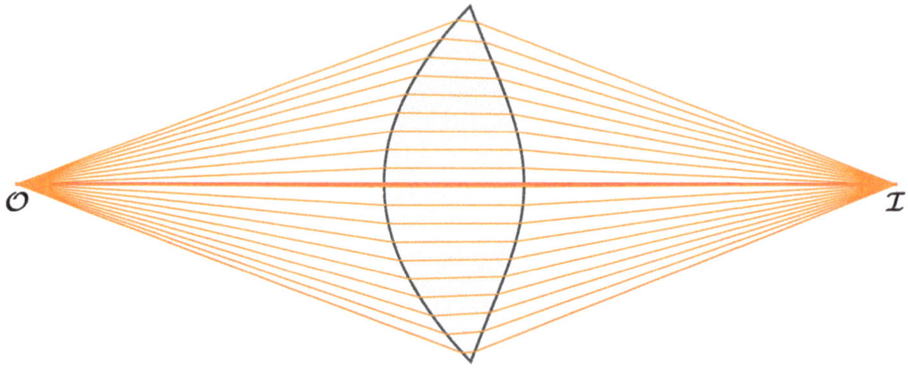


**Figure 16.3.** Design specifications: $n = 1.5$, $t_a = -30$ mm, $t = 10$ mm, $t_b = 30$ mm and $M = 1$.

## 16.3 Infinite object finite image aplanatic singlet

In the case when the object is far away from the Abbe sine condition it takes the following form,

$$\frac{r_a}{\sin\beta} = M = \text{const.} \tag{16.3}$$

If we want to design a lens such that $\alpha = 0$, we need to compute the limit $t_a \to -\infty$ through equations of the collector of chapter 4 and then express the Abbe sine condition, equation (16.3), in terms of them.

### 16.3.1 The stigmatic collector equations

The lens free from spherical aberration or an object very far away is now well known, we just recall them from chapter 4. The second surface of the on-axis stigmatic collector is given by

$$\begin{cases} \lim\limits_{t_a\to-\infty}(z_b) = z_a + \left[\lim\limits_{t_a\to-\infty}(\vartheta)\right]\left[\lim\limits_{t_a\to-\infty}(\wp_z)\right], \\ \lim\limits_{t_a\to-\infty}(r_b) = r_a + \left[\lim\limits_{t_a\to-\infty}(\vartheta)\right]\left[\lim\limits_{t_a\to-\infty}(\wp_r)\right], \end{cases} \tag{4.42}$$

where,

$$\lim_{t_a\to-\infty}(\vartheta) = \frac{-\beta \pm \sqrt{\beta^2 - (1-n^2)\left\{r_a^2 + \tau^2 - \left[\lim\limits_{t_a\to-\infty}(f)\right]^2\right\}}}{(1-n^2)}, \tag{4.40}$$

where,

$$\beta \equiv \left\{\left[\lim_{t_a \to -\infty}(f)\right]n + r_a\left[\lim_{t_a \to -\infty}(\wp_r)\right] + \tau\left[\lim_{t_a \to -\infty}(\wp_z)\right]\right\}, \tag{4.41}$$

$$\lim_{t_a \to -\infty} f = nt + t_b - z_a, \tag{4.39}$$

the cosine directors are,

$$\lim_{t_a \to -\infty} \wp_r = -\frac{(z_a')\left(n\sqrt{(z_a')^2 + 1}\sqrt{\frac{(n^2 - 1)(z_a')^2 + n^2}{n^2((z_a')^2 + 1)}} - 1\right)}{n((z_a')^2 + 1)}, \tag{4.37}$$

and,

$$\lim_{t_a \to -\infty} \wp_z = \frac{\sqrt{\frac{(n^2 - 1)(z_a')^2 + n^2}{n^2((z_a')^2 + 1)}}}{\sqrt{(z_a')^2 + 1}} + \frac{(z_a')^2}{n(z_a')^2 + n}. \tag{4.38}$$

### 16.3.2 The aplanatic condition of the collector

The next step is to express the Abbe sine condition, equation (16.3), when the object is at minus infinity. Using simple trigonometric relations we can express equation (16.3) as,

$$\boxed{r_a = M\frac{\left[\lim_{t_a \to -\infty}(r_b)\right]}{\sqrt{\left[\lim_{t_a \to -\infty}(r_b)\right]^2 + (z_b - t_b - t)^2}}.} \tag{16.4}$$

equation (16.4) is also a nonlinear differential equation which can be solved only numerically. The procedure to solve it is similar to the procedure to solve (16.2). An example of a lens that meets the condition of equation (16.4) is presented in figure 16.4.

## 16.4 Conclusion

In this chapter, we derived the Abbe sine conditions for exact stigmatic lenses for finite objects and infinite objects and in both cases with a real finite image. The conditions are expressed in equations (16.2) and (16.4). To design the mentioned lenses, equations (16.2) and (16.4) should be solved numerically, since they are nonlinear differential equations. The degree to which a ray satisfies the Abbe sine condition is proportional to the precision of the numerical method to solve the mentioned nonlinear differential equations.

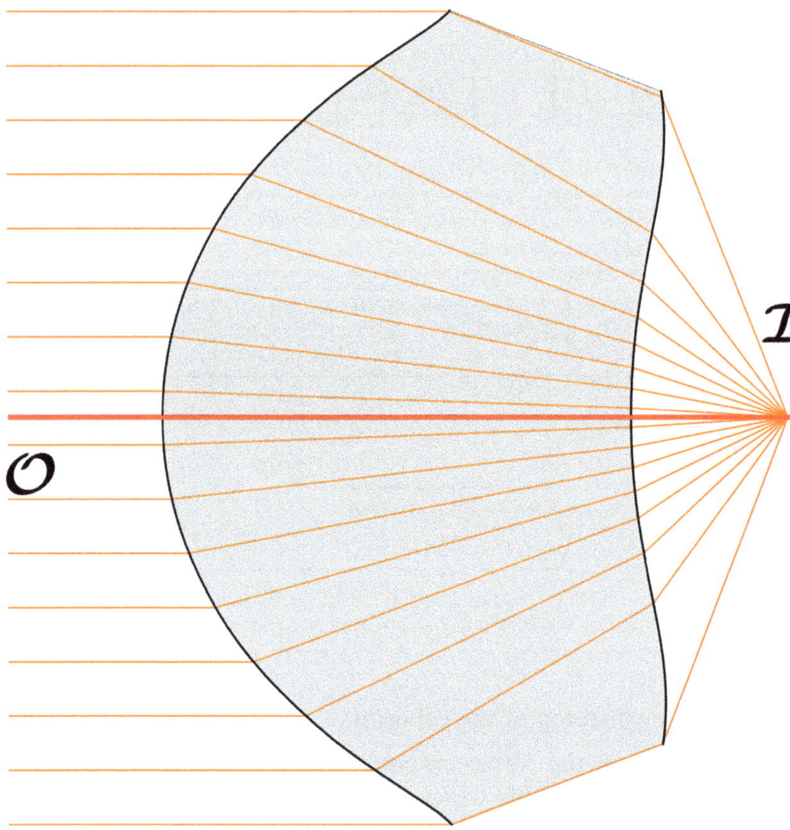


**Figure 16.4.** Design specifications: $n = 2$, $t_a \to -\infty$ mm, $t = 6$ mm, $t_b = 2$ mm and $M = 1.2$.

# Further reading

Abbe E 1881 On the estimation of aperture in the microscope *J. R. Microsc. Soc.* **1** 388–423

Born M and Wolf E 2013 Principles of Optics: Electromagnetic Theory of Propagation *Interference and Diffraction of Light* (Amsterdam: Elsevier)

Braat J J M 1997 Abbe sine condition and related imaging conditions in geometrical optics *5th Int. Topical Meeting on Education and Training in Optics* ed C H F Velzel vol 3190 (Bellingham, WA: International Society for Optics and Photonics, SPIE) pp 59–64

Chaves J 2016 *Introduction to Nonimaging Optics* 2nd edn (Boca Raton, FL: CRC Press)

Courtial J, Oxburgh S and Tyc T 2015 Direct stigmatic imaging with curved surfaces *J. Opt. Soc. Am.* A **32** 478–81

Duerr F, Benítez P, Miñano J C, Meuret Y and Thienpont H 2012 Analytic design method for optimal imaging: coupling three ray sets using two free-form lens profiles *Opt. Express* **20** 5576–85

Elazhary T T, Zhou P, Zhao C and Burge J H 2015 Generalized sine condition *Appl. Opt.* **54** 5037–49

Gimenez-Benitez P, Miñano J C, Blen J, Arroyo R, Chaves J, Dross O, Hernández M and Falicoff W 2004 Simultaneous multiple surface optical design method in three dimensions *Opt. Eng.* **43** 1489–503

González-Acuña R G 2021 Exact equations to design aplanatic sequential optical systems *Appl. Opt.* **60** 9263–8

González-Acuña R G 2022 Design of a pair of aplanatic mirrors *Appl. Opt.* **61** 1982–6

González-Acuña R G and Chaparro-Romo H A 2018 General formula for bi-aspheric singlet lens design free of spherical aberration *Appl. Opt.* **57** 9341–5

González-Acuña R G and Guitiérrez-Vega J C 2018 Generalization of the axicon shape: the gaxicon *J. Opt. Soc. Am.* A **35** 1915–8

González-Acuña R G and Gutiérrez-Vega J C 2019a General formula of the refractive telescope design free spherical aberration *Novel Optical Systems, Methods, and Applications XXII* vol 11105 (Bellingham, WA: International Society for Optics and Photonics)

González-Acuña R G and Gutiérrez-Vega J C 2019b Analytic formulation of a refractive-reflective telescope free of spherical aberration *Opt. Eng.* **58** 085105

González-Acuña R G and Gutiérrez-Vega J C 2019c General formula for aspheric collimator lens design free of spherical aberration *Current Developments in Lens Design and Optical Engineering XX* vol 11104 (Bellingham, WA: International Society for Optics and Photonics)

González-Acuña R G and Gutiérrez-Vega J C 2020 Analytic design of a spherochromatic singlet *J. Opt. Soc. Am.* A **37** 149–53

González-Acuña R G and Gutiérrez-Vega J C 2021 Exact equations to design a stigmatic singlet that meets the Herschel's condition *Opt. Commun.* **485** 126727

González-Acuña R G, Avendaño-Alejo M and Gutiérrez-Vega J C 2019a Singlet lens for generating aberration-free patterns on deformed surfaces *J. Opt. Soc. Am.* A **36** 925–9

González-Acuña R G, Chaparro-Romo H A and Gutiérrez-Vega J C 2019b General formula to design freeform singlet free of spherical aberration and astigmatism *Appl. Opt.* **58** 1010–5

González-Acuña R G, Chaparro-Romo H A and Gutiérrez-Vega J C 2020 *Analytical Lens Design* (Bristol: IOP Publishing)

González-Acuña R G, Chaparro-Romo H A and Gutiérrez-Vega J C 2021 Exact equations for stigmatic singlet design meeting the Abbe sine condition *Opt. Commun.* **479** 126415

Gross H, Singer W, Totzeck M, Blechinger F and Achtner B 2005 *Handbook of Optical Systems* vol 1 (New York: Wiley Online Library)

Hazra L N, Han Y and Delisle C 1993 Curved kinoform lenses for stigmatic imaging of axial objects *Appl. Opt.* **32** 4775–84

Herzberger M 1935 On the fundamental optical invariant, the optical tetrality principle, and on the new development of Gaussian optics based on this law *J. Opt. Soc. Am.* **25** 295–304

Herzberger M 1936 A new theory of optical image formation *J. Opt. Soc. Am.* **26** 197–204

Herzberger M 1939 Normal systems with two caustic lines *J. Opt. Soc. Am.* **29** 392–4

Herzberger M 1940 Normal systems with two caustic lines *J. Opt. Soc. Am.* **30** 307–8

Herzberger M 1943 Direct methods in geometrical optics *Trans. Am. Math. Soc.* **53** 218–29

Herzberger M 1951 Some remarks on ray tracing *J. Opt. Soc. Am.* **41** 805–7

Herzberger M 1948 Image error theory for finite aperture and field I. The image of a point; geometry of the wave surface *J. Opt. Soc. Am.* **38** 736–8

Herzberger M 1958 Modern geometrical optics *Phys. Today* **12** 50–2

Herzberger M 1963 Some recent ideas in the field of geometrical optics *J. Opt. Soc. Am.* **53** 661–71

Hsueh C-C, Elazhary T, Nakano M and Sasian J 2011 Closed-form sag solutions for Cartesian oval surfaces *J. Opt.* **40** 168–75
Kingslake R and Johnson R B 2009 *Lens Design Fundamentals* (New York: Academic)
Luneburg R K and Herzberger M 1964 *Mathematical Theory of Optics* (Berkeley, CA: University of California Press)
Malacara-Hernández D and Malacara-Hernández Z 2016 *Handbook of Optical Design* (Boca Raton, FL: CRC Press)
Mansuripur M 1998 Abbe's sine condition *Opt. Photonics News* **9** 56–60
Masters B R 2007 Ernst Abbe and the foundation of scientific microscopes *Opt. Photonics News* **18** 18–23
Miñano J C, Mohedano R and Benítez P 2007 Nonimaging optics *The Optics Encyclopedia* (New York: Wiley) pp 1–34
Miñano J C, Benítez P, Lin W, Infante J, Muñoz F and Santamaría A 2009 An application of the SMS method for imaging designs *Opt. Express* **17** 24036–44
Shibuya M 1992 Exact sine condition in the presence of spherical aberration *Appl. Opt.* **31** 2206–10
Silva-Lora A and Torres R 2020 Explicit Cartesian oval as a superconic surface for stigmatic imaging optical systems with real or virtual source or image *Proc. R. Soc. Sect.* A **476** 20190894
Stavroudis O N and Fronczek R C 1978 Generalized ray tracing and the caustic surface *Opt. Laser Technol.* **10** 185–91
Stavroudis O 2012 The Optics of Rays *Wavefronts, and Caustics* vol 38 (Amsterdam: Elsevier)
Stavroudis O N 1972 Some consequences of Herzberger's fundamental optical invariant *J. Opt. Soc. Am.* **62** 59–63
Stavroudis O N 2006 *The Mathematics of Geometrical and Physical Optics: The k-Function and Its Ramifications* (New York: Wiley)
Stavroudis O N and Fronczek R C 1976 Caustic surfaces and the structure of the geometrical image *J. Opt. Soc. Am.* **66** 795–800
Stavroudis O N, Fronczek R C and Chang R-S 1978 Geometry of the half-symmetric image *J. Opt. Soc. Am.* **68** 739–42
Valencia-Estrada J C, Pereira-Ghirghi M V, Malacara-Hernández Z and Chaparro-Romo H A 2017 Aspheric coefficients of deformation for a Cartesian oval surface *J. Opt.* **46** 100–7
Velzel C H F 2014 *A Course in Lens Design* vol 183 (Berlin: Springer)
Volkmann H 1966 Ernst Abbe and his work *Appl. Opt.* **5** 1720–31

**IOP** Publishing

Analytical Lens Design (Second Edition)

**Rafael G González-Acuña**

# Chapter 17

## Abbe aplanatic optical systems

We present the exact differential equations to design an aplanatic sequential optical system, a system that is free of spherical aberration and linear coma. We get the exact set of equations from the Fermat principle and the Abbe sine condition. We solve the mentioned set of equations numerically and present an illustrative example.

### 17.1 Introduction

The analysis presented in chapter 16, can be extended to an arbitrary number of lenses in order to get robust aplanatic systems.

To achieve that goal, the Abbe sine condition should be generalized for an arbitrary number of lenses. Therefore, we are going to use some of the results presented in chapters 11 and 12. However, in this chapter, we are only going to deal with the case when the object is far away and the image is finite and real. Thus we focus on generalizing the Abbe sine condition for an arbitrary number of lenses when the object is far away, in the notation used in chapters 11 and 12, we have,

$$r_1 = M \sin \theta_{N+1}. \tag{17.1}$$

The last expression relates the height $r_1$ in which a ray strikes the entrance pupil with the output ray angle $\theta_{N+1}$. $M$ is a constant related to magnification, see figure 17.1. $M$ is the magnification and $N$ is a constant integer related to the number of surfaces of the optical system. To hold for equation (17.1) the optical path length of every ray should be constant, which is why we are going to use the results of chapters 11 and 12.

Therefore, the goal of this chapter expresses equation (17.1) in terms of the equations presented in chapters 11 and 12 in order to obtain a nonlinear differential equation where if we solve it we get surfaces of an aplanatic system.

doi:10.1088/978-0-7503-5774-6ch17  

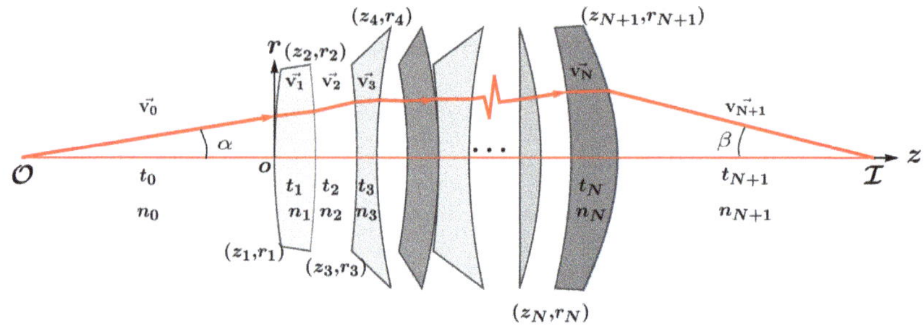


**Figure 17.1.** Ray tracing of an aplanatic optical system. The coordinate system is placed at the vertex of the first surface. $n_0$, $n_1$, $n_2$,…,$n_{N+1}$ are the refraction indexes in each region of the system. $t_1$, $t_2$,…,$t_N$ are the axial thicknesses of the elements that compose the system. The distance from the image $\mathcal{O}$ to the vertex of the first surfaces $t_0$ and the distance from the image $\mathcal{I}$ to the vertex of the last surface $t_{N+1}$.

## 17.2 Mathematical model

Let's retain the notation present in chapters 11 and 12. Let $n_0$, $n_1$, $n_2$,…,$n_{N+1}$ be the refraction indices of the elements that compose the optical system, where $n_0$ and $n_{N+1}$ are the refraction indices at the object space and image space, respectively. Let $t_1$, $t_2$,…,$t_N$ be the axial thicknesses of the elements that compose the optical system, see figure 17.1. Then, let $t_0$ be the distance from the object to the first surface of the system and let $t_{N+1}$ be the distance from the last surface of the system to the image. As mentioned in the introduction of this chapter, the object is considered far away so $t_0 \to -\infty$. Since we are considering that from the beginning and for clarity we are not going to express the limit in every parameter. Remember, we do the same thing for the on-axis stigmatic telescope of chapter 12.

Well, going back with the notation, let $(z_1, r_1)$, $(z_2, r_2)$,…,$(z_{N+1}, r_{N+1})$ be the surfaces. Note that the coordinate system is placed at the vertex of the first surface. The only independent variable is $r_1$, this means that all the other surfaces are in terms of $r_1$. Observe that $r_1$ is the height $r_1$ at which a ray strikes the entrance pupil.

Please observe that in this system there are $N + 1$ refractive surfaces and $N$ thick elements. So, let be $r_i$ is the radial coordinate at the $i$th interface. With this in mind and recalling the results of chapter 11, we can express the last surface as

$$\boxed{\begin{cases} z_{N+1} = z_N + \vartheta_N \wp_{z,N}, \\ r_{N+1} = r_N + \vartheta_N \wp_{r,N}, \end{cases}} \tag{11.36}$$

where,

$$\vartheta_N = \frac{-[f n_N + n_{N+1}^2 (r_N \wp_{r,N} + \tau \wp_{z,N})]}{(n_{N+1}^2 - n_N^2)} \pm \frac{\sqrt{[f n_N + n_{N+1}^2 (r_N \wp_{r,N} + \tau \wp_{z,N})]^2 - (n_{N+1}^2 - n_N^2)[n_{N+1}^2 (r_N^2 + \tau^2) - f^2]}}{(n_{N+1}^2 - n_N^2)}, \tag{11.35}$$

$\wp_{r,N}$ and $\wp_{z,N}$ are given by

$$\wp_{r,N} = \frac{n_{N-1}r_N'[(r_N - r_{N-1})r_N' + (z_N - z_{N-1})z_N']}{n_N\sqrt{(r_{N-1} - r_N)^2 + (z_{N-1} - z_N)^2}(r_N'^2 + z_N'^2)} - \frac{z_N'\sqrt{1 - \frac{n_{N-1}^2[(r_N - r_{N-1})r_N' + (z_N - z_{N-1})z_N']^2}{n_N^2[(r_{N-1} - r_N)^2 + (z_{N-1} - z_N)^2](r_N'^2 + z_N'^2)}}}{\sqrt{r_N'^2 + z_N'^2}}, \tag{11.17}$$

and,

$$\wp_{z,N} = \frac{n_{N-1}z_N'[(r_N - r_{N-1})r_N' + (z_N - z_{N-1})z_N']}{n_N\sqrt{(r_{N-1} - r_N)^2 + (z_{N-1} - z_1)^2}(r_N'^2 + z_N'^2)} + \frac{r_N'\sqrt{1 - \frac{n_{N-1}^2[(r_N - r_{N-1})r_N' + (z_N - z_{N-1})z_N']^2}{n_N^2[(r_{N-1} - r_N)^2 + (z_{N-1} - z_N)^2](r_N'^2 + z_N'^2)}}}{\sqrt{r_N'^2 + z_N'^2}}, \tag{11.18}$$

where $z_i' = \mathrm{d}z_i(r_1)/\mathrm{d}r_1$ and $r_i' = \mathrm{d}r_i(r_1)/\mathrm{d}r_1$ are the derivatives of $z_i$ and $r_i$ with respect to $r_1$. So, $z_N' = \mathrm{d}z_N(r_1)/\mathrm{d}r_1$ and $r_N' = \mathrm{d}r_N(r_1)/\mathrm{d}r_1$. The cosine directors $\wp_{r,i}$ and $\wp_{z,i}$ between consecutive surfaces are,

$$\wp_{r,i} = \frac{n_{i-1}r_i'[(r_i - r_{i-1})r_i' + (z_i - z_{i-1})z_i']}{n_i\sqrt{(r_{i-1} - r_i)^2 + (z_{i-1} - z_i)^2}(r_i'^2 + z_i'^2)} - \frac{z_i'\sqrt{1 - \frac{n_{i-1}^2[(r_i - r_{i-1})r_i' + (z_i - z_{i-1})z_i']^2}{n_i^2[(r_{i-1} - r_i)^2 + (z_{i-1} - z_i)^2](r_i'^2 + z_i'^2)}}}{\sqrt{r_i'^2 + z_i'^2}}, \tag{11.15}$$

and,

$$\wp_{z,i} = \frac{n_{i-1}z_i'[(r_i - r_{i-1})r_i' + (z_i - z_{i-1})z_i']}{n_i\sqrt{(r_{i-1} - r_i)^2 + (z_{i-1} - z_1)^2}(r_i'^2 + z_i'^2)} + \frac{r_i'\sqrt{1 - \frac{n_{i-1}^2[(r_i - r_{i-1})r_i' + (z_i - z_{i-1})z_i']^2}{n_i^2[(r_{i-1} - r_i)^2 + (z_{i-1} - z_i)^2](r_i'^2 + z_i'^2)}}}{\sqrt{r_i'^2 + z_i'^2}}. \tag{11.16}$$

Finally, for a finite object $f$ is,

$$f \equiv -t_0 n_0 + \sum_{i=1}^{N+1} n_i t_i - n_0\sqrt{r_1^2 + (z_1 - t_0)^2} - \sum_{i=0}^{N-2} n_{i+1}\sqrt{(r_{i+2} - r_{i+1})^2 + (z_{i+2} - z_{i+1})^2}. \tag{11.26}$$

But we are going to use $f$ for a far away object, so we compute the limit when $t_0 \to -\infty$ in the last expression,

$$f \equiv -z_1 n_0 + \sum_{i=1}^{N+1} n_i t_i - \sum_{i=0}^{N-2} n_{i+1}\sqrt{(r_{i+2} - r_{i+1})^2 + (z_{i+2} - z_{i+1})^2}, \tag{17.2}$$

and, $\tau$ is,

$$\tau \equiv z_N - \sum_{i=1}^{N+1} t_i. \tag{11.27}$$

With the above mathematical expression, we have the first part of the problem. The second one is to express the Abbe sine condition in terms of them. Analyzing figure 17.1 and using basic trigonometric relationships, we can express equation (17.1) as,

$$r_1 = \frac{Mn_{N+1}}{n_1}\frac{r_{N+1}}{\sqrt{r_{N+1}^2 + (z_{N+1} - \sum_{i=1}^{N+1} t_i)^2}}. \tag{17.3}$$

Equation (17.3) is the exact equation that the stigmatic lens needs to satisfy such that it meets the Abbe condition. Note that the free parameters are all the previous surfaces to the last surface. We set the first surface such that each ray meets the Abbe sine condition. Observe that equation (11.36) is expressed in terms of the first surface $z_1$ and its derivative $z_1'$, with the boundary condition $z_1(0) = 0$. Therefore, equation (17.3) is a differential equation that can be solved numerically. All surfaces between the first and last surfaces are still free parameters. This is the value of equation (17.3) and equation (11.36) because we can set the equations such that the lens is free from spherical aberration and coma, and then we can set all the other surfaces such that the system is free from other aberrations.

## 17.3 An illustrative example

In this section, we present an aplanatic telephoto with a half field of view 30° designed with only two aspheres at the beginning and at the end of the system. The wavelength of this example is 660 nm. The rest of the surfaces are spheres. The first surface is an asphere designed by numerically solving equation (17.3) and the second asphere, the last surface, is given by equation (11.36). All the other parameters are given by the user and can be found in table 17.1. The ray tracing of this example is presented in figure 17.2. In figure 17.3 are the spot diagrams of the example presented in figure 17.2. The spot diagram is the front view of the image plane being intersected by the rays. This is a valuable tool to identify the possible aberrations present in an optical system briefly. The different aberrations present different shapes of the spot diagram, it can work as a starting point for identifying the aberrations, if the field is on-axis and it is free of the spherical aberration the spot diagram is a single point. In general, if the system is stigmatic the spot diagram is just a point. In this case, the spot diagram is not a point since the solution of (17.3) and equation (11.36) is numerical. But the solution is good because it is diffracted limited. A design can be considered diffracted limited when all the rays in its spot

**Table 17.1.** The $M$ constant of the Abbe sine condition is 0.95.

| Surface | Profile/Radius [mm] | Refraction index | Thickness [mm] |
|---|---|---|---|
| 0 | Object | $n_0 = 1$ | $t_0 \to -\infty$ |
| 1 | $z_1$ = solution of equation (17.3) | $n_1 = 1.6195$ | $t_1 = 0.0494$ |
| 2 | $R_2 = 0.564\ 68$ | $n_2 = 1$ | $t_2 = 0.1279$ |
| 3 | $R_3 = 1.4622$ | $n_3 = 1.1.6197$ | $t_3 = 0.048$ |
| 4 | $R_4 = 0.652\ 61$ | $n_4 = 1$ | $t_4 = 0.4505$ |
| 5 | $R_5 = 1.100\ 72$ | $n_5 = 1.6591$ | $t_5 = 0.1653$ |
| 6 | $R_6 = -1.880\ 76$ | $n_6 = 1$ | $t_6 = 0.0053$ |
| 7 | $R_7 = 0.513\ 51$ | $n_7 = 1.4623$ | $t_7 = 0.068$ |
| 8 | $R_8 = 1.7307$ | $n_8 = 1.6922$ | $t_8 = 0.1597$ |
| 9 | $R_9 = -1.0523$ | $n_9 = 1.6922$ | $t_9 = 0.0395$ |
| 10 | $R_{10} = 0.560\ 92$ | $n_{10} = 1$ | $t_{10} = 0.0448$ |
| 11 | $R_{11} = 2.191\ 06$ | $n_{11} = 1.6667$ | $t_{11} = 0.0319$ |
| 12 | $R_{12} = 1.412\ 83$ | $n_{12} = 1.6742$ | $t_{12} = 0.1121$ |
| 13 | $z_4$ = equation (11.36) | $n_{13} = 1$ | $t_{13} = 1.2034$ |
| 14 | Image | Not apply | Not apply |

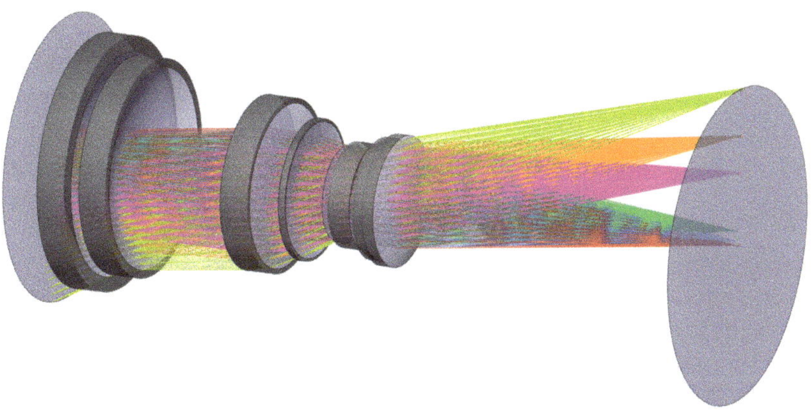

**Figure 17.2.** Ray tracing of the optical system of table 17.1.

diagram are inside the Airy disk. The diameter of the maximum brightness on the diffraction pattern is called Airy disk. The diffraction happens in an optical element due to the wavelets interfering with each other, so a circular aperture will present an energy distribution around the focal region with a Bessel function shape, in other means a maximum brightness surrounded by rings with different intensities.

In order to compute the ray tracing and the spot diagrams, the numerical solution of equation (17.3) and equation (11.36) has been exported as a point cloud to the commercial software Quadoa Optical CAD.

The spot diagram of the aplanatic telephoto is presented in the figure 17.3. The spot diagrams shown correspond to the fields with 0°, 1° 3°, 15°, 22° and 33°. Note

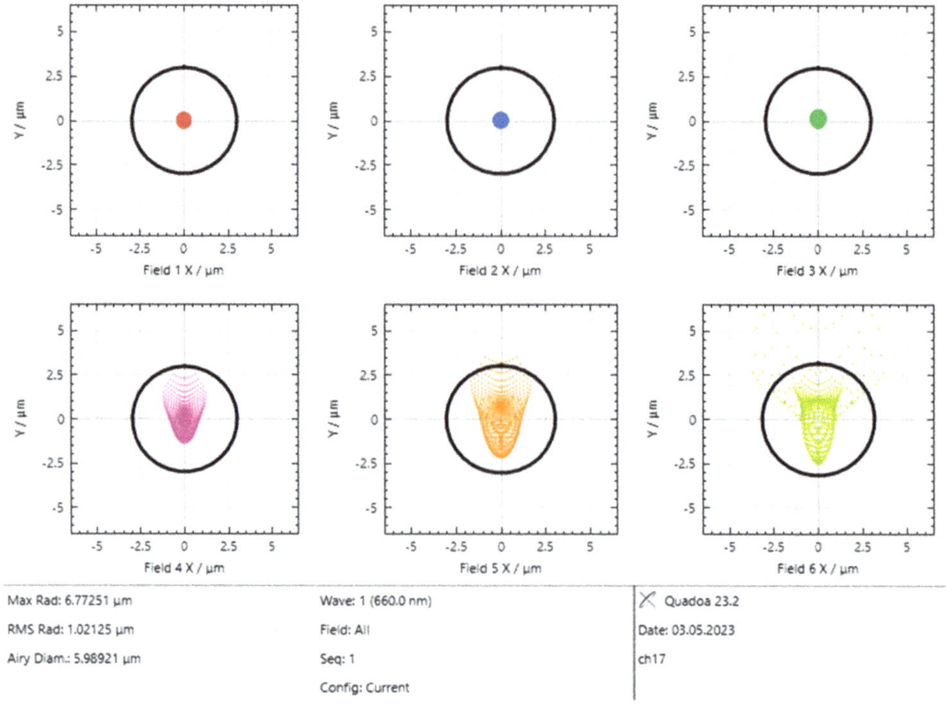


**Figure 17.3.** Spot diagrams of the system of table 17.1 for two objects. The first object is on the axis and the second has a height of 2 mm. The ray tracing of the aplanatic telephoto is shown in figure 17.2.

that most of them are inside the Airy disk. Thus, these results confirm what equations (11.36) and (17.3) predict, a system free from spherical aberration and free from coma. Observe that the system is free from the mentioned aberrations despite the solution of equations (11.36) and (17.3) being an approximation since it is numeric. In general, the system behaves well depsite not considering, in general, astigmatism, distortion and field curvature. However, this is just a particular case, from the solution of equations (11.36) and (17.3) we only should expect that the system is free from spherical aberration and free from coma. However, it is important to remark, that if the number of surfaces is reduced, the performance of the optical system designed with equations (17.3) and (11.36) regarding the spherical aberration and the linear coma will remain the same.

The result is as predicted by the theory that equations (11.36) and (17.3) are the exact equations to design an aplanatic system. It is important to remark that equation (11.36) is the general equation for a sequential stigmatic optical system but also it is unique since it is deviation shows it. This can be confirmed by observing the steps presented in chapter 11. Then, for a given input there will be only one last surface that made the system stigmatic. It is important to see that the whole system works together such that the optical path length is constant for all the rays, thus systems designed with equations (11.36) contemplate all the possible stigmatic optical systems with radial symmetry.

The method proposed in this chapter can be very useful for optical designers because they can focus on other aberrations like astigmatism, field curvature and distortion. Because by imposing the restrictions presented in this manuscript they can design aplanatic systems with the faculty of the solutions of equations (17.3) and (11.36). Therefore, they can focus on selecting the proper intermediate surfaces such that the other aberrations are minimized. This selection can be obtained using optimization techniques that are commonly found in commercial optical design software.

## 17.4 Conclusions

In this chapter, we presented the generalization of the method presented in the previous chapter. This generalization consists in considering an arbitrary number of surfaces instead of a single lens.

To achieve that we used equations (17.3) and (11.36). Equation (17.3) is the generalization of the Abbe sine condition for an arbitrary number of refractive surfaces. Equation (11.36) expresses the last surface of a radially symmetric optical system such that this system is free from spherical aberration. Once we obtained both equations we solved them using a numerical procedure just like in the previous chapter. The method presented here can be useful for an initial design free from spherical aberration and coma. These designs can be next optimized using commercial optical design software in order to remove other aberrations.

Finally, the procedure presented in this chapter is only focused on the case when the object is far away and the image is finite and real. Other cases can be easily generalized from the results presented in this chapter. These generalizations are left as an exercise for the reader.

## Further reading


Quadoa Optical CAD 2022 *Quadoa Optical CAD Software Manual V* (Berlin: Quadoa Optical Systems) https://www.quadoa.com

Abbe E 1881 On the estimation of aperture in the microscope *J. R. Microsc. Soc.* **1** 388–423

Born M and Wolf E 2013 *Principles of Optics: Electromagnetic Theory of Propagation Interference and Diffraction of Light* (Amsterdam: Elsevier)

Braat J J M 1997 Abbe sine condition and related imaging conditions in geometrical optics *5th Int. Topical Meeting on Education and Training in Optics* ed C H F Velzel vol 3190 (Bellingham, WA: International Society for Optics and Photonics, SPIE) pp 59–64

Chaves J 2016 *Introduction to Nonimaging Optics* 2nd edn (Boca Raton, FL: CRC Press)

Courtial J, Oxburgh S and Tyc T 2015 Direct stigmatic imaging with curved surfaces *J. Opt. Soc. Am.* A **32** 478–81

Doskolovich L L, Bykov D A, Greisukh G I, Strelkov Y S and Bezus E A 2021 Designing stigmatic lenses with minimal Fresnel losses *J. Opt. Soc. Am.* A **38** 855–61

Duerr F, Benítez P, Miñano J C, Meuret Y and Thienpont H 2012 Analytic design method for optimal imaging: coupling three ray sets using two free-form lens profiles *Opt. Express* **20** 5576–85

Elazhary T T, Zhou P, Zhao C and Burge J H 2015 Generalized sine condition *Appl. Opt.* **54** 5037–49

Gimenez-Benitez P, Miñano J C, Blen J, Arroyo R, Chaves J, Dross O, Hernández M and Falicoff W 2004 Simultaneous multiple surface optical design method in three dimensions *Opt. Eng.* **43** 1489–503

González-Acuña R G 2021 Exact equations to design aplanatic sequential optical systems *Appl. Opt.* **60** 9263–8

González-Acuña R G 2022 Design of a pair of aplanatic mirrors *Appl. Opt.* **61** 1982–6

González-Acuña R G and Chaparro-Romo H A 2018 General formula for bi-aspheric singlet lens design free of spherical aberration *Appl. Opt.* **57** 9341–5

González-Acuña R G and Chaparro-Romo H A 2020 *Stigmatic Optics* (Bristol: IOP Publishing)

González-Acuña R G and Guitiérrez-Vega J C 2018 Generalization of the axicon shape: the gaxicon *J. Opt. Soc. Am.* A **35** 1915–8

González-Acuña R G and Gutiérrez-Vega J C 2019a General formula of the refractive telescope design free spherical aberration *Novel Optical Systems, Methods, and Applications XXII* vol 11105 (Bellingham, WA: International Society for Optics and Photonics)

González-Acuña R G and Gutiérrez-Vega J C 2019b Analytic formulation of a refractive-reflective telescope free of spherical aberration *Opt. Eng.* **58** 085105

González-Acuña R G and Gutiérrez-Vega J C 2019c General formula for aspheric collimator lens design free of spherical aberration *Current Developments in Lens Design and Optical Engineering XX* vol 11104 (Bellingham, WA: International Society for Optics and Photonics)

González-Acuña R G and Gutiérrez-Vega J C 2019d General formula to eliminate spherical aberration produced by an arbitrary number of lenses *Opt. Eng.* **58** 085106

González-Acuña R G and Gutiérrez-Vega J C 2020 Analytic design of a spherochromatic singlet *J. Opt. Soc. Am.* A **37** 149–53

González-Acuña R G and Gutiérrez-Vega J C 2021 Exact equations to design a stigmatic singlet that meets the Herschel's condition *Opt. Commun.* **485** 126727

González-Acuña R G, Avendaño-Alejo M and Gutiérrez-Vega J C 2019a Singlet lens for generating aberration-free patterns on deformed surfaces *J. Opt. Soc. Am.* A **36** 925–9

González-Acuña R G, Chaparro-Romo H A and Gutiérrez-Vega J C 2019b General formula to design freeform singlet free of spherical aberration and astigmatism *Appl. Opt.* **58** 1010–5

González-Acuña R G, Chaparro-Romo H A and Gutiérrez-Vega J C 2020a Analytic solution of the eikonal for a stigmatic singlet lens *Phys. Scr.* **95** 085201

González-Acuña R G, Chaparro-Romo H A and Gutiérrez-Vega J C 2020b *Analytical Lens Design* (Bristol: IOP Publishing)

González-Acuña R G, Chaparro-Romo H A and Gutiérrez-Vega J C 2021 Exact equations for stigmatic singlet design meeting the Abbe sine condition *Opt. Commun.* **479** 126415

Gross H, Singer W, Totzeck M, Blechinger F and Achtner B 2005 *Handbook of Optical Systems* vol 1 (New York: Wiley Online Library)

Hazra L N, Han Y and Delisle C 1993 Curved kinoform lenses for stigmatic imaging of axial objects *Appl. Opt.* **32** 4775–84

Hopkins H H 1946 Herschel's condition *Proc. Phys. Soc.* **58** 100–5

Hsueh C-C, Elazhary T, Nakano M and Sasian J 2011 Closed-form sag solutions for Cartesian oval surfaces *J. Opt.* **40** 168–75

Kingslake R and Johnson R B 2009 *Lens Design Fundamentals* (New York: Academic)

Luneburg R K and Herzberger M 1964 *Mathematical Theory of Optics* (Berkeley, CA: University of California Press)

Malacara-Hernández D and Malacara-Hernández Z 2016 *Handbook of Optical Design* (Boca Raton, FL: CRC Press)

Mansuripur M 1998 Abbe's sine condition *Opt. Photonics News* **9** 56–60

Masters B R 2007 Ernst Abbe and the foundation of scientific microscopes *Opt. Photonics News* **18** 18–23

Miñano J C, Mohedano R and Benítez P 2007 Nonimaging optics *The Optics Encyclopedia* (New York: Wiley) pp 1–34

Miñano J C, Benítez P, Lin W, Infante J, Muñoz F and Santamaría A 2009 An application of the SMS method for imaging designs *Opt. Express* **17** 24036–44

Shibuya M 1992 Exact sine condition in the presence of spherical aberration *Appl. Opt.* **31** 2206–10

Silva-Lora A and Torres R 2020a Aplanatism in stigmatic optical systems *Opt. Lett.* **45** 6390–3

Silva-Lora A and Torres R 2020b Explicit Cartesian oval as a superconic surface for stigmatic imaging optical systems with real or virtual source or image *Proc. R. Soc.* A **476** 20190894

Silva-Lora A and Torres R 2020c Superconical aplanatic ovoid singlet lenses *J. Opt. Soc. Am.* A **37** 1155–65

Silva-Lora A and Torres R 2021 Rigorously aplanatic Descartes ovoids *J. Opt. Soc. Am.* A **38** 1160–9

Stavroudis O N 2006 *The Mathematics of Geometrical and Physical Optics: The k-Function and Its Ramifications* (New York: Wiley)

Steward G C 1927 On Herschel's condition and the optical cosine law *Math. Proc. Camb. Philos. Soc.* **23** 703–12

Valencia-Estrada J C, Flores-Hernández R B and Malacara-Hernández D 2015 Singlet lenses free of all orders of spherical aberration *Proc. R. Soc.* A **471** 20140608

Valencia-Estrada J C, Pereira-Ghirghi M V, Malacara-Hernández Z and Chaparro-Romo H A 2017 Aspheric coefficients of deformation for a Cartesian oval surface *J. Opt.* **46** 100–7

Velzel C H F 2014 *A Course in Lens Design* vol 183 (Berlin: Springer)

Volkmann H 1966 Ernst Abbe and his work *Appl. Opt.* **5** 1720–31

Wassermann G D and Wolf E 1949 On the theory of aplanatic aspheric systems *Proc. Phys. Soc.* B **62** 2

# Part VI

## Stigmatic mirror systems

**IOP** Publishing

Analytical Lens Design (Second Edition)

**Rafael G González-Acuña**

# Chapter 18

# The set of all possible pairs of stigmatic mirrors

This chapter presents the combination of all possible mirrors that form a stigmatic system. The equations that describe the mentioned mirrors are closed-form. The derivation of the mentioned equations is presented step by step. Several illustrative examples are shown.

## 18.1 Introduction

So far in this treatise, we have just mainly focused on the refractive surfaces, with the exception of the telescope of chapter 12 and the conic mirrors in the introduction of the book. But now it is time to move forward in the direction of the reflective surface. This is the last part of the treatise to express the ideas that are contained in this part, we need to understand all the ideas presented in the previous parts.

The part consists of four chapters in which the set of mirrors that conforms to stigmatic systems is explored. In this initial chapter, we focus on the smallest set possible, the pair of mirrors that form a stigmatic system. In this chapter, we set the foundations for the following chapters. The next chapter is a generalization of the results shown in this chapter, but if presented directly may cause confusion, so let's build these models step by step.

## 18.2 Mathematical model

As mentioned in the introduction the goal of this chapter is to get the pair of mirrors for a stigmatic optical system. This mirror can be placed in any place of the Cartesian coordinate system, as well as the object and image points. Due to these initial constraints it is expected to have freeform geometries that describe the mirrors.

A diagram of the stigmatic pair of mirrors is shown in figure 18.1. Our goal is to express a set of all the possible pairs of stigmatic mirrors. So let $(x_a, y_a, z_a)$ be the first mirror, where $z_a$ is the profile of the mirror, it is a function that depends on $x_a$ and $y_a$, thus $z_a(x_a, y_a)$.$x_a$ and $y_a$ are the $x$ and $y$ components, respectively; and they are the

doi:10.1088/978-0-7503-5774-6ch18  

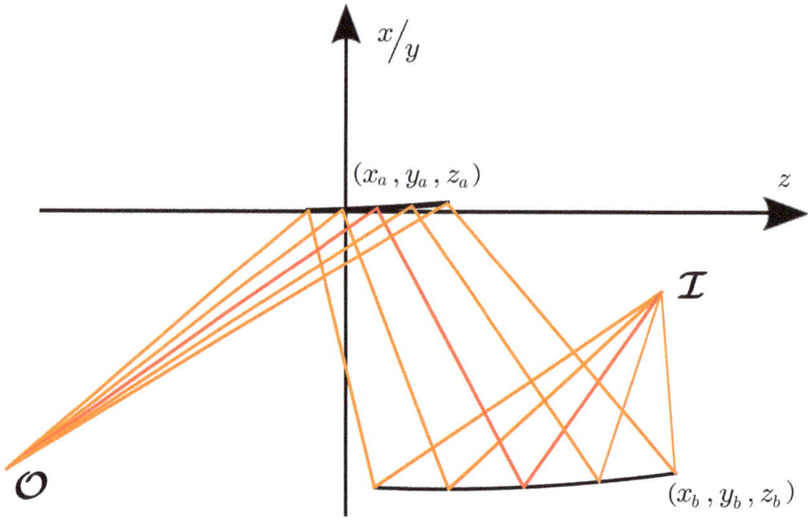


**Figure 18.1.** Diagram of two stigmatic mirrors. The first mirror is denoted by $(x_a, y_a, z_a)$ and the second mirror is given by $(x_b, y_b, z_b)$. From the vertex of the first mirror to the vertex of the second mirror, there is a gap with the size $\tau = (x_\tau^2 + y_\tau^2 + z_\tau^2)^{1/2}$, where $x_\tau$, $y_\tau$, and $z_\tau$ are the $x$, $y$ and $z$ distances between the vertexes.

only independent variables of the model. This is just like the model of the freeform lens presented in previous chapters.

Let $(x_b, y_b, z_b)$ be the second mirror, where $x_b$, $y_b$ and $z_b$ are the components in $x$, $y$ and $z$ directions, respectively. Without loss of generalization let us express the second mirror in terms of the first mirror such that the pair is stigmatic like $x_b(x_a, y_a)$, $y_b(x_a, y_a)$ and $z_b(x_a, y_a)$ are functions of $x_a$ and $y_a$. Now observe that by the principle of reciprocity of light propagation, without loss of generality, we assume that the shape of the first mirror is given and the shape of the second mirror is unknown. Thus we can choose the second mirror to be the unknown of the model, but what must the shape of the second mirror be such that the system is stigmatic, given the shape of the first mirror? It will be useful to set the origin of the Cartesian coordinate at the vertex of the first surface ($x_a = 0$, $y_a = 0$, $z_a = 0$). With the same notation let $(x_b(0, 0), y_b(0, 0), z_b(0, 0))$ be the vertex of the second surface.

The object is placed at a distance $(x_o, y_o, z_o)$ from the origin of the vertex of the first mirror. The image is placed at a distance $(x_i, y_i, z_i)$ from the origin of the vertex of the second mirror. Let $t = \sqrt{x_t^2 + y_t^2 + z_t^2}$ be the distance between the vertices of both mirrors.

### 18.2.1 Snell's law

In order to achieve our expectations let's start with the Snell law, equation (18.1), at the surface of the first mirror,

$$\boxed{\vec{\mathbf{v}}_2 = \left(\vec{\mathbf{v}}_1 - \left(\vec{\mathbf{n}}_a \cdot \vec{\mathbf{v}}_1\right)\vec{\mathbf{n}}_a\right) - \vec{\mathbf{n}}_a\sqrt{1 + (\vec{\mathbf{n}}_a \wedge \vec{\mathbf{v}}_1)^2} \quad \text{for } \vec{\mathbf{v}}_2, \vec{\mathbf{v}}_1, \vec{\mathbf{n}}_a \in \mathbb{R}^3,} \tag{18.1}$$

where $\vec{\mathbf{v}}_1$ is the unit vector of the incident ray; it travels from the object to the first surface. $\vec{\mathbf{v}}_2$ is the unit vector of the refracted ray; it moves inside the lens. Finally, $\vec{\mathbf{n}}_a$ is the normal vector of the first surface, see figure 18.1.

The unit vectors related to the refraction at the first surface are the following,

$$\begin{aligned}\vec{\mathbf{v}}_1 &= \frac{(x_a - x_o)\vec{\mathbf{e}}_1 + (y_a - y_o)\vec{\mathbf{e}}_2 + (z_a - z_o)\vec{\mathbf{e}}_3}{\sqrt{(x_a - x_o)^2 + (y_a - y_o)^2 + (z_a - x_o)^2}},\\ \vec{\mathbf{v}}_2 &= \frac{(x_b - x_a)\vec{\mathbf{e}}_1 + (y_b - y_a)\vec{\mathbf{e}}_2 + (z_b - z_a)\vec{\mathbf{e}}_3}{\sqrt{(x_b - x_a)^2 + (y_b - y_a)^2 + (z_b - z_a)^2}},\\ \vec{\mathbf{n}}_a &= \frac{z_{a_x}\vec{\mathbf{e}}_1 + z_{a_y}\vec{\mathbf{e}}_2 - \vec{\mathbf{e}}_3}{\sqrt{z_{a_x}^2 + z_{a_y}^2 + 1}},\end{aligned} \tag{18.2}$$

where, $z_{a_x} \equiv \partial_x z_a$ and $z_{a_y} \equiv \partial_y z_a$ are the partial derivatives of $z_a(x_a, y_a)$ with respect to $x_a$ and $y_a$.

Let's replace the unit vectors of equation (18.2) in equation (18.1). The procedure is easy but long, so be patient. Let's start with $(\vec{\mathbf{n}}_a \wedge \vec{\mathbf{v}}_1)$, which is the term inside the square root of equation (18.1),

$$\begin{aligned}(\vec{\mathbf{n}}_a \wedge \vec{\mathbf{v}}_1) =& \frac{\left[z_{a_x}(y_a - y_o) + z_{a_y}(x_o - x_a)\right]\left(\vec{\mathbf{e}}_1 \wedge \vec{\mathbf{e}}_2\right)}{\sqrt{z_{a_x}^2 + z_{a_y}^2 + 1}\sqrt{(z_o - z_a)^2 + (x_a - x_o)^2 + (y_a - y_o)^2}}\\ &- \frac{[z_{a_x}(z_a - z_o) + x_a - x_o](\vec{\mathbf{e}}_3 \wedge \vec{\mathbf{e}}_1)}{\sqrt{z_{a_x}^2 + z_{a_y}^2 + 1}\sqrt{(z_o - z_a)^2 + (x_a - x_o)^2 + (y_a - y_o)^2}}\\ &+ \frac{\left[z_{a_y}(z_a - z_o) + y_a - y_o\right]\left(\vec{\mathbf{e}}_2 \wedge \vec{\mathbf{e}}_3\right)}{\sqrt{z_{a_x}^2 + z_{a_y}^2 + 1}\sqrt{(z_o - z_a)^2 + (x_a - x_o)^2 + (y_a - y_o)^2}}\\ =& \frac{\left[z_{a_x}(y_a - y_o) + z_{a_y}(x_o - x_a)\right]\vec{\mathbf{e}}_3}{\sqrt{z_{a_x}^2 + z_{a_y}^2 + 1}\sqrt{(z_o - z_a)^2 + (x_a - x_o)^2 + (y_a - y_o)^2}}\\ &- \frac{[z_{a_x}(z_a - z_o) + x_a - x_o]\vec{\mathbf{e}}_2}{\sqrt{z_{a_x}^2 + z_{a_y}^2 + 1}\sqrt{(z_o - z_a)^2 + (x_a - x_o)^2 + (y_a - y_o)^2}}\\ &+ \frac{\left[z_{a_y}(z_a - z_o) + y_a - y_o\right]\vec{\mathbf{e}}_1}{\sqrt{z_{a_x}^2 + z_{a_y}^2 + 1}\sqrt{(z_o - z_a)^2 + (x_a - x_o)^2 + (y_a - y_o)^2}}.\end{aligned} \tag{18.3}$$

Squaring, $(\vec{\mathbf{n}}_a \wedge \vec{\mathbf{v}}_1) \cdot (\vec{\mathbf{n}}_a \wedge \vec{\mathbf{v}}_1) = -(\vec{\mathbf{n}}_a \wedge \vec{\mathbf{v}}_1)^2$,

$$\begin{aligned}&(\vec{\mathbf{n}}_a \wedge \vec{\mathbf{v}}_1)^2 = -\\ &\frac{\left(z_{a_x}(y_a - y_o) + z_{a_y}(x_o - x_a)\right)^2 + \left(z_{a_x}(z_o - z_a) - x_a + x_o\right)^2 + \left(z_{a_y}(z_a - z_o) + y_a - y_o\right)^2}{\left(z_{a_x}^2 + z_{a_y}^2 + 1\right)\left((x_a - x_o)^2 + (y_a - y_o)^2 + (z_a - z_o)^2\right)}.\end{aligned} \tag{18.4}$$

So, the square root term in equation (18.2) is,

$$\sqrt{1-(\vec{\mathbf{n}}_a \wedge \vec{\mathbf{v}}_1)^2} = \sqrt{1-\frac{\left(z_{a_x}(y_a-y_o)+z_{a_y}(x_o-x_a)\right)^2 + (z_{a_x}(z_o-z_a)-x_a+x_o)^2 + \left(z_{a_y}(z_a-z_o)+y_a-y_o\right)^2}{\left(z_{a_x}^2+z_{a_y}^2+1\right)((x_a-x_o)^2+(y_a-y_o)^2+(z_a-z_o)^2)}}. \tag{18.5}$$

The next step is the term $(\vec{\mathbf{n}}_a \cdot \vec{\mathbf{v}}_1)\vec{\mathbf{n}}_a$,

$$\begin{aligned}(\vec{\mathbf{n}}_a \cdot \vec{\mathbf{v}}_1)\vec{\mathbf{n}}_a &= \left[\frac{(x_a-x_o)z_{a_x}+(y_a-y_o)z_{a_y}-(z_a-z_o)}{\sqrt{(z_o-z_a)^2+(x_a-x_o)^2+(y_a-y_o)^2}\sqrt{z_{a_x}^2+z_{a_y}^2+1}}\right]\frac{z_{a_x}\overline{\mathbf{e}}_1+z_{a_y}\overline{\mathbf{e}}_2-\overline{\mathbf{e}}_3}{\sqrt{z_{a_x}^2+z_{a_y}^2+1}},\\
&= \frac{\left[(x_a-x_o)z_{a_x}+(y_a-y_o)z_{a_y}-(z_a-z_o)\right]}{\sqrt{(z_o-z_a)^2+(x_a-x_o)^2+(y_a-y_o)^2}}\frac{z_{a_x}\overline{\mathbf{e}}_1+z_{a_y}\overline{\mathbf{e}}_2-\overline{\mathbf{e}}_3}{\left(z_{a_x}^2+z_{a_y}^2+1\right)}\\
&= \frac{z_{a_x}\left[(x_a-x_o)z_{a_x}+(y_a-y_o)z_{a_y}-(z_a-z_o)\right]\overline{\mathbf{e}}_1}{\sqrt{(z_o-z_a)^2+(x_a-x_o)^2+(y_a-y_o)^2}\left(z_{a_x}^2+z_{a_y}^2+1\right)}\\
&+ \frac{z_{a_y}\left[(x_a-x_o)z_{a_x}+(y_a-y_o)z_{a_y}-(z_a-z_o)\right]\overline{\mathbf{e}}_2}{\sqrt{(z_o-z_a)^2+(x_a-x_o)^2+(y_a-y_o)^2}\left(z_{a_x}^2+z_{a_y}^2+1\right)}\\
&- \frac{\left[(x_a-x_o)z_{a_x}+(y_a-y_o)z_{a_y}-(z_a-z_o)\right]\overline{\mathbf{e}}_3}{\sqrt{(z_o-z_a)^2+(x_a-x_o)^2+(y_a-y_o)^2}\left(z_{a_x}^2+z_{a_y}^2+1\right)}.\end{aligned} \tag{18.6}$$

Then let's see the following term, $\vec{\mathbf{v}}_1 - (\vec{\mathbf{n}}_a \cdot \vec{\mathbf{v}}_1)\vec{\mathbf{n}}_a$,

$$\begin{aligned}\vec{\mathbf{v}}_1 - (\vec{\mathbf{n}}_a \cdot \vec{\mathbf{v}}_1)\vec{\mathbf{n}}_a &=\\
&- \frac{z_{a_x}\left[(x_a-x_o)z_{a_x}+(y_a-y_o)z_{a_y}-(z_a-z_o)\right]\overline{\mathbf{e}}_1}{\sqrt{(z_o-z_a)^2+(x_a-x_o)^2+(y_a-y_o)^2}\left(z_{a_x}^2+z_{a_y}^2+1\right)} + \frac{x_a\overline{\mathbf{e}}_1}{\sqrt{x_a^2+y_a^2+(z_a-x_o)^2}}\\
&- \frac{z_{a_y}\left[(x_a-x_o)z_{a_x}+(y_a-y_o)z_{a_y}-(z_a-z_o)\right]\overline{\mathbf{e}}_2}{\sqrt{(z_o-z_a)^2+(x_a-x_o)^2+(y_a-y_o)^2}\left(z_{a_x}^2+z_{a_y}^2+1\right)} + \frac{y_a\overline{\mathbf{e}}_2}{\sqrt{x_a^2+y_a^2+(z_a-x_o)^2}}\\
&+ \frac{\left[(x_a-x_o)z_{a_x}+(y_a-y_o)z_{a_y}-(z_a-z_o)\right]\overline{\mathbf{e}}_3}{\sqrt{(z_o-z_a)^2+(x_a-x_o)^2+(y_a-y_o)^2}\left(z_{a_x}^2+z_{a_y}^2+1\right)} + \frac{(z_a-x_o)\overline{\mathbf{e}}_3}{\sqrt{x_a^2+y_a^2+(z_a-x_o)^2}}.\end{aligned}$$

Simplifying,

$$\begin{aligned}\vec{\mathbf{v}}_1 - (\vec{\mathbf{n}}_a \cdot \vec{\mathbf{v}}_1)\vec{\mathbf{n}}_a = & \frac{z_{a_x}\left(z_{a_y}(y_o - y_a) + z_a - z_o\right) + x_a\left(z_{a_y}^2 + 1\right) - \left(z_{a_y}^2 + 1\right)x_o}{\left(z_{a_x}^2 + z_{a_y}^2 + 1\right)\sqrt{(x_a - x_o)^2 + (y_a - y_o)^2 + (z_a - z_o)^2}}\vec{\mathbf{e}}_1 \\ & + \frac{z_{a_y}\left(z_{a_x}(x_o - x_a) + z_a - z_o\right) + y_a\left(z_{a_x}^2 + 1\right) - \left(z_{a_x}^2 + 1\right)y_o}{\left(z_{a_x}^2 + z_{a_y}^2 + 1\right)\sqrt{(x_a - x_o)^2 + (y_a - y_o)^2 + (z_a - z_o)^2}}\vec{\mathbf{e}}_2 \\ & + \frac{(z_a - z_o)\left(z_{a_x}^2 + z_{a_y}^2\right) + x_a z_{a_x} + z_{a_y}(y_a - y_o) - z_{a_x}x_o}{\left(z_{a_x}^2 + z_{a_y}^2 + 1\right)\sqrt{(x_a - x_o)^2 + (y_a - y_o)^2 + (z_a - z_o)^2}}\vec{\mathbf{e}}_3.\end{aligned} \tag{18.7}$$

Replacing equations (18.5) and (18.7) in equation (18.1),

$$\begin{aligned}\frac{(x_b - x_a)}{\vartheta} = & \frac{z_{a_x}\left(z_{a_y}(y_o - y_a) + z_a - z_o\right) + x_a\left(z_{a_y}^2 + 1\right) - \left(z_{a_y}^2 + 1\right)x_o}{\left(z_{a_x}^2 + z_{a_y}^2 + 1\right)\sqrt{(x_a - x_o)^2 + (y_a - y_o)^2 + (z_a - z_o)^2}} \\ & + \frac{z_{a_x}\sqrt{1 - \dfrac{\left(z_{a_x}(y_a - y_o) + z_{a_y}(x_o - x_a)\right)^2 + (z_{a_x}(z_o - z_a) - x_a + x_o)^2 + \left(z_{a_y}(z_a - z_o) + y_a - y_o\right)^2}{\left(z_{a_x}^2 + z_{a_y}^2 + 1\right)((x_a - x_o)^2 + (y_a - y_o)^2 + (z_a - z_o)^2)}}}{\sqrt{z_{a_x}^2 + z_{a_y}^2 + 1}},\end{aligned} \tag{18.8}$$

in $y$,

$$\begin{aligned}\frac{(y_b - y_a)}{\vartheta} = & \frac{z_{a_y}\left(z_{a_x}(x_o - x_a) + z_a - z_o\right) + y_a\left(z_{a_x}^2 + 1\right) - \left(z_{a_x}^2 + 1\right)y_o}{\left(z_{a_x}^2 + z_{a_y}^2 + 1\right)\sqrt{(x_a - x_o)^2 + (y_a - y_o)^2 + (z_a - z_o)^2}} \\ & + \frac{z_{a_y}\sqrt{1 - \dfrac{\left(z_{a_x}(y_a - y_o) + z_{a_y}(x_o - x_a)\right)^2 + \left(z_{a_x}(z_o - z_a) - x_a + x_o\right)^2 + \left(z_{a_y}(z_a - z_o) + y_a - y_o\right)^2}{\left(z_{a_x}^2 + z_{a_y}^2 + 1\right)\left((x_a - x_o)^2 + (y_a - y_o)^2 + (z_a - z_o)^2\right)}}}{\sqrt{z_{a_x}^2 + z_{a_y}^2 + 1}},\end{aligned} \tag{18.9}$$

and, in $z$,

$$\begin{aligned}\frac{(z_b - z_a)}{\vartheta} = & \frac{(z_a - z_o)\left(z_{a_x}^2 + z_{a_y}^2\right) + x_a z_{a_x} + z_{a_y}(y_a - y_o) - z_{a_x}x_o}{\left(z_{a_x}^2 + z_{a_y}^2 + 1\right)\sqrt{(x_a - x_o)^2 + (y_a - y_o)^2 + (z_a - z_o)^2}} \\ & - \frac{\sqrt{1 - \dfrac{\left(z_{a_x}(y_a - y_o) + z_{a_y}(x_o - x_a)\right)^2 + (z_{a_x}(z_o - z_a) - x_a + x_o)^2 + \left(z_{a_y}(z_a - z_o) + y_a - y_o\right)^2}{\left(z_{a_x}^2 + z_{a_y}^2 + 1\right)((x_a - x_o)^2 + (y_a - y_o)^2 + (z_a - z_o)^2)}}}{\sqrt{z_{a_x}^2 + z_{a_y}^2 + 1}},\end{aligned} \tag{18.10}$$

where,

$$\vartheta \equiv \sqrt{(x_b - x_a)^2 + (y_b - y_a)^2 + (z_b - z_a)^2}. \tag{18.11}$$

The right side of equations (18.8), (18.9) and (18.10) only depends on known parameters; in reality, on the right side are the cosine directors of the refracted ray.

$$\begin{aligned}\wp_x \equiv &\frac{z_{a_x}\left(z_{a_y}(y_o - y_a) + z_a - z_o\right) + x_a\left(z_{a_y}^2 + 1\right) - \left(z_{a_y}^2 + 1\right)x_o}{\left(z_{a_x}^2 + z_{a_y}^2 + 1\right)\sqrt{(x_a - x_o)^2 + (y_a - y_o)^2 + (z_a - z_o)^2}} \\ &+ \frac{z_{a_x}\sqrt{1 - \frac{\left(z_{a_x}(y_a - y_o) + z_{a_y}(x_o - x_a)\right)^2 + (z_{a_x}(z_o - z_a) - x_a + x_o)^2 + \left(z_{a_y}(z_a - z_o) + y_a - y_o\right)^2}{\left(z_{a_x}^2 + z_{a_y}^2 + 1\right)((x_a - x_o)^2 + (y_a - y_o)^2 + (z_a - z_o)^2)}}}{\sqrt{z_{a_x}^2 + z_{a_y}^2 + 1}},\end{aligned} \tag{18.12}$$

$$\begin{aligned}\wp_y \equiv &\frac{z_{a_y}(z_{a_x}(x_o - x_a) + z_a - z_o) + y_a\left(z_{a_x}^2 + 1\right) - \left(z_{a_x}^2 + 1\right)y_o}{\left(z_{a_x}^2 + z_{a_y}^2 + 1\right)\sqrt{(x_a - x_o)^2 + (y_a - y_o)^2 + (z_a - z_o)^2}} \\ &+ \frac{z_{a_y}\sqrt{1 - \frac{\left(z_{a_x}(y_a - y_o) + z_{a_y}(x_o - x_a)\right)^2 + (z_{a_x}(z_o - z_a) - x_a + x_o)^2 + \left(z_{a_y}(z_a - z_o) + y_a - y_o\right)^2}{\left(z_{a_x}^2 + z_{a_y}^2 + 1\right)((x_a - x_o)^2 + (y_a - y_o)^2 + (z_a - z_o)^2)}}}{\sqrt{z_{a_x}^2 + z_{a_y}^2 + 1}},\end{aligned} \tag{18.13}$$

and,

$$\begin{aligned}\wp_z \equiv &\frac{(z_a - z_o)\left(z_{a_x}^2 + z_{a_y}^2\right) + x_a z_{a_x} + z_{a_y}(y_a - y_o) - z_{a_x}x_o}{\left(z_{a_x}^2 + z_{a_y}^2 + 1\right)\sqrt{(x_a - x_o)^2 + (y_a - y_o)^2 + (z_a - z_o)^2}} \\ &- \frac{\sqrt{1 - \frac{\left(z_{a_x}(y_a - y_o) + z_{a_y}(x_o - x_a)\right)^2 + (z_{a_x}(z_o - z_a) - x_a + x_o)^2 + \left(z_{a_y}(z_a - z_o) + y_a - y_o\right)^2}{\left(z_{a_x}^2 + z_{a_y}^2 + 1\right)((x_a - x_o)^2 + (y_a - y_o)^2 + (z_a - z_o)^2)}}}{\sqrt{z_{a_x}^2 + z_{a_y}^2 + 1}}.\end{aligned} \tag{18.14}$$

Observe that the following relation always meets $\wp_x^2 + \wp_y^2 + \wp_z^2 = 1$. Equaling the cosine directors of equations (18.12), (18.13) and (18.14) with (18.8), (18.9) and (18.10), we have,

$$\frac{x_b - x_a}{\vartheta} = \wp_x, \tag{18.15}$$

$$\frac{y_b - y_a}{\vartheta} = \wp_y, \tag{18.16}$$

and,

$$\frac{z_b - z_a}{\vartheta} = \wp_z. \tag{18.17}$$

Solving for $x_b$, $y_b$ and $z_b$, we get the structure of the second surface of the singlet.

$$\begin{cases} x_b = x_a + \vartheta\wp_x, \\ y_b = y_a + \vartheta\wp_y, \\ z_b = z_a + \vartheta\wp_z. \end{cases} \tag{18.18}$$

The next step and final step is to find $\vartheta$ in terms of the know parameters. We mean the first surface, the distance $t$, the cosine directors $\wp_x$, $\wp_y$ and $\wp_z$ and the distances from the object and image points (18.18).

## 18.3 Fermat principle

Equation (18.18) is the structure of the solution, for which we only need to express $\vartheta$ in terms of the other known parameters instead of the unknowns $x_b$, $y_b$, and $z_b$. So let's recall the Fermat principle to solve for $\vartheta$ on it. Remember that the Fermat principle implies that the optical path for all rays in the beam must be constant. We equalize the optical path of the ray that passes through the vertices of the mirror and the optical path of any other ray, equation (18.19),

$$\begin{aligned} &\sqrt{x_o^2 + y_o^2 + z_o^2} + \sqrt{x_\tau^2 + y_\tau^2 + z_\tau^2} + \sqrt{x_i^2 + y_i^2 + z_i^2} \\ &= \sqrt{(x_a - x_o)^2 + (y_a - y_o)^2 + (z_a - z_o)^2} \\ &+ \sqrt{(x_b - x_a)^2 + (y_b - y_a)^2 + (z_b - z_a)^2} \\ &\quad + \sqrt{(x_b - r_i - r_\tau)^2 + (y_b - r_i - r_\tau)^2 + (z_b - z_i - z_\tau)^2}. \end{aligned} \tag{18.19}$$

Let's define the following parameters,

$$\varsigma \equiv x_a - x_i - x_\tau, \quad \xi \equiv y_a - y_i - y_\tau, \quad \zeta \equiv z_a - z_i - z_\tau, \tag{18.20}$$

and,

$$\begin{aligned} f \equiv &\sqrt{x_o^2 + y_o^2 + z_o^2} + \sqrt{x_\tau^2 + y_\tau^2 + z_\tau^2} + \sqrt{x_i^2 + y_i^2 + z_i^2} \\ &- \sqrt{(x_a - x_o)^2 + (y_a - y_o)^2 + (z_a - z_o)^2}. \end{aligned} \tag{18.21}$$

Replacing,

$$f = \vartheta + \sqrt{(\vartheta\wp_x + \varsigma)^2 + (\vartheta\wp_y + \xi)^2 + (\vartheta\wp_z + \zeta)^2}. \tag{18.22}$$

Squaring,

$$(f - \vartheta)^2 = (\vartheta \wp_x + \varsigma)^2 + \left(\vartheta \wp_y + \xi\right)^2 + (\vartheta \wp_z + \zeta)^2. \tag{18.23}$$

Expanding,

$$\begin{aligned} f^2 - 2f\vartheta + \vartheta^2 = \varsigma^2 + 2\varsigma\vartheta\wp_x + \vartheta^2\wp_x^2 + \xi^2 \\ + 2\xi\vartheta\wp_x + \vartheta^2\wp_y^2 + \zeta^2 + 2\zeta\vartheta\wp_x + \vartheta^2\wp_z^2, \end{aligned} \tag{18.24}$$

simplifying,

$$f^2 - 2f\vartheta = \varsigma^2 + 2\varsigma\vartheta\wp_x + \xi^2 + 2\xi\vartheta\wp_x + \zeta^2 + 2\zeta\vartheta\wp_x. \tag{18.25}$$

Solving for $\vartheta$,

$$\vartheta = \frac{f^2 - \zeta^2 - \xi^2 - \varsigma^2}{2\left(f + \varsigma\wp_x + \xi\wp_y + \zeta\wp_z\right)}. \tag{18.26}$$

Therefore, the second mirror can be expressed by equation (18.27),

$$\begin{bmatrix} x_b \\ y_b \\ z_b \end{bmatrix} = \begin{bmatrix} x_a \\ y_a \\ z_a \end{bmatrix} + \frac{f^2 - \zeta^2 - \xi^2 - \varsigma^2}{2\left(f + \varsigma\wp_x + \xi\wp_y + \zeta\wp_z\right)} \begin{bmatrix} \wp_x \\ \wp_y \\ \wp_z \end{bmatrix}. \tag{18.27}$$

Equation (18.27) is the most important result in the paper because it expresses the shape of the second mirror in such a way that the pair of mirrors becomes stigmatic. Note that the first mirror has been taken as a free parameter or given by the user, the shape of the second mirror only depends on the given parameters, the positions of the mirrors, the object and the image. This does not affect the generality of the solution since by the principle of reciprocity of light the object and the image can be interchanged. This is possible because the system is stigmatic. Thus the pair of mirrors work as a team, they form a perfect image for a given object at $(x_o, y_o, z_o)$.

Equation (18.27) has been obtained with no iterations nor paraxial approximations. The deduction of equation (18.27) is totally analytical and it is a closed-form equation. Thus, equation (18.27) is as rigorous as the mathematical equations in most of the equations in this treatise. It is important to remark that equation (18.27) meets the uniqueness theorem since for a given configuration with a known shape of the first surface there is only one unique shape for the second surface determined by equation (18.27) such that the two-mirror system is stigmatic and equation (18.27) contains all the possible pairs of mirrors that form a stigmatic system.

## 18.4 Gallery

In this section, we present several stigmatic pairs of mirrors designed directly using equation (18.27). The examples of the gallery are presented in figures 18.2–18.5. The specification of each example is presented in the caption of its respective image.

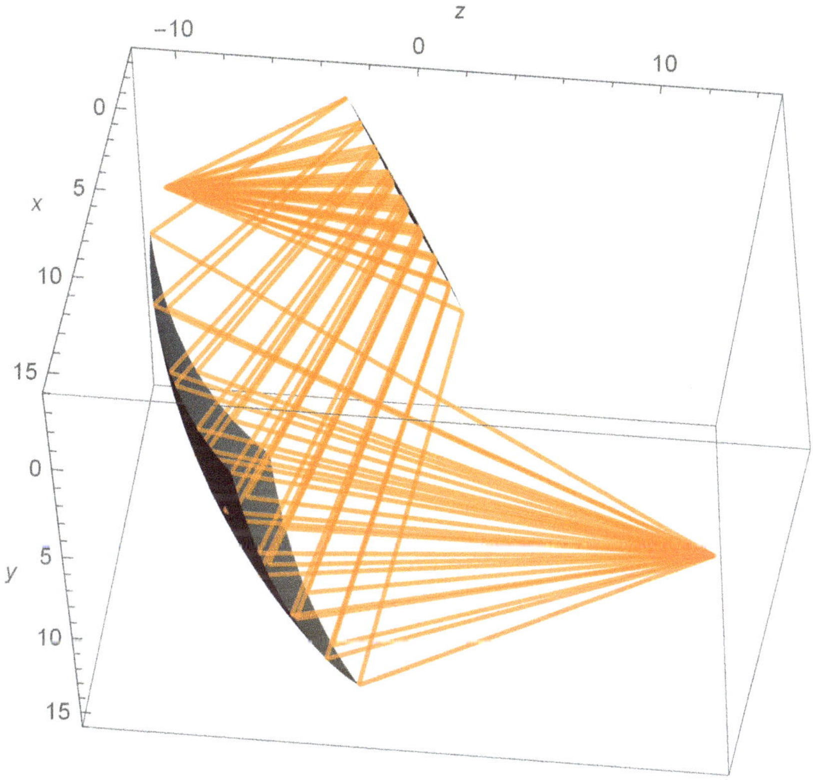


**Figure 18.2.** Design specifications: the first surface is given by $z_a = 0.45(x_a + y_a) - \left(x_a^2 + y_a^2\right)/150$, the second surface is equation (18.27), $x_o = 0$ mm, $y_o = 0$ mm, $z_o = -10$ mm, $x_i = 0$ mm, $y_i = 0$ mm, $z_i = 20$ mm and $\tau = 40$ mm.

## 18.5 Stigmatic collector

In order to get a stigmatic collector made by a pair of mirrors, we need to compute the limit when $z_o \to -\infty$ over equations (18.12), (18.13), (18.14) and (18.21) and replace them (18.27). So let's compute the limits,

$$\lim_{z_o \to -\infty} (\wp x) = \lim_{z_o \to -\infty} \left[ \frac{z_{a_x}\left(z_{a_y}(y_o - y_a) + z_a - z_o\right) + x_a\left(z_{a_y}^2 + 1\right) - \left(z_{a_y}^2 + 1\right)x_o}{\left(z_{a_x}^2 + z_{a_y}^2 + 1\right)\sqrt{(x_a - x_o)^2 + (y_a - y_o)^2 + (z_a - z_o)^2}} + \frac{z_{a_x}\sqrt{1 - \frac{\left(z_{a_x}(y_a - y_o) + z_{a_y}(x_o - x_a)\right)^2 + \left(z_{a_x}(z_o - z_a) - x_a + x_o\right)^2 + \left(z_{a_y}(z_a - z_o) + y_a - y_o\right)^2}{\left(z_{a_x}^2 + z_{a_y}^2 + 1\right)\left((x_a - x_o)^2 + (y_a - y_o)^2 + (z_a - z_o)^2\right)}}}{\sqrt{z_{a_x}^2 + z_{a_y}^2 + 1}} \right], \tag{18.28}$$

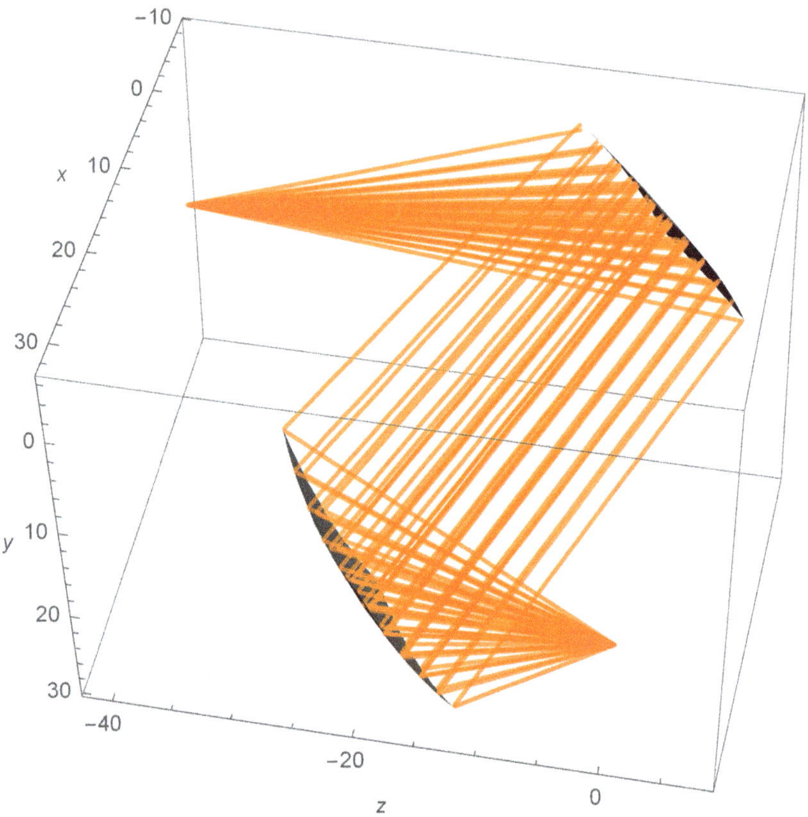


**Figure 18.3.** Design specifications: the first surface is given by $z_a = 0.75(x_a + y_a) - \left(x_a^2 + 4y_a^2\right)/150$, the second surface is equation (18.27), $x_o = 1$ mm, $y_o = 4$ mm, $z_o = -40$ mm, $x_i = 0$ mm, $y_i = 0$ mm, $z_i = 20$ mm and $\tau = 40$ mm.

computing the limit,

$$\lim_{z_o \to -\infty} (\wp_x) = \frac{2z_{a_x}}{z_{a_x}^2 + z_{a_y}^2 + 1} \tag{18.29}$$

for the $y$ direction we have,

$$\lim_{z_o \to -\infty} (\wp_y) = \left[ \frac{z_{a_y}\left(z_{a_x}(x_o - x_a) + z_a - z_o\right) + y_a\left(z_{a_x}^2 + 1\right) - \left(z_{a_x}^2 + 1\right)y_o}{\left(z_{a_x}^2 + z_{a_y}^2 + 1\right)\sqrt{(x_a - x_o)^2 + (y_a - y_o)^2 + (z_a - z_o)^2}} + \frac{z_{a_y}\sqrt{1 - \dfrac{\left(z_{a_x}(y_a - y_o) + z_{a_y}(x_o - x_a)\right)^2 + \left(z_{a_x}(z_o - z_a) - x_a + x_o\right)^2 + \left(z_{a_y}(z_a - z_o) + y_a - y_o\right)^2}{\left(z_{a_x}^2 + z_{a_y}^2 + 1\right)\left((x_a - x_o)^2 + (y_a - y_o)^2 + (z_a - z_o)^2\right)}}}{\sqrt{z_{a_x}^2 + z_{a_y}^2 + 1}} \right], \tag{18.30}$$

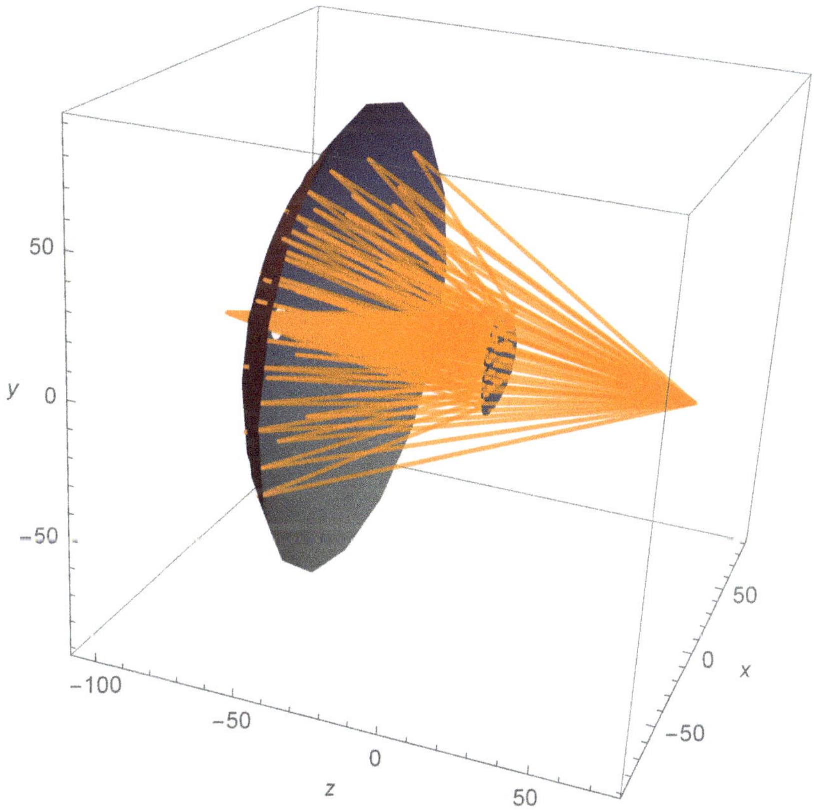


**Figure 18.4.** Design specifications: the first surface is given by $z_a = 70 - \sqrt{70^2 - x_a^2 - (y_a/2)^2}$, the second surface is equation (18.27), $x_o = 0$ mm, $y_o = 0$ mm, $z_o \to -\infty$ mm, $x_i = 0$ mm, $y_i = 0$ mm, $z_i = 20$ mm and $\tau = 20$ mm.

after the computation we get,

$$\lim_{z_o \to -\infty} (\wp_y) = \frac{2z_{a_y}}{z_{a_x}^2 + z_{a_y}^2 + 1}, \tag{18.31}$$

in $z$ direction we have,

$$\lim_{z_o \to -\infty} (\wp_z) = \left[ \frac{(z_a - z_o)\left(z_{a_x}^2 + z_{a_y}^2\right) + x_a z_{a_x} + z_{a_y}(y_a - y_o) - z_{a_x} x_o}{\left(z_{a_x}^2 + z_{a_y}^2 + 1\right)\sqrt{(x_a - x_o)^2 + (y_a - y_o)^2 + (z_a - z_o)^2}} - \frac{\sqrt{1 - \frac{\left(z_{a_x}(y_a - y_o) + z_{a_y}(x_o - x_a)\right)^2 + \left(z_{a_x}(z_o - z_a) - x_a + x_o\right)^2 + \left(z_{a_y}(z_a - z_o) + y_a - y_o\right)^2}{\left(z_{a_x}^2 + z_{a_y}^2 + 1\right)\left((x_a - x_o)^2 + (y_a - y_o)^2 + (z_a - z_o)^2\right)}}}{\sqrt{z_{a_x}^2 + z_{a_y}^2 + 1}} \right]. \tag{18.32}$$

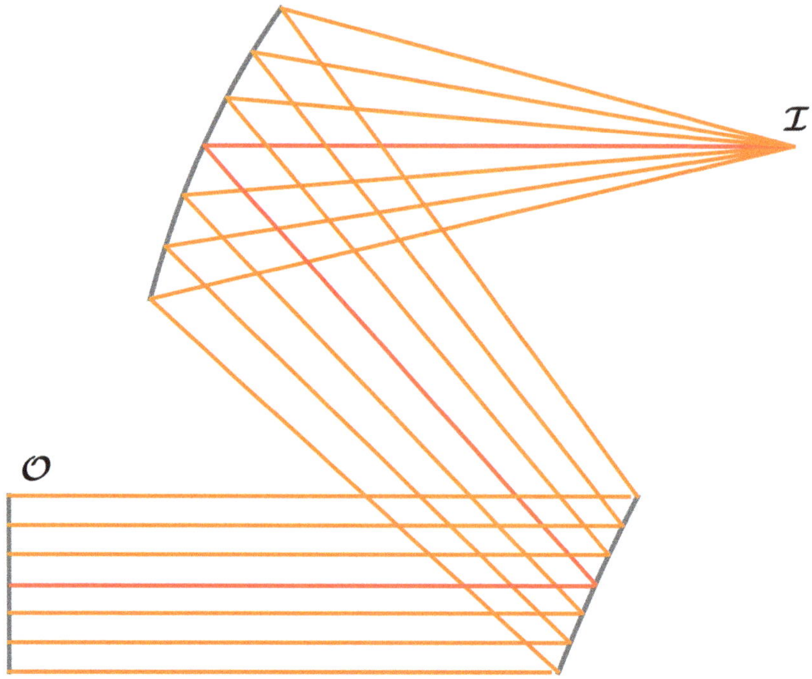


**Figure 18.5.** Design specifications: the first surface is given by $z_a = -\sqrt{50^2 - x_a^2} + 0.45x_a + 50$, the second surface is equation (18.36), $x_o = 0$ mm, $y_o = 0$ mm, $z_o \to -\infty$ mm, $x_i = 0$ mm, $y_i = 0$ mm, $z_i = 20$ mm and $\tau = 20$ mm.

simplifying,

$$\lim_{z_o \to -\infty} (\wp_z) = \frac{z_{a_x}^2 + z_{a_y}^2 - 1}{z_{a_x}^2 + z_{a_y}^2 + 1}, \tag{18.33}$$

and for $f$ we have,

$$\lim_{z_o \to -\infty} (f) \equiv z_a + \sqrt{x_i^2 + y_i^2 + z_i^2} + \sqrt{x_\tau^2 + y_\tau^2 + z_\tau^2}. \tag{18.34}$$

Therefore, $\vartheta$ over the limit when $z_o \to -\infty$ is,

$$\lim_{z_o \to -\infty} (\vartheta) = \frac{\left[\lim_{z_o \to -\infty} (f)\right]^2 - \zeta^2 - \xi^2 - \varsigma^2}{2\left[\lim_{z_o \to -\infty} (f) + \varsigma \lim_{z_o \to -\infty} (\wp_x) + \xi \lim_{z_o \to -\infty} (\wp_y) + \zeta \lim_{z_o \to -\infty} (\wp_z)\right]}. \tag{18.35}$$

Then, we can use equation (18.27) with the parameters of equations (18.29), (18.31), (18.33), (18.34), (18.35) and test it. This leads to,

$$\begin{cases} \lim_{t_a \to -\infty} (x_b) = x_a + \left[ \lim_{t_a \to -\infty} (\vartheta) \right] \left[ \lim_{t_a \to -\infty} (\wp_x) \right], \\ \lim_{t_a \to -\infty} (y_b) = y_a + \left[ \lim_{t_a \to -\infty} (\vartheta) \right] \left[ \lim_{t_a \to -\infty} (\wp_y) \right], \\ \lim_{t_a \to -\infty} (z_b) = z_a + \left[ \lim_{t_a \to -\infty} (\vartheta) \right] \left[ \lim_{t_a \to -\infty} (\wp_z) \right]. \end{cases} \tag{18.36}$$

An example of a stigmatic collector made by a pair of mirrors can be seen in figure 18.2.

## 18.6 Conclusion

In this chapter, we get the exact equations to design a pair of mirrors that forms a stigmatic system. These equations are closed-form with no need for paraxial approximations. Given the shape of the first mirror, equation (18.27) uniquely describes the shape of the second mirror and guarantees that the pair of mirrors is a stigmatic system. We also showed examples of this stigmatic pair and the performance was as predicted by the theory.

## Further reading


Quadoa Optical CAD 2022 *Quadoa Optical CAD Software Manual V* (Berlin: Quadoa Optical Systems) https://www.quadoa.com

Abbe E 1881 On the estimation of aperture in the microscope *J. R. Microsc. Soc.* **1** 388–423

Beier M, Scheiding S, Gebhardt A, Loose R, Risse S, Eberhardt R and Tünnermann A 2013 Fabrication of high precision metallic freeform mirrors with magnetorheological finishing (MRF) *Optifab 2013* vol 8884 (Bellingham, WA: International Society for Optics and Photonics)

Born M and Wolf E 1959 Principles of Optics: Electromagnetic theory of Propagation *Interference and Diffraction of Light* (New York: Pergamon)

Born M and Wolf E 2013 Principles of Optics: Electromagnetic Theory of Propagation *Interference and Diffraction of Light* (Amsterdam: Elsevier)

Braat J J M 1997a Abbe sine condition and related imaging conditions in geometrical optics *5th Int. Topical Meeting on Education and Training in Optics* ed C H F Velzel vol 3190 (Bellingham, WA: International Society for Optics and Photonics, SPIE) pp 59–64

Challita Z, Agócs T, Hugot E, Jaskó A, Kroes G, Taylor W, Miller C, Schnetler H, Venema L and Mosoni L 2014 Design and development of a freeform active mirror for an astronomy application *Opt. Eng.* **53** 031311

Chaves J 2016 *Introduction to Nonimaging Optics* 2nd edn (Boca Raton, FL: CRC Press)

Courtial J, Oxburgh S and Tyc T 2015 Direct stigmatic imaging with curved surfaces *J. Opt. Soc. Am.* A **32** 478–81

Doskolovich L L, Bezus E A, Moiseev M A, Bykov D A and Kazanskiy N L 2016 Analytical source-target mapping method for the design of freeform mirrors generating prescribed 2D intensity distributions *Opt. Express* **24** 10962–71

Doskolovich L L, Bykov D A, Andreeva K V and Kazanskiy N L 2018 Design of an axisymmetrical refractive optical element generating required illuminance distribution and wavefront *J. Opt. Soc. Am.* A **35** 1949–53

Doskolovich L L, Bykov D A, Andreev E S, Byzov E V, Moiseev M A, Bezus E A and Kazanskiy N L 2020 Design and fabrication of freeform mirrors generating prescribed far-field irradiance distributions *Appl. Opt.* **59** 5006–12

Doskolovich L L, Bykov D A, Bezus E A and Greisukh G I 2021a Design of a stigmatic lens implementing a required ray mapping *Appl. Opt.* **60** 9138–45

Doskolovich L L, Bykov D A, Greisukh G I, Strelkov Y S and Bezus E A 2021b Designing stigmatic lenses with minimal Fresnel losses *J. Opt. Soc. Am.* A **38** 855–61

Duerr F, Benítez P, Miñano J C, Meuret Y and Thienpont H 2012 Analytic design method for optimal imaging: coupling three ray sets using two free-form lens profiles *Opt. Express* **20** 5576–85

Elazhary T T, Zhou P, Zhao C and Burge J H 2015 Generalized sine condition *Appl. Opt.* **54** 5037–49

Gimenez-Benitez P, Miñano J C, Blen J, Arroyo R, Chaves J, Dross O, Hernández M and Falicoff W 2004 Simultaneous multiple surface optical design method in three dimensions *Opt. Eng.* **43** 1489–503

González-Acuña R G and Chaparro-Romo H A 2018 General formula for bi-aspheric singlet lens design free of spherical aberration *Appl. Opt.* **57** 9341–5

González-Acuña R G and Chaparro-Romo H A 2020 *Stigmatic Optics* (Bristol: IOP Publishing)

González-Acuña R G and Guitiérrez-Vega J C 2018 Generalization of the axicon shape: the gaxicon *J. Opt. Soc. Am.* A **35** 1915–8

González-Acuña R G and Gutiérrez-Vega J C 2019a General formula of the refractive telescope design free spherical aberration *Novel Optical Systems, Methods, and Applications XXII* vol 11105 (Bellingham, WA: International Society for Optics and Photonics)

González-Acuña R G and Gutiérrez-Vega J C 2019b Analytic formulation of a refractive-reflective telescope free of spherical aberration *Opt. Eng.* **58** 085105

González-Acuña R G and Gutiérrez-Vega J C 2019c General formula for aspheric collimator lens design free of spherical aberration *Current Developments in Lens Design and Optical Engineering XX* vol 11104 (Bellingham, WA: International Society for Optics and Photonics)

González-Acuña R G and Gutiérrez-Vega J C 2020 Analytic design of a spherochromatic singlet *J. Opt. Soc. Am.* A **37** 149–53

González-Acuña R G, Avendaño-Alejo M and Gutiérrez-Vega J C 2019a Singlet lens for generating aberration-free patterns on deformed surfaces *J. Opt. Soc. Am.* A **36** 925–9

González-Acuña R G, Chaparro-Romo H A and Gutiérrez-Vega J C 2019b General formula to design freeform singlet free of spherical aberration and astigmatism *Appl. Opt.* **58** 1010–5

González-Acuña R G, Chaparro-Romo H A and Gutiérrez-Vega J C 2020a Analytic solution of the eikonal for a stigmatic singlet lens *Phys. Scr.* **95** 085201

González-Acuña R G, Chaparro-Romo H A and Gutiérrez-Vega J C 2020b Exact equations for stigmatic singlet design meeting the Abbe sine condition *Opt. Commun.* **479** 126415

González-Acuña R G, Chaparro-Romo H A and Gutiérrez-Vega J C 2020c *Analytical Lens Design* (Bristol: IOP Publishing)

González-Acuña R G, Borne J and Thibault S 2021 On the diffraction of a high-NA aplanatic and stigmatic singlet *J. Opt. Soc. Am.* A **38** 1332–8

Gonzalez-Acuna R G, Chaparro-Romo H A and Gutierrez-Vega J C 2021 Exact equations for stigmatic singlet design meeting the Abbe sine condition *Opt. Commun.* **479** 126415

Gross H, Singer W, Totzeck M, Blechinger F and Achtner B 2005 *Handbook of Optical Systems* vol 1 (New York: Wiley Online Library)

Hazra L N, Han Y and Delisle C 1993 Curved kinoform lenses for stigmatic imaging of axial objects *Appl. Opt.* **32** 4775–84

Head A K 1960 Aplanatic lenses of high-refractive index *J. Opt. Soc. Am.* **50** 922

Head A K 1959 Aplanatic lenses *Proc. Phys. Soc.* **74** 731

Hopkins H H 1946 Herschel's condition *Proc. Phys. Soc.* **58** 100–5

Hsueh C-C, Elazhary T, Nakano M and Sasian J 2011 Closed-form sag solutions for Cartesian oval surfaces *J. Opt.* **40** 168–75

Hui X, Liu J, Wan Y and Lin H 2017 Realization of uniform and collimated light distribution in a single freeform-Fresnel double surface LED lens *Appl. Opt.* **56** 4561–5

Ji H, Zhu Z, Tan H, Shan Y, Tan W and Ma D 2021 Design of a high-throughput telescope based on scanning an off-axis three-mirror anastigmat system *Appl. Opt.* **60** 2817–23

Kingslake R and Johnson R B 2009 *Lens Design Fundamentals* (New York: Academic)

Lerman G M and Levy U 2008 Effect of radial polarization and apodization on spot size under tight focusing conditions *Opt. Express* **16** 4567–81

Liu J (ed) 2018 *Elliptical Mirrors* (Bristol: IOP Publishing) pp 2053–563

Luneburg R K and Herzberger M 1964 *Mathematical Theory of Optics* (Berkeley, CA: University of California Press)

Malacara-Hernández D and Malacara-Hernández Z 2016 *Handbook of Optical Design* (Boca Raton, FL: CRC Press)

Mansuripur M 1998 Abbe's sine condition *Opt. Photonics News* **9** 56–60

Mashaal H, Feuermann D and Gordon J M 2015 New types of refractive-reflective aplanats for maximal flux concentration and collimation *Opt. Express* **23** A1541–8

Mashaal H, Feuermann D and Gordon J M 2016 Aplanatic lenses revisited: the full landscape *Appl. Opt.* **55** 2537–42

Masters B R 2007 Ernst Abbe and the foundation of scientific microscopes *Opt. Photonics News* **18** 18–23

Meng Q, Wang W, Ma H and Dong J 2014 Easy-aligned off-axis three-mirror system with wide field of view using freeform surface based on integration of primary and tertiary mirror *Appl. Opt.* **53** 3028–34

Meng Q, Wang H, Wang K, Wang Y, Ji Z and Wang D 2016 Off-axis three-mirror freeform telescope with a large linear field of view based on an integration mirror *Appl. Opt.* **55** 8962–70

Meng Q, Wang H, Liang W, Yan Z and Wang B 2019 Design of off-axis three-mirror systems with ultrawide field of view based on an expansion process of surface freeform and field of view *Appl. Opt.* **58** 609–15

Miñano J C, Mohedano R and Benítez P 2007 Nonimaging optics *The Optics Encyclopedia* (New York: Wiley) pp 1–34

Miñano J C, Benítez P, Lin W, Infante J, Muñoz F and Santamaría A 2009 An application of the SMS method for imaging designs *Opt. Express* **17** 24036–44

Shibuya M 1992 Exact sine condition in the presence of spherical aberration *Appl. Opt.* **31** 2206–10

Silva-Lora A and Torres R 2020a Aplanatism in stigmatic optical systems *Opt. Lett.* **45** 6390–3

Silva-Lora A and Torres R 2020b Explicit Cartesian oval as a superconic surface for stigmatic imaging optical systems with real or virtual source or image *Proc. R. Soc.* A **476** 20190894

Silva-Lora A and Torres R 2020c Superconical aplanatic ovoid singlet lenses *J. Opt. Soc. Am.* A **37** 1155–65

Silva-Lora A and Torres R 2021 Rigorously aplanatic Descartes ovoids *J. Opt. Soc. Am.* A **38** 1160–9

Stavroudis O N 2006 *The Mathematics of Geometrical and Physical Optics: The k-Function and Its Ramifications* (New York: Wiley)

Steward G C 1927 On Herschel's condition and the optical cosine law *Math. Proc. Camb. Philos. Soc.* **23** 703–12

Strojnik M and Kirk M S 2017 Telescopes *Fundamentals and Basic Optical Instruments* (Boca Raton, FL: CRC Press) pp 325–74

Turk A S 2006 Analysis of aperture illumination and edge rolling effects for parabolic reflector antenna design *AEU—Int. J. Electron. Commun.* **60** 257–66

Valencia-Estrada J C, Flores-Hernández R B and Malacara-Hernández D 2015 Singlet lenses free of all orders of spherical aberration *Proc. Math. Phys. Eng. Sci.* **471** 20140608

Valencia-Estrada J C, Pereira-Ghirghi M V, Malacara-Hernández Z and Chaparro-Romo H A 2017 Aspheric coefficients of deformation for a Cartesian oval surface *J. Opt.* **46** 100–7

Vaskas E M 1957 Note on the Wasserman-Wolf method for designing aspheric surfaces *J. Opt. Soc. Am.* **47** 669–70

Velzel C H F 2014 *A Course in Lens Design* vol 183 (Berlin: Springer)

Wassermann G D and Wolf E 1949 On the theory of aplanatic aspheric systems *Proc. Phys. Soc. Sect.* B **62** 2–8

Zhen Y-k and Ye Z 2007 Illumination system with compound parabolic retro-reflector for single LCOS panel projection display *J. Zhejiang Univ. Sci.* **8** 2021–6

**IOP** Publishing

Analytical Lens Design (Second Edition)

**Rafael G González-Acuña**

# Chapter 19

# Design of a pair of aplanatic mirrors

This chapter presents a mathematical model to design a pair of mirrors such that they form an aplanatic system. Thus, this system is free from spherical aberration and coma. The equations that represent this system come from the equations presented in the last chapter and the Abbe sine condition. They form a system of differential equations that is solved with numerical methods.

## 19.1 Introduction

Aplanatic optical systems have the particularity of being free from spherical aberration and coma for objects near the focus. Therefore, for a system to be aplanatic first it needs to be stigmatic. Stigmatism of the particularity of an optical system where given a point-object, all the rays that emerge from it converge in a single point-image, once the mentioned rays have passed the optical system. The other condition for a system to be aplanatic is that it obeys the Abbe sine condition, which is the following relation,

$$\sin\theta_a = M\,\sin\theta_b, \tag{19.1}$$

where $M$ is a constant, $\theta_a$ and $\theta_b$ are the input and output angles of the light rays, respectively. The angles are measured from the optical axis. If the object is considered to be far away from the optical system Abbe's sine condition turns to,

$$r_a = M\,\sin\theta_b, \tag{19.2}$$

where $r_a$ is the height with respect to the optical axis, where the input light ray strikes at the first surface of the optical system.

In chapter 16 we have studied the Abbe sine condition for singlets and in chapter 17 the Abbe sine condition for more complex optical systems. Here in this chapter, we are

doi:10.1088/978-0-7503-5774-6ch19  

going now to study the Abbe sine condition for a pair of mirrors. Therefore, in this chapter, we are going to use some of the results of chapters 16 and 17.

## 19.2 Mathematical model

A diagram of the pair of aplanatic mirrors is presented in figure 19.1. Let $(r_a, z_a)$ be the shape of the first mirror, and let $(r_b, z_b)$ be the shape of the second mirror. The profiles of the first and second mirrors are given by $z_a$ and $z_b$, respectively. The radial components of the mirrors are $r_a$ for the first and $r_b$ for the second. The goal is to get a single differential equation and we should define only a single independent variable. This differential equation comes from the Abbe sine condition. Thus, let $r_a$ be the only independent variable of the model. As consequence the parameters $z_a$, $z_b$ and $r_b$ depend only on $r_a$, in other means $z_a(r_a)$, $z_b(r_a)$ and $r_b(r_a)$.

Let $(r_a = 0, z_a(r_a = 0) = 0)$ be the origin of the Cartesian coordinate system. By notation, we call this point the vertex of the first mirror. Then, let $(r_b(r_a = 0), z_b(r_a = 0))$ be the vertex of the second mirror. The gap from the object to the vertex's first surface is $z_o \to -\infty$. The gap between the vertexes is denoted by $t \equiv \sqrt{r_t^2 + z_t^2}$. The gap between the vertex of the second mirror and the image is $z_i$.

So, as we have seen in earlier chapters, in order to get an aplanatic system, we need to find $z_a(r_a)$, $z_b(r_a)$ and $r_b(r_a)$ such that the optical system is stigmatic and obeys the Abbe since condition. This implies that $z_a(r_a)$, $z_b(r_a)$ and $r_b(r_a)$ are the unknowns of our system. We have three unknowns but we also have three equations: Snell's law, the Fermat principle, and the Abbe sine condition. The goal solves the system conformed by three equations and three unknowns.

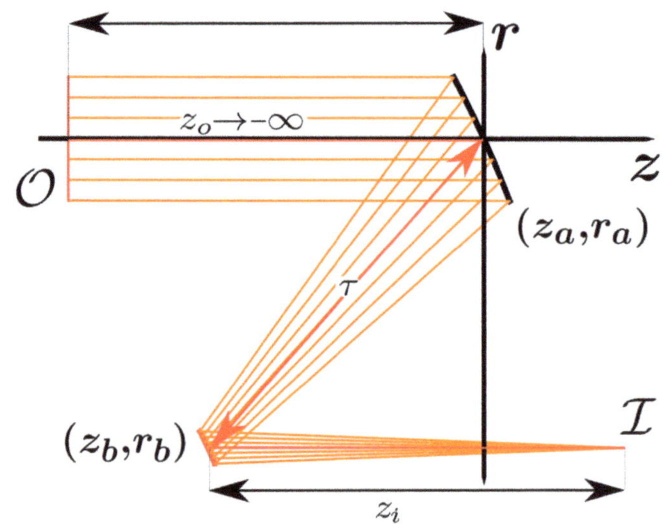


**Figure 19.1.** Diagram of the pair of aplanatic mirrors. The first mirror is given by $[r_a, z_a]$ and the second mirror is $[r_b, z_b]$. The gap between the vertices of the mirrors is $t \equiv \sqrt{r_t^2 + z_t^2}$. The space between the object and the first mirror is $z_o \to -\infty$ and the gap between the second mirror and the image is $z_i$.

Let's begin with the first equation, Snell's law at the first surface,

$$\vec{\mathbf{v}}_2 = (\vec{\mathbf{v}}_1 - (\vec{\mathbf{n}}_a \cdot \vec{\mathbf{v}}_1)\vec{\mathbf{n}}_a) - \vec{\mathbf{n}}_a\sqrt{1 + (\vec{\mathbf{n}}_a \wedge \vec{\mathbf{v}}_1)^2}, \tag{19.3}$$

where $\vec{\mathbf{v}}_1$ is the unit vector of the incident ray, $\vec{\mathbf{v}}_2$ is unit vector of the refracted ray, and $\vec{\mathbf{n}}_a$ is the normal vector,

$$\begin{aligned} &\vec{\mathbf{v}}_1 = \vec{\mathbf{e}}_2, \quad \vec{\mathbf{v}}_2 = \wp_r\vec{\mathbf{e}}_1 + \wp_z\vec{\mathbf{e}}_2 = \frac{(r_b - r_a)\vec{\mathbf{e}}_1 + (z_b - z_a)\vec{\mathbf{e}}_2}{\sqrt{(r_b - r_a)^2 + (z_b - z_a)^2}}, \\ &\vec{\mathbf{n}}_a = \frac{z_a'\vec{\mathbf{e}}_1 - \vec{\mathbf{e}}_2}{\sqrt{1 + z_a'^2}}, \end{aligned} \tag{19.4}$$

where $z_a'$ is the derivative of $z_a$ with respect to $r_a$, and $\wp_r$ and $\wp_z$ are the cosine directors of the ray between the mirrors. Note that we can place the unit vectors of equation (19.4) into equation (19.3) and get all the mathematical processes to get the following expression,

$$r_b = r_a + \vartheta\wp_r, \quad z_b = z_a + \vartheta\wp_z, \tag{19.5}$$

where, $\vartheta$ is the distance between mirrors, equation (19.6),

$$\vartheta \equiv \sqrt{(r_b - r_a)^2 + (z_b - z_a)^2}. \tag{19.6}$$

But, also we can recall the cosine directors of the stigmatic collector, chapter 4,

$$\wp_r = -\frac{z_a'\left(n\sqrt{(z_a')^2 + 1}\sqrt{\frac{(n^2 - 1)(z_a')^2 + n^2}{n^2((z_a')^2 + 1)}} - 1\right)}{n[(z_a')^2 + 1]}, \tag{4.37}$$

and,

$$\wp_z = \frac{\sqrt{\frac{(n^2 - 1)(z_a')^2 + n^2}{n^2[(z_a')^2 + 1]}}}{\sqrt{(z_a')^2 + 1}} + \frac{(z_a')^2}{n(z_a')^2 + n}, \tag{4.38}$$

and computed when $n = -1$ over them. Instead all the mathematical processes place equation (19.4) into equation (19.3). Thus, by doing that we get,

$$\wp_r = \frac{2z_a'}{z_a'^2 + 1}, \quad \wp_z = 1 - \frac{2}{z_a'^2 + 1}. \tag{19.7}$$

Note that $\wp_r^2 + \wp_z^2 = 1$.

Now let's focus on our second equation the Fermat principle. The Fermat principle predicts that if an optical system is stigmatic it means that the optical path length of all its rays is constant. Comparing the optical path length of the ray that passes through the vertex and any other ray we end with equation (19.8),

$$
\begin{aligned}
-z_o + t + z_i = & \sqrt{(r_a)^2 + (z_a - z_o)^2} + \sqrt{(r_b - r_a)^2 + (z_b - z_a)^2} \\
& + \sqrt{(r_b - r_t)^2 + (z_b - z_i - z_t)^2}\,.
\end{aligned}
\tag{19.8}
$$

Note that equation (19.8) considers a finite object, to recover the parallelism of the input rays we shall compute the limit when $z_o \to -\infty$ over equation (19.8),

$$
-z_a + z_i + t = \sqrt{(r_b - r_a)^2 + (z_b - z_a)^2} + \sqrt{(r_b - r_t)^2 + (z_b - z_i - z_t)^2}\,. \tag{19.9}
$$

Let's define the following parameters,

$$
f = -z_a + z_i + t, \quad \xi = r_a - r_t, \quad \zeta = z_a - z_i - z_t, \tag{19.10}
$$

Equations (19.11) turn to,

$$
f = \vartheta + \sqrt{(\xi + \vartheta\wp_r)^2 + (\zeta + \vartheta\wp_z)^2}\,, \tag{19.11}
$$

squaring,

$$
(f - \vartheta)^2 = (\xi + \vartheta\wp_r)^2 + (\zeta + \vartheta\wp_z)^2, \tag{19.12}
$$

expanding,

$$
f^2 - 2f\vartheta + \vartheta^2 = \xi^2 + 2\xi\vartheta\wp_r + (\xi\vartheta\wp_r)^2 + \zeta^2 + 2\zeta\vartheta\wp_z + (\zeta\vartheta\wp_z)^2, \tag{19.13}
$$

simplifying,

$$
f^2 - 2f\vartheta = \xi^2 + 2\xi\vartheta\wp_r + \zeta^2 + 2\zeta\vartheta\wp_z, \tag{19.14}
$$

and solving for $\vartheta$,

$$
\vartheta = \frac{-\zeta^2 - \xi^2 + f^2}{2\left(\xi\wp_r + \zeta\wp_z + f\right)}. \tag{19.15}
$$

Replacing equation (19.7), equation (19.15) and equation (19.10) in equation (19.5), we get,

$$
\begin{aligned}
\begin{bmatrix} r_b \\ z_b \end{bmatrix} = & \begin{bmatrix} r_a \\ z_a \end{bmatrix} \\
& - \frac{(z_a'^2 + 1)[(r_a - r_t)^2 - (t - z_t)(-2z_a + 2z_i + t + z_t)]}{2[(t - z_t)z_a'^2 + 2(r_a - r_t)z_a' - 2z_a + 2z_i + t + z_t]} \begin{bmatrix} \dfrac{2z_a'}{z_a'^2 + 1} \\ 1 - \dfrac{2}{z_a'^2 + 1} \end{bmatrix}.
\end{aligned}
\tag{19.16}
$$

Equation (19.16) expresses the second mirror such that the system is stigmatic. The deduction of equation (19.16) at this time is completely analytical, no approximation nor iteration has been implemented yet. Observe that in equation (19.16) there is still the unknown $z_a(r_a)$. $z_a(r_a)$ can be such that the system is aplanatic. To do that we need to call our third equation the Abbe sine condition. For a faraway object, the Abbe sine condition in terms of both mirrors is,

$$\boxed{r_a = M \frac{r_b}{\sqrt{(r_b - r_t)^2 + (z_b - z_i - z_t)^2}}.} \tag{19.17}$$

Equation (19.17) is in fact a differential equation because equation (19.17) depends on equation (19.16) which depends on $z_a(r_a)$ and also on $z_a'(r_a)$. The boundary values of equation (19.17) are given by the vertex of $z_a(r_a)$, $(r_a = 0,\ z_a(r_a = 0) = 0)$ of the first mirror. Equation (19.17) is not so complicated to solve numerically since it only has an independent variable $r_a$.

Once the solution of equation (19.17) is known then the second mirror can be directly computed with equation (19.16). Equation (19.16) guarantees that the system is stigmatic and equation (19.17) guarantees that the system is aplanatic. Thus, no further optimization or iteration is needed in order to reduce the coma and the spherical aberration.

## 19.3 Illustrative example

Figure 19.2 presents the ray tracing of an illustrative example designed with equation (19.17) and equation (19.16). The ray tracing has been done with Quadoa Optical CAD. The surfaces had been uploaded to the software as point clouds. The example is a telescope which consists of two aspheric mirrors directly obtained from equation (19.17) and equation (19.16) with no other optimization nor iteration.

Once the data of the mirrors is uploaded as a point cloud it is converted to an aspherical surface by the software. Specifications of the design before are $z_o \to -\infty$, $t = 10$ mm, and $z_i = 7$ mm.

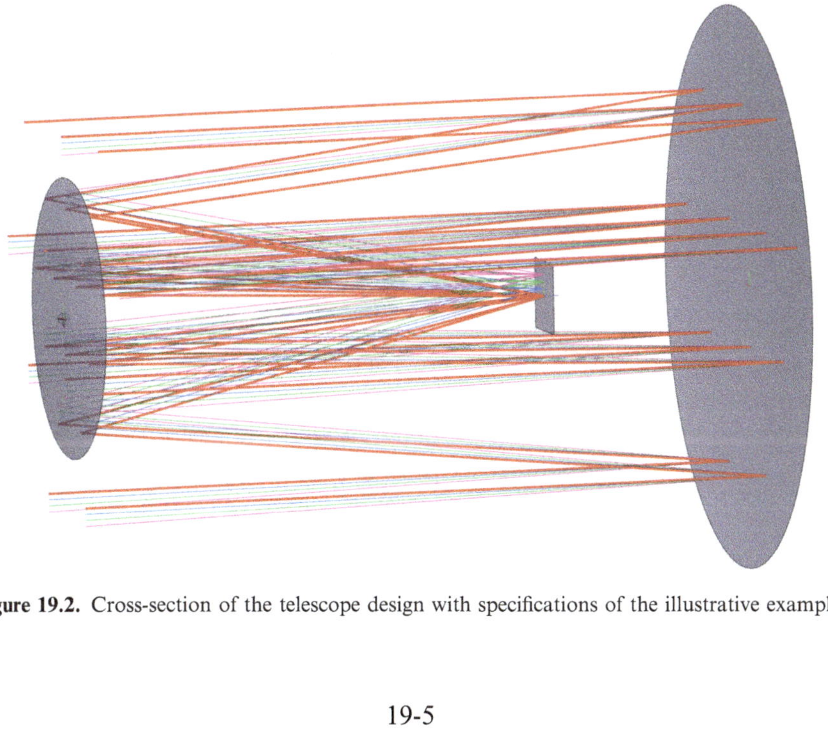

**Figure 19.2.** Cross-section of the telescope design with specifications of the illustrative example.

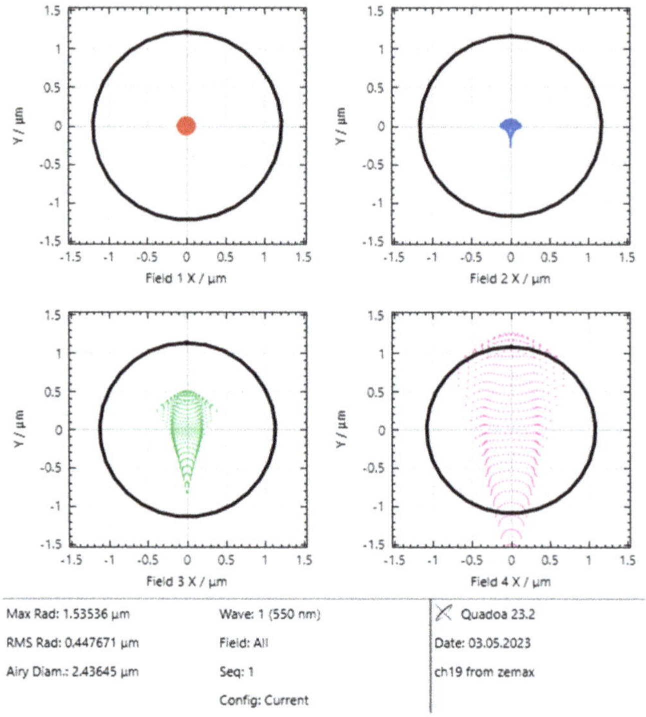


**Figure 19.3.** Spot diagram of the telescope design with specifications of illustrative example.

The results are as predicted by the theory, the telescope designed with equation (19.17) and equation (19.16) is diffracted limited for fields field of view 3°. In figure 19.3 we present the spot diagram for fields with the following half field of view: 0°, 0.5°, 1° and 1.5°. The first three fields are diffracted limited while the last one at 0.5° is nearly diffracted limited.

The method presented here only uses two equations that are not solved simultaneously. First equation (19.17) is solved and then equation (19.16) is fed with the data of the solution of equation (19.17). This procedure is a lower computational cost than methods that consider more independent variables.

## 19.4 Conclusions

In this chapter, a differential equation (19.17) has been presented. Its solutions gives information on the shape of the first mirror of a pair of mirrors that forms an aplanatic system. With the shape of the first mirror and equation (19.16) the shape of the second mirror can be computed. The deduction of getting equation (19.16) and equation (19.17) is entirely analytic. But, unfortunately, equation (19.17) has only a numeric solution so far. An illustrative example of the designs obtained from equation (19.16) and equation (19.17) has been presented and exported to Quadoa Optical CAD.

## Further reading


Quadoa Optical CAD 2022 *Quadoa Optical CAD Software Manual V* (Berlin: Quadoa Optical Systems) https://www. quadoa.com

Abbe E 1991 On the estimation of aperture in the microscope *J. R. Microsc. Soc.* **1** 388–423

Born M and Wolf E 2013 *Principles of Optics: Electromagnetic Theory of Propagation Interference and Diffraction of Light* (Amsterdam: Elsevier)

Braat J J M 1997 Abbe sine condition and related imaging conditions in geometrical optics *5th Int. Topical Meeting on Education and Training in Optics* ed C H F Velzel vol 3190 (Bellingham, WA: International Society for Optics and Photonics, SPIE) pp 59–64

Chaves J 2016 *Introduction to Nonimaging Optics* 2nd edn (Boca Raton, FL: CRC Press)

Courtial J, Oxburgh S and Tyc T 2015 Direct stigmatic imaging with curved surfaces *J. Opt. Soc. Am.* A **32** 478–81

Doskolovich L L, Bykov D A, Bezus E A and Greisukh G I 2021a Design of a stigmatic lens implementing a required ray mapping *Appl. Opt.* **60** 9138–45

Doskolovich L L, Bykov D A, Greisukh G I, Strelkov Y S and Bezus E A 2021b Designing stigmatic lenses with minimal Fresnel losses *J. Opt. Soc. Am.* A **38** 855–61

Duerr F, Benitez P, Miñano J C, Meuret Y and Thienpont H 2012 Analytic design method for optimal imaging: coupling three ray sets using two free-form lens profiles *Opt. Express* **20** 5576–85

Elazhary T T, Zhou P, Zhao C and Burge J H 2015 Generalized sine condition *Appl. Opt.* **54** 5037–49

Gimenez-Benitez P, Miñano J C, Blen J, Arroyo R, Chaves J, Dross O, Hernández M and Falicoff W 2004 Simultaneous multiple surface optical design method in three dimensions *Opt. Eng.* **43** 1489–503

González-Acuña R G 2021 Exact equations to design aplanatic sequential optical systems *Appl. Opt.* **60** 9263–8

González-Acuña R G and Chaparro-Romo H A 2018 General formula for bi-aspheric singlet lens design free of spherical aberration *Appl. Opt.* **57** 9341–5

González-Acuña R G and Chaparro-Romo H A 2020 *Stigmatic Optics* (Bristol: IOP Publishing)

González-Acuña R G and Guitiérrez-Vega J C 2018 Generalization of the axicon shape: the gaxicon *J. Opt. Soc. Am.* A **35** 1915–8

González-Acuña R G and Gutiérrez-Vega J C 2019a General formula of the refractive telescope design free spherical aberration *Novel Optical Systems, Methods, and Applications XXII* vol 11105 (Bellingham, WA: International Society for Optics and Photonics)

González-Acuña R G and Gutiérrez-Vega J C 2019b Analytic formulation of a refractive-reflective telescope free of spherical aberration *Opt. Eng.* **58** 085105

González-Acuña R G and Gutiérrez-Vega J C 2019d General formula for aspheric collimator lens design free of spherical aberration *Current Developments in Lens Design and Optical Engineering XX* vol 11104 (Bellingham, WA: International Society for Optics and Photonics)

González-Acuña R G and Gutiérrez-Vega J C 2019e General formula to eliminate spherical aberration produced by an arbitrary number of lenses *Opt. Eng.* **58** 085106

González-Acuña R G and Gutiérrez-Vega J C 2020 Analytic design of a spherochromatic singlet *J. Opt. Soc. Am.* A **37** 149–53

González-Acuña R G and Gutiérrez-Vega J C 2021 Exact equations to design a stigmatic singlet that meets the Herschel's condition *Opt. Commun.* **485** 126727

González-Acuña R G, Avendaño-Alejo M and Gutiérrez-Vega J C 2019a Singlet lens for generating aberration-free patterns on deformed surfaces *J. Opt. Soc. Am.* A **36** 925–9

González-Acuña R G, Chaparro-Romo H A and Gutiérrez-Vega J C 2019b General formula to design freeform singlet free of spherical aberration and astigmatism *Appl. Opt.* **58** 1010–5

González-Acuña R G, Chaparro-Romo H A and Gutiérrez-Vega J C 2020a Analytic solution of the eikonal for a stigmatic singlet lens *Phys. Scr.* **95** 085201

González-Acuña R G, Chaparro-Romo H A and Gutiérrez-Vega J C 2020b Exact equations for stigmatic singlet design meeting the Abbe sine condition *Opt. Commun.* **479** 126415

González-Acuña R G, Chaparro-Romo H A and Gutiérrez-Vega J C 2020c *Analytical Lens Design* (Bristol: IOP Publishing)

Gross H 2005 *Handbook of Optical Systems* (New York: Wiley Online Library)

Hazra L N, Han Y and Delisle C 1993 Curved kinoform lenses for stigmatic imaging of axial objects *Appl. Opt.* **32** 4775–84

Hopkins H H 1946 Herschel's condition *Proc. Phys. Soc.* **58** 100–5

Hsueh C-C, Elazhary T, Nakano M and Sasian J 2011 Closed-form sag solutions for Cartesian oval surfaces *J. Opt.* **40** 168–75

Kingslake R and Johnson R B 2009a *Lens Design Fundamentals* (New York: Academic)

Loveday N 1981 A folded Newtonian telescope with dual focal lengths *Sky Telesc.* **61** 545

Luneburg R K and Herzberger M 1964 *Mathematical Theory of Optics* (Berkeley, CA: University of California Press)

Malacara-Hernández D and Malacara-Hernández Z 2016 *Handbook of Optical Design* (Boca Raton, FL: CRC Press)

Mansuripur M 1998 Abbe's sine condition *Opt. Photonics News* **9** 56–60

Mashaal H, Feuermann D and Gordon J M 2016 Aplanatic lenses revisited: the full landscape *Appl. Opt.* **55** 2537–42

Masters B R 2007 Ernst Abbe and the foundation of scientific microscopes *Opt. Photonics News* **18** 18–23

Miñano J C, Mohedano R and Benítez P 2007 Nonimaging optics *The Optics Encyclopedia* (New York: Wiley) pp 1–34

Miñano J C, Benítez P, Lin W, Infante J, Muñoz F and Santamaría A 2009 An application of the SMS method for imaging designs *Opt. Express* **17** 24036–44

Rumsey N J 1969 Telescopic system utilizing three axially aligned substantially hyperbolic mirrors *US Patent* 3460886

Rutten H and van Venrooij M 1989 *Telescope Optics* (Richmond, VA: Willman-Bell)

Schroeder D J 1999 *Astronomical Optics* (Amsterdam: Elsevier)

Schwarzschild K 1905 Untersuchungen zur geometrischen Optik, II; Theorie der Spiegelteleskope *Abh. der Konigl Ges. der Wiss. Zu Gottingen, Math.-Phys. Klasse* **9**

Shibuya M 1992 Exact sine condition in the presence of spherical aberration *Appl. Opt.* **31** 2206–10

Silva-Lora A and Torres R 2020a Aplanatism in stigmatic optical systems *Opt. Lett.* **45** 6390–3

Silva-Lora A and Torres R 2020b Explicit Cartesian oval as a superconic surface for stigmatic imaging optical systems with real or virtual source or image *Proc. R. Soc.* A **476** 20190894

Silva-Lora A and Torres R 2020c Superconical aplanatic ovoid singlet lenses *J. Opt. Soc. Am.* A **37** 1155–65

Silva-Lora A and Torres R 2021 Rigorously aplanatic Descartes ovoids *J. Opt. Soc. Am.* A **38** 1160–9

Stavroudis O N 2006 *The Mathematics of Geometrical and Physical Optics: The k-Function and Its Ramifications* (New York: Wiley)

Steward G C 1927 On Herschel's condition and the optical cosine law *Math. Proc. Camb. Philos. Soc.* **23** 703–12

Valencia-Estrada J C, Flores-Hernández R B and Malacara-Hernández D 2015 Singlet lenses free of all orders of spherical aberration *Proc. R. Soc.* A **471** 20140608

Valencia-Estrada J C, Pereira-Ghirghi M V, Malacara-Hernández Z and Chaparro-Romo H A 2017 Aspheric coefficients of deformation for a Cartesian oval surface *J. Opt.* **46** 100–7

Velzel C H F 2014 *A Course in Lens Design* vol 183 (Berlin: Springer)

Volkmann H 1966 Ernst Abbe and his work *Appl. Opt.* **5** 1720–31

Wassermann G D and Wolf E 1949 On the theory of aplanatic aspheric systems *Proc. Phys. Soc. Sect.* B **62** 2

Wilson R N 1996 *Reflecting Telescope Optics II* (Berlin: Springer Science & Business Media)

**IOP** Publishing

Analytical Lens Design (Second Edition)

**Rafael G González-Acuña**

# Chapter 20

## The stigmatic three-freeform-mirror system

An innovative method is presented for designing three-freeform-mirror systems from scratch. In this technique, an initial set-up is obtained directly from the set of stigmatic three-freeform-mirror systems, before optimization. To reduce aberration caused by additional fields, deformation coefficients are added to each surface. Results of the test show that the method is accurate.

## 20.1 Introduction

In this chapter, we introduce a new method that in part follows the strategy presented in chapter 18. The method presented in chapter 18 is only for pairs of stigmatic mirrors, but if we add an extra mirror the mathematical complexity grows dramatically. In this chapter, we are going to deal with the stigmatic three-freeform-mirror system.

Therefore, we have to consider in this chapter just three mirrors and see how the mathematical calculations are done. Finally, the next chapter generalizes the system for an arbitrary number of mirrors. The idea is that this book can be useful to the reader and that he or she read it step by step, chapter by chapter, as the ideas grow and become more complex. If we just present the results of each chapter in a direct way, the purpose of the book will be lost. That is not just to show meteorology but to teach it.

## 20.2 Mathematical model

A representation of the three-freeform-mirror stigmatic system is shown in figure 20.1. Let $(x_a, y_a, z_a)$, $(x_b, y_b, z_b)$ and $(x_c, y_c, z_c)$, be the first, second and third mirrors, respectively. Let subscripts $a$, $b$, and $c$ be sight of correspondence related to the first, second and third mirrors, respectively. Let $(x_a = 0,\ y_a = 0,\ z_a(0, 0) = 0)$ be the origin of the coordinate system, which in fact is placed at the center of the first mirror. Let $x_a$, $y_a$ be the only independent variables, so the other parameters

doi:10.1088/978-0-7503-5774-6ch20

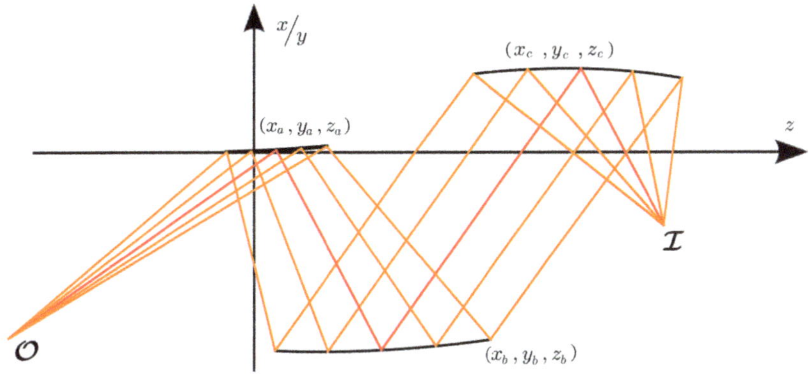


**Figure 20.1.** Diagram of the three-freeform-mirror stigmatic system. The first mirror is given by $(x_a, y_a, z_a)$, the second mirror is described by $(x_b, y_b, z_b)$ and the last mirror is $(x_c, y_c, z_c)$. The origin of the Cartesian coordinate system is placed at the center of the first mirror $(x_a = 0, y_a = 0, z_a(0, 0) = 0)$.

like $z_a$, $x_b$, $y_b$, $z_b$, $x_c$, $y_c$, and $z_c$ only depend on $x_a$, $y_a$, in other means $z_a(x_a, y_a)$, $x_b(thex_a, y_a)$, $y_b(x_a, y_a)$, $z_b(x_a, y_a)$, $x_c(x_a, y_a)$, $y_c(x_a, y_a)$, and $z_c(x_a, y_a)$. The goal is to know $x_c$, $y_c$, and $z_c$ in terms of the first and second surfaces which are given by the user. We are going to work with the case when the object is at minus infinity, but for the moment consider that the object is placed at $(x_o, y_o, z_o)$. The image is always considered finite and real.

The gap from the center of the first mirror $(x_a = 0, y_a = 0, z_a(0, 0) = 0)$ to the center of the second mirror $(x_b(0, 0), y_b(0, 0), z_b(0, 0))$ is $\sqrt{x_\kappa^2 + y_\kappa^2 + z_\kappa^2}$. The distance from the center of the second mirror $(x_b(0, 0), y_b(0, 0), z_b(0, y0))$ to the center of the third mirror black $(x_c(0, 0), y_c(0, 0), z_b(0, 0))$ is $\sqrt{x_\tau^2 + y_\tau^2 + z_\tau^2}$. The gap between the center of the last mirror and the image is $\sqrt{x_i^2 + y_i^2 + z_i^2}$.

### 20.2.1 Snell's law

In order to have the expression of the last mirror such that the system is stigmatic we need to apply the Snell law at the the surface of the first mirror and the surface of the second mirror. We recall Snell's law at the first mirror, equation (20.1),

$$\vec{\mathbf{v}}_1 = [\vec{\mathbf{v}}_0 - (\vec{\mathbf{n}}_a \cdot \vec{\mathbf{v}}_0)\vec{\mathbf{n}}_a] - \vec{\mathbf{n}}_a\sqrt{1 + (\vec{\mathbf{n}}_a \wedge \mathbf{v}_0)^2}, \tag{20.1}$$

where $\vec{\mathbf{v}}_0$ is the unit vector with the direction of the striking ray, $\vec{\mathbf{n}}_a$ is the normal vector of the first mirror and $\vec{\mathbf{v}}_1$ is the reflected ray,

$$\vec{\mathbf{v}}_0 = [0, 0, 1], \quad \vec{\mathbf{n}}_a = \frac{[z_{a_x}, z_{a_y}, -1]}{\sqrt{(z_{a_x})^2 + (z_{a_y})^2 + 1}}, \quad \vec{\mathbf{v}}_1 = [\varrho_x, \varrho_y, \varrho_z], \tag{20.2}$$

where $z_{a_x}$ and $z_{a_y}$ are the deviates of $z_a$ with respect to $x_a$ and $y_a$. $\varrho_x$, $\varrho_y$, $\varrho_z$ are the cosine directors of the reflected ray at the first surface. We replace equation (20.2) in equation (20.1). Now we can replace the unit vector of Snell's law at the first mirror and do all the mathematics, or we just can recall (9.36), (9.37), (9.38), and apply some limit over them. We are going to do the second option. Notice that both options give the same result. Recalling (9.36), that we call with our notation, $\varrho_x$,

$$\varrho_x = \frac{z_{a_x}\left(\sqrt{(n^2 - 1)(z_{a_x}^2 + z_{a_y}^2) + n^2} - 1\right)}{n\left(z_{a_x}^2 + z_{a_y}^2 + 1\right)}, \tag{9.36}$$

taking negative square root and setting $n = -1$, of the last expression,

$$\varrho_x = \frac{2z_{a_x}}{(z_{a_x})^2 + (z_{a_y})^2 + 1}. \tag{20.3}$$

Equation (9.37) with the notation of $\varrho_y$ looks like,

$$\varrho_y = \frac{z_{a_y}\left(\sqrt{(n^2 - 1)(z_{a_x}^2 + z_{a_y}^2) + n^2} - 1\right)}{n\left(z_{a_x}^2 + z_{a_y}^2 + 1\right)}, \tag{9.37}$$

computing the square root and let $n = -1$ be,

$$\varrho_y = \frac{2z_{a_y}}{(z_{a_x})^2 + (z_{a_y})^2 + 1}, \tag{20.4}$$

Equation (9.38) for us is given written as,

$$\varrho_z = \frac{z_{a_x}^2 + z_{a_y}^2 + \sqrt{(n^2 - 1)(z_{a_x}^2 + z_{a_y}^2) + n^2}}{n\left(z_{a_x}^2 + z_{a_y}^2 + 1\right)}. \tag{9.38}$$

Finally, repeating the process of the negative square root and $n = -1$, we have,

$$\varrho_z = 1 - \frac{2}{(z_{a_x})^2 + (z_{a_y})^2 + 1}. \tag{20.5}$$

The next step is to compute Snell's law at the second mirror with equation (20.6),

$$\vec{\mathbf{v}}_2 = [\vec{\mathbf{v}}_1 - (\vec{\mathbf{n}}_b \cdot \vec{\mathbf{v}}_1)\vec{\mathbf{n}}_b] - \vec{\mathbf{n}}_b\sqrt{1 - (\vec{\mathbf{n}}_b \times \vec{\mathbf{v}}_1)^2}, \tag{20.6}$$

where,

$$\vec{\mathbf{v}}_1 = [\varrho_x, \varrho_y, \varrho_z], \quad \vec{\mathbf{v}}_2 = [\wp_x, \wp_y, \wp_z], \quad \vec{\mathbf{n}}_b = [\Phi_x, \Phi_y, \Phi_z], \tag{20.7}$$

where the components of the normal vector $\vec{\mathbf{n}}_b$ of the second surface are given by,

$$\Phi_x = \frac{\partial_x y_b \partial_y z_b - \partial_x z_b \partial_y y_b}{\sqrt{(\partial_x x_b \partial_y y_b - \partial_x y_b \partial_y x_b)^2 + (\partial_x z_b \partial_y x_b - \partial_x x_b \partial_y z_b)^2 + (\partial_x y_b \partial_y z_b - \partial_x z_b \partial_y y_b)^2}}, \tag{20.8}$$

$$\Phi_y = \frac{\partial_x z_b \partial_y x_b - \partial_x x_b \partial_y z_b}{\sqrt{(\partial_x x_b \partial_y y_b - \partial_x y_b \partial_y x_b)^2 + (\partial_x z_b \partial_y x_b - \partial_x x_b \partial_y z_b)^2 + (\partial_x y_b \partial_y z_b - \partial_x z_b \partial_y y_b)^2}}, \quad (20.9)$$

$$\Phi_z = \frac{\partial_x x_b \partial_y y_b - \partial_x y_b \partial_y x_b}{\sqrt{(\partial_x x_b \partial_y y_b - \partial_x y_b \partial_y x_b)^2 + (\partial_x z_b \partial_y x_b - \partial_x x_b \partial_y z_b)^2 + (\partial_x y_b \partial_y z_b - \partial_x z_b \partial_y y_b)^2}}. \quad (20.10)$$

Note that the components of the normal vector $\vec{\mathbf{n}}_b$ are more complicated than the components of the normal vector $\vec{\mathbf{n}}_a$ of the first mirror since the first mirror is in explicit form and the second mirror is in a parametric form. So remember that if a surface is expressed in a parameter form its normal vector is given by the following expression,

Well, let's focus once more on our task. Replacing equation (20.7) in equation (20.6),

$$\begin{aligned} \wp_x = {} & \left[\varrho_x\left(\Phi_y^2 + \Phi_z^2\right) - \varrho_y \Phi_x \Phi_y - \varrho_z \Phi_x \Phi_z\right] \\ & + \Phi_x \sqrt{1 - \left[(\varrho_y \Phi_x - \varrho_x \Phi_y)^2 + (\varrho_z \Phi_x - \varrho_x \Phi_z)^2 + (\varrho_z \Phi_y - \varrho_y \Phi_z)^2\right]}, \end{aligned} \quad (20.11)$$

$$\begin{aligned} \wp_y = {} & \left[\quad \varrho_y(\Phi_x^2 + \Phi_z^2) + \Phi_y(\varrho_x \Phi_x + \varrho_z \Phi_z)\right] \\ & + \Phi_y \sqrt{1 - \left[(\varrho_y \Phi_x - \varrho_x \Phi_y)^2 + (\varrho_z \Phi_x - \varrho_x \Phi_z)^2 + (\varrho_z \Phi_y - \varrho_y \Phi_z)^2\right]}, \end{aligned} \quad (20.12)$$

$$\begin{aligned} \wp_z = {} & \left[-\varrho_z\left(\Phi_x^2 + \Phi_y^2\right) + \Phi_z(\varrho_x \Phi_x + \varrho_y \Phi_y)\right] \\ & + \Phi_z \sqrt{1 - \left[(\varrho_y \Phi_x - \varrho_x \Phi_y)^2 + (\varrho_z \Phi_x - \varrho_x \Phi_z)^2 + (\varrho_z \Phi_y - \varrho_y \Phi_z)^2\right]}. \end{aligned} \quad (20.13)$$

We now have the expression for the cosine directors reflected by the first and second surfaces. With them in the next section and the Fermat principle we are going to obtain the expression desired for the third surface.

### 20.2.2 Optical path and solution

Let's do this by a notation that the reference ray is the ray that passes through all the centers of the surface, from the object point to the image point. If the system is stigmatic, the optical path length for all rays should be constant,

$$\begin{aligned} & \sqrt{x_o^2 + y_o^2 + z_o^2} + \sqrt{x_\kappa^2 + y_\kappa^2 + z_\kappa^2} + \sqrt{x_\tau^2 + y_\tau^2 + z_\tau^2} + \sqrt{x_i^2 + y_i^2 + z_i^2} \\ & = \sqrt{(x_b - x_a)^2 + (y_b - y_a)^2 + (z_b - z_a)^2} + \sqrt{(x_a - x_o)^2 + (y_a - y_o)^2 + (z_a - z_o)^2} \\ & \qquad + \sqrt{(x_c - x_b)^2 + (y_c - y_b)^2 + (z_c - z_b)^2} \\ & \qquad + \sqrt{(x_c - x_i - x_\kappa - x_\tau)^2 + (y_c - y_i - y_\kappa - y_\tau)^2 + (z_c - z_i - z_\tau)^2}. \end{aligned} \quad (20.14)$$

The unknowns of equation (20.14) are $x_c$, $y_c$, $z_c$, to solve of them we need to replace the first and second surface in terms of the independent variables $x_a$, $y_a$. We have done that already expressing the cosine directors of the reflected rays at the second and first surfaces. So, for the second surface, we have,

$$x_b = x_a + \phi \varrho_x, \quad y_b = y_a + \phi \varrho_y, \quad z_b = z_a + \phi \varrho_z, \tag{20.15}$$

where, $\varrho_x$, $\varrho_y$, and $\varrho_z$ are the cosine directors of the ray reflected by the first mirror and $\phi$ is the distance from the first mirror to the second,

$$\phi = \sqrt{(x_b - x_a)^2 + (y_b - y_a)^2 + (z_b - z_a)^2}. \tag{20.16}$$

Then, the third surface can be also expressed in terms of the cosine directors of the ray reflected by the second mirror,

$$x_c = x_b + \vartheta \wp_x, \quad y_c = y_b + \vartheta \wp_y, \quad z_c = z_b + \vartheta \wp_z. \tag{20.17}$$

where, $\wp_x$, $\wp_y$ and $\wp_z$ are the cosine directors of the ray reflected by the second surface. $\vartheta$ is the distance from the second surface to the last one,

$$\vartheta = \sqrt{(x_c - x_b)^2 + (y_c - y_b)^2 + (z_c - z_b)^2}. \tag{20.18}$$

The idea is to express $\vartheta$ in terms of the known parameters. So, If equations (20.15), (20.16), (20.17), (20.18) are replaced in equations (20.14) the limit when $z_o \to -\infty$ is computed over it, the comparison of the optical path lengths is given by

$$f = \sqrt{(\zeta + \vartheta \wp_z)^2 + (\chi + \vartheta \wp_x)^2 + \left(\upsilon + \vartheta \wp_y\right)^2} + \vartheta, \tag{20.19}$$

where $f$ is given by,

$$\begin{aligned} f \equiv & -\sqrt{(x_a - x_b)^2 + (y_a - y_b)^2 + (z_a - z_b)^2} \\ & - z_a + \sqrt{x_i^2 + y_i^2 + z_i^2} + \sqrt{x_\kappa^2 + y_\kappa^2 + z_\kappa^2} + \sqrt{x_\tau^2 + y_\tau^2 + z_\tau^2}, \end{aligned} \tag{20.20}$$

and, $\chi$, $\upsilon$, $\zeta$ are defined as,

$$\chi \equiv x_b - x_i - x_\kappa - x_\tau, \quad \upsilon \equiv y_b - y_i - y_\kappa - y_\tau, \quad \zeta \equiv z_b - z_i - z_\kappa - z_\tau. \tag{20.21}$$

Thus, we go back to equation (20.19),

$$(f - \vartheta)^2 = (\zeta + \vartheta \wp_z)^2 + (\chi + \vartheta \wp_x)^2 + \left(\upsilon + \vartheta \wp_y\right)^2, \tag{20.22}$$

then, expanding

$$\begin{aligned} f^2 - 2f\vartheta + \vartheta^2 = & \wp_z^2 \vartheta^2 + 2\wp_z \vartheta \zeta + \zeta^2 + \wp_x^2 \vartheta^2 \\ & + 2\wp_x \vartheta \chi + \chi^2 + \wp_y^2 \vartheta^2 + 2\wp_y \vartheta \upsilon + \upsilon^2, \end{aligned} \tag{20.23}$$

simplifying,

$$f^2 - 2f\vartheta = 2\wp_z \vartheta \zeta + \zeta^2 + 2\wp_x \vartheta \chi + \chi^2 + 2\wp_y \vartheta \upsilon + \upsilon^2. \tag{20.24}$$

Solving for $\vartheta$,

$$\vartheta = \frac{-\zeta^2 - \upsilon^2 - \chi^2 + f^2}{2\left(\chi \wp_x + \upsilon \wp_y + \zeta \wp_z + f\right)}. \tag{20.25}$$

Replacing equation (20.25) in equation (20.17), we have,

$$\begin{bmatrix} x_c \\ y_c \\ z_c \end{bmatrix} = \begin{bmatrix} x_b \\ y_b \\ z_b \end{bmatrix} + \frac{-\zeta^2 - \upsilon^2 - \chi^2 + f^2}{2\left(\chi \wp_x + \upsilon \wp_y + \zeta \wp_z + f\right)} \begin{bmatrix} \wp_x \\ \wp_y \\ \wp_z \end{bmatrix}. \tag{20.26}$$

Equation (20.26) is the most important expression in this chapter. It rigorously expresses the form of the last mirror such that the system is stigmatic. It is important to remark that equation (20.26) is also expressed in terms of the cosine directors of the reflected rays at the first and second surfaces. Observe that in the proof of equation (20.26) there is only a single solution, (20.26). From this we can interpret that there is a uniqueness theorem related to equation (20.26).

The uniqueness theorem should not be surprising since all the stigmatic systems analytically obtained had their respective uniqueness theorem, so far it is as we have shown in this book. Since the first and second surfaces are given, the set of mirrors computed by equation (20.26) contains all the possible stigmatic three-freeform-mirror systems.

## 20.3 Illustrative example

The illustrative example presented in this section is a three-mirrors stigmatic system with $y_\kappa = y_\tau = 0$. The system that directly computes with equation (20.26) is presented in figure 20.2. The first gray circle is the stop of the system and it has a diameter of 7 mm. Note that the first and the second surfaces are not confocal. In this illustrative design, the first two surfaces are spheres, their radii are −87.88 mm, 285.16 mm, respectively. The last surface is highlighted in pink, it is computed with equation (20.26). The analysis of this design is done in Quadoa optical CAD. The last surface is uploaded to Quadoa optical CAD as a point cloud. The ray tracing of this design is presented in figure 20.2 and the spot diagram is shown in figure 20.3. As expected, the image is stigmatic. The disk that surrounds the spot is the Airy disk.

Now the design presented in figure 20.2, can be seen as a good starting point for optimization of a more complex design that considers off-axis objects. The off-axis objects have the field of view of 8.6° × 6.5°. For this example that we are going to optimize, the image screen size is considered to be 3 × 4 mm. Next, the starting design of figure 20.2 is optimized to reduce the aberrations of the off-axis fields. At this point, the surfaces of the initial design are considered to be bi-dimensional polynomials. The details of the optimization are out of the scope of this treatise, when the optimization starts, the analytical lens design ends. But the optimization done for this example follows the standard paradigm in lens design.

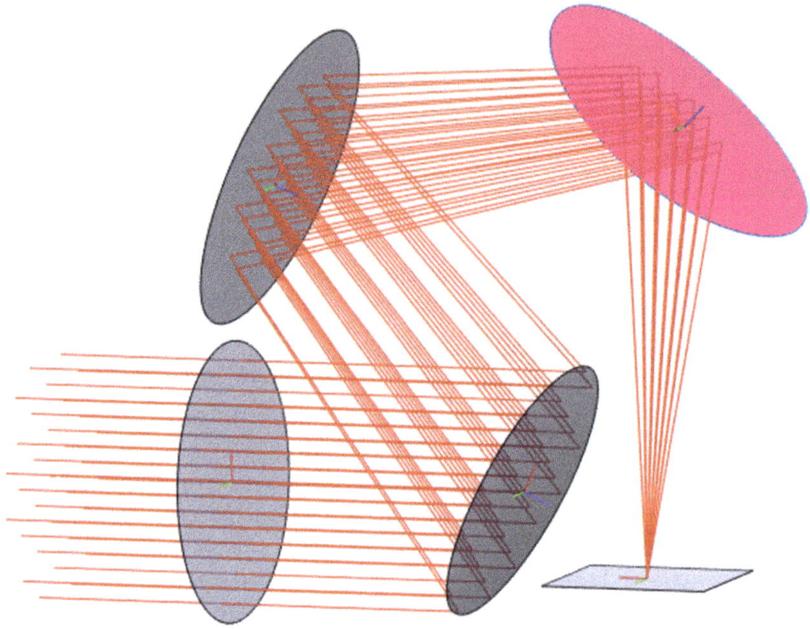

**Figure 20.2.** Ray tracing of the illustrative example before optimization.

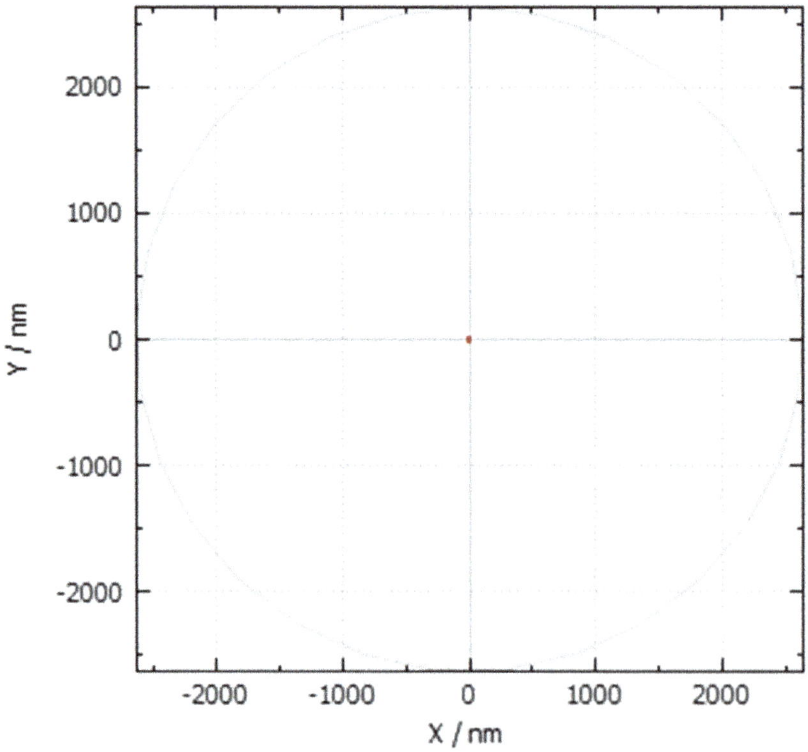


**Figure 20.3.** Spot diagram of the illustrative example before optimization.

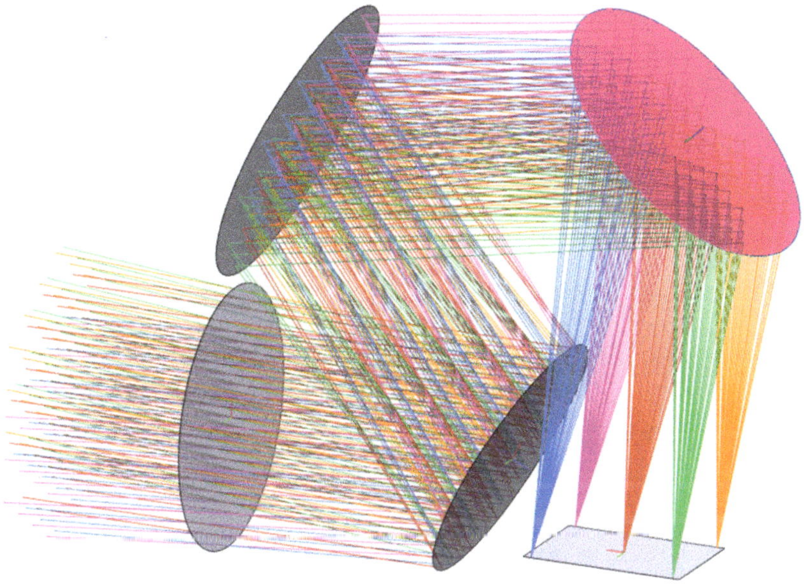

**Figure 20.4.** Ray tracing of the illustrative example after optimization.

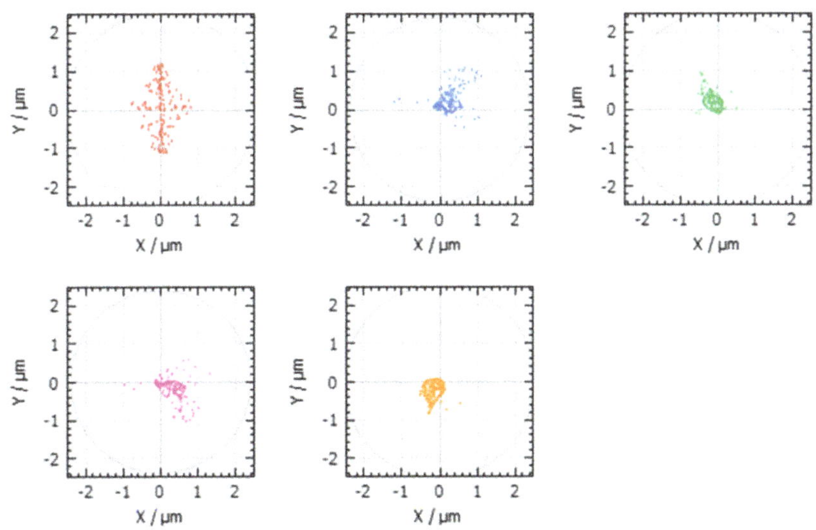


**Figure 20.5.** Spot diagram of illustrative example after optimization, *left* ray tracing, *right* spot diagram.

The ray tracing of the design after the optimization is presented in figure 20.4. The spot diagram of the mentioned design is shown in figure 20.5. The spot diagram confirms that the system is nearly diffracted limited, despite the central image not being stigmatic. The central image is not stigmatic anymore because the

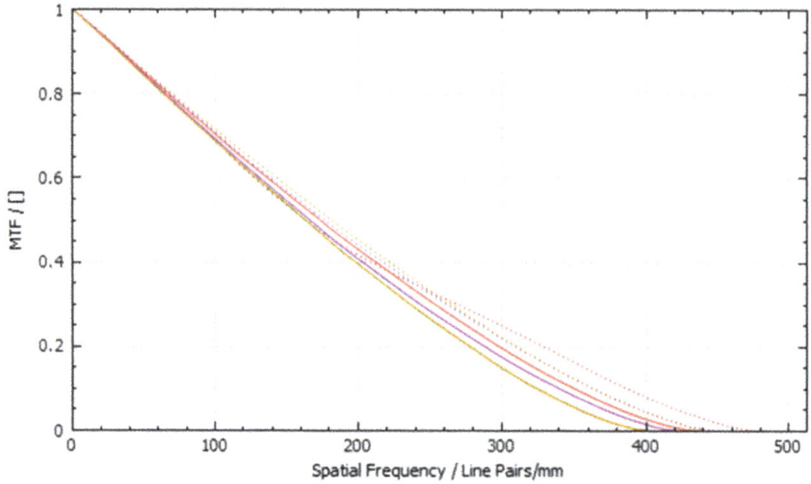


**Figure 20.6.** Modulus transfer function and the illustrative example after optimization, *left* ray tracing, *right* spot diagram.

optimization process balanced the off-axis aberrations. In figure 20.6 is the modulation transfer function (MTF). The MTF is the mathematical description of the comparison between the detail contrast at the edges of an object and the detail contrast of its pictorial representation. It is a useful tool to measure the quality of an optical system and from figure 20.6 it is clear that the system is near to being diffracted limited.

## 20.4 Conclusion

We present a novel method for designing three-freeform-mirror systems from scratch in this chapter. By optimizing an element from the set of stigmatic three-freeform-mirror systems, the method is based on the set of all possible stigmatic systems. In addition to the initial set-up, other parameters such as the shapes and positions of the first and second mirrors, and the positions of the object and image are also calculated by equation. In order to reduce aberration caused by additional fields, deformation coefficients are added to each surface after the initial set-up is properly selected. Mirror, decenter, and tilt distances are not modified during optimization.

We tested the method on a system in figures 20.2 and 20.3, and the diffracted limited system in both cases produced the same result, despite the large field of view and the constraints on the dimensions. The results were as expected by the theory because equation (20.26) represents the set of all stigmatic three-freeform-mirror systems, therefore it inherits all the mathematical properties of the mentioned systems.

## References

Alvarado-Martínez J J, Granados-Agustín F S, Montiel S V Y, Vázquez-Villa A and Cornejo-Rodríguez A A 2022 Design of a compact off-axis four-mirror objective system for a thermal camera *Appl. Opt.* **61** A43–9

Bauer A and Rolland J P 2021 Roadmap for the unobscured three-mirror freeform design space *Opt. Express* **29** 26736–44

Bauer A, Schiesser E M and Rolland J P 2018 Starting geometry creation and design method for freeform optics *Nat. Commun.* **9** 1–11

Caron J and Bäumer S 2021 Progress in freeform mirror design for space applications *Int. Conf. on Sace Optics-ICSO 2020* vol 11852 (Bellingham, WA: International Society for Optics and Photonics), 118521S

Duerr F and Thienpont H 2021 Freeform imaging systems: Fermat's principle unlocks 'first time right' design *Light Sci. Appl.* **10** 1–12

González-Acuña R G 2021 Two-mirror system for tunable apodization *Appl. Opt.* **60** 10756–60

González-Acuña R G 2022 Design of a pair of aplanatic mirrors *Appl. Opt.* **61** 1982–6

González-Acuña R G 2022 Set of all possible stigmatic pairs of mirrors *Appl. Opt.* **61** 2513–7

González-Acuna R G 2023 Power set of mirror-based non-symmetric stigmatic optical systems *Appl. Opt.* **62** 536–40

González-Acuña R G and Chaparro-Romo H A 2020 *Stigmatic Optics* (Bristol: IOP Publishing)

González-Acuna R G, Chaparro-Romo H A and Gutiérrez-Vega J C 2020 *Analytical Lens Design* (Bristol: IOP Publishing)

Narasimhan B, Benitez P, Miñano J C, Chaves J, Grabovickic D, Nikolic M and Infante J 2015 Design of three freeform mirror aplanat *Novel Optical Systems Design and Optimization XVIII* vol 9579 (Bellingham, WA: SPIE) pp 125–35

Nie Y, Shafer D R, Ottevaere H, Thienpont H and Duerr F 2021 Automated freeform imaging system design with generalized ray tracing and simultaneous multi-surface analytic calculation *Opt. Express* **29** 17227–45

Optic Studio 2019 *Optic Studio User Manual 19.4* (Bellevue, WA: Zemax LLC)

Papa J C, Howard J M and Rolland J P 2018 Three-mirror freeform imagers *Optical Design and Engineering VII* vol 10690 (Bellingham, WA: International Society for Optics and Photonics) p 106901D

Papa J C, Howard J M and Rolland J P 2019 Automatic solution space exploration for freeform optical design *Freeform Optics* (Washington, DC: Optical Society of America) p FM4B–1

Quadoa Optical CAD 2022 *Quadoa Optical CAD Software Manual V* (Berlin: Quadoa Optical Systems) https://www. quadoa.com

Reshidko D and Sasian J 2018 Method for the design of nonaxially symmetric optical systems using free-form surfaces *Opt. Eng.* **57** 101704

Rolland J P, Davies M A, Suleski T J, Evans C, Bauer A, Lambropoulos J C and Falaggis K 2021 Freeform optics for imaging *Optica* **8** 161–76

Sasián J 2019 Method of confocal mirror design *Opt. Eng.* **58** 015101

Yang T, Zhu J, Wu X and Jin G 2015 Direct design of freeform surfaces and freeform imaging systems with a point-by-point three-dimensional construction-iteration method *Opt. Exp.* **23** 10233–46

Zhang B, Jin G and Zhu J 2021 Towards automatic freeform optics design: coarse and fine search of the three-mirror solution space *Light Sci. Appl.* **10** 1–11

**IOP** Publishing

Analytical Lens Design (Second Edition)

**Rafael G González-Acuña**

# Chapter 21

# The power set of mirror-based stigmatic optical systems

This chapter shows the set of all possible stigmatic systems made by mirrors. The construction of this set is entirely analytical. The proof is directly obtained from the Fermat principle. The importance of this set is that its properties are shared by all the possible stigmatic systems made by mirrors. With a practical example of optical design, the mentioned set is tested and the results are as expected.

## 21.1 Introduction

This is the last chapter of this treatise and in it we are going to focus on the most general expression we can get from a stigmatic optical system made of mirrors. So, in other words, we are going to generalize the results presented in the last three chapters as much as possible.

The generalization will consist of an arbitrary number of reflective surfaces, located anywhere in the space and in any orientation. These surfaces are free parameters given by the user and the system has a last surface that makes the system stigmatic for a given point object and a point image. This system is so general that the equations that describe it can be seen as the equations that describe the power set of mirror-based stigmatic optical systems. The power set of mirror-based stigmatic optical systems is defined as the set that contains all the possible sets of mirrors that form a stigmatic system, the introduced definition is inspired by the power set definition of general topology.

## 21.2 Mathematical model

Figure 21.1 shows the system power set of mirror-based stigmatic optical systems. The system consists of an initial number of $N$ mirrors of any shape, orientation, and location. The final mirror forms a stigmatic image for a single field given all the previous $N$ mirrors. Therefore, $N \in \mathbb{N}$ is the total number of mirrors, and $N + 1$ is

 

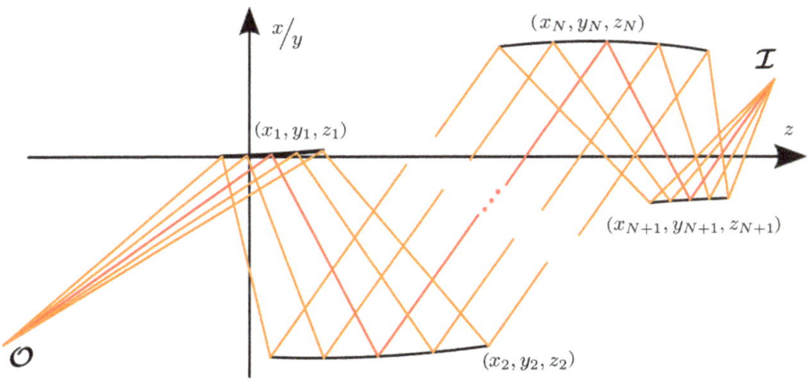


**Figure 21.1.** Diagram of a stigmatic system composed by an arbitrary number of mirrors. The coordinate system is located at the vertex of the first mirror, $(x_1, y_1, z_1)$. The rest of the mirrors have also arbitrary locations and orientations. The last mirror is such that system is stigmatic.

the total number of mirrors. First, the designer gives the first mirrors, while the last one is unknown. Let $(x_1, y_1, z_1)$ be the first mirror, where $z_1$ is a function of $x_1$ and $y_1$, so $z_1(x_1, y_1)$. Using independent variables $x_1$ and $y_1$, the rest of the mirrors can be expressed as parametric functions. Thus the $i$th mirror is given by $(x_i(x_1, y_1), y_i(x_1, y_1), z_i(x_1, y_1))$ and it's a function of $x_1$ and $y_1$. The coordinate system is located at the vertex of the first mirror ($x_1 = 0$, $y_1 = 0$, $z_1 = 0$).

Let $\vartheta_i$ be the distance that a ray needs to travel between consecutive mirrors,

$$\vartheta_i \equiv \sqrt{(x_{i+1} - x_i)^2 + (y_{i+1} - y_i)^2 + (z_{i+1} - z_i)^2}. \tag{21.1}$$

Without losing generality, each surface can be expressed as,

$$x_{i+1} = x_i + \vartheta_i \wp_{x,i}, \quad y_{i+1} = y_i + \vartheta_i \wp_{y,i}, \quad z_{i+1} = z_i + \vartheta_i \wp_{z,i}, \tag{21.2}$$

where $\wp_{x,i}$, $\wp_{y,i}$, and $\wp_{z,i}$ are the cosine directors of the reflected ray at surface $(x_i, y_i, z_i)$ in the $x$, $y$ and $z$, directions, respectively. So, $\wp_{x,\,i}^2 + \wp_{y,\,i}^2 + \wp_{z,\,i}^2 = 1$. The cosine directors can be obtained for a surface using the Snell law that we used in chapter 20. This procedure is left as an exercise to the reader. Here we are going to focus on the optical path and the Fermat principle.

An expression for the last mirror can be obtained by applying the Fermat principle. The Fermat principle predicts that a stigmatic system has a constant optical path for all rays. Thus, we compare the optical path of the reference ray and any other ray,

$$\begin{aligned} -t_0 + \sum_{i=1}^{N+1} t_i = \sqrt{x_1^2 + y_1^2 + (z_1 - t_0)^2} \\ + \sum_{i=0}^{N-1} \sqrt{(x_{i+2} - x_{i+1})^2 + (y_{i+2} - y_{i+1})^2 + (z_{i+2} - z_{i+1})^2} \\ + \sqrt{\left(x_{N+1} - \sum_{i=1}^{N+1} t_{x,i}\right)^2 + \left(y_{N+1} - \sum_{i=1}^{N+1} t_{y,i}\right)^2 + \left(z_{N+1} - \sum_{i=1}^{N+1} t_{z,i}\right)^2}, \end{aligned} \tag{21.3}$$

where $t_0$ is the distance from the object to the first mirror and $t_i$ is the distance that the reference ray needs to travel between consecutive mirrors,

$$t_i \equiv \sqrt{(t_{x,i+1} - t_{x,i})^2 + (t_{y,i+1} - t_{y,i})^2 + (t_{z,i+1} - t_{z,i})^2}\,, \tag{21.4}$$

where $t_{\omega,i}$ is the $\omega$ coordinate where the reference ray touches the $i$th mirror, $\omega \in \{x, y, z\}$. Thus, the left side equation (21.3) is the optical path length of the reference ray and the right side is the optical path length of any other ray. If $\vartheta_N$ is known, the last surface can be expressed in the form of equation (21.2).

Replacing equations (21.1) and (21.2) in equation (21.3) and solving for $\vartheta_N$,

$$\vartheta_N = \frac{-\left(x_N - \sum_{i=1}^{N+1} t_{x,\,i}\right)^2 - \left(y_N - \sum_{i=1}^{N+1} t_{y,\,i}\right)^2 - \left(z_N - \sum_{i=1}^{N+1} t_{z,\,i}\right)^2 + f^2}{2\left[\left(x_N - \sum_{i=1}^{N+1} t_{x,\,i}\right)\wp_{x,\,N} + \left(x_N - \sum_{i=1}^{N+1} t_{x,\,i}\right)\wp_{y,\,N} + \left(x_N - \sum_{i=1}^{N+1} t_{x,\,i}\right)\wp_{z,\,N} + f\right]}, \tag{21.5}$$

where $f$ for a finite object is given by,

$$\begin{aligned} f \equiv -t_0 + \sum_{i=1}^{N+1} t_i - \sqrt{x_1^2 + y_1^2 + (z_1 - t_0)^2} \\ - \sum_{i=0}^{N-2} \sqrt{(x_{i+2} - x_{i+1})^2 + (y_{i+2} - y_{i+1})^2 + (z_{i+2} - z_{i+1})^2}\,. \end{aligned} \tag{21.6}$$

If the object is not finite, $t_0 \to -\infty$, then $f$ becomes,

$$f \equiv -z_0 + \sum_{i=1}^{N+1} t_i - \sum_{i=0}^{N-2} \sqrt{(x_{i+2} - x_{i+1})^2 + (y_{i+2} - y_{i+1})^2 + (z_{i+2} - z_{i+1})^2}\,. \tag{21.7}$$

Finally, replacing equation (21.1) in equation (21.2), with $i = N$, the shape of the last surface is obtained,

$$\begin{aligned} \begin{bmatrix} x_{N+1} \\ y_{N+1} \\ z_{N+1} \end{bmatrix} = \begin{bmatrix} x_N \\ y_N \\ z_N \end{bmatrix} \\ + \frac{-\left(x_N - \sum_{i=1}^{N+1} t_{x,i}\right)^2 - \left(y_N - \sum_{i=1}^{N+1} t_{y,i}\right)^2 - \left(z_N - \sum_{i=1}^{N+1} t_{z,i}\right)^2 + f^2}{2\left[\left(x_N - \sum_{i=1}^{N+1} t_{x,i}\right)\wp_{x,N} + \left(y_N - \sum_{i=1}^{N+1} t_{y,i}\right)\wp_{y,N} + \left(z_N - \sum_{i=1}^{N+1} t_{z,i}\right)\wp_{z,N} + f\right]} \begin{bmatrix} \wp_{x,N} \\ \wp_{y,N} \\ \wp_{z,N} \end{bmatrix}. \end{aligned} \tag{21.8}$$

Equation (21.8) is the most important equation in this chapter. It describes how the last surface of a mirror-base system should be in order that the system becomes stigmatic. The mirror-base system can be composed of an arbitrary number of mirrors, with different positions and orientations.

Observe that the set of systems designed with equation (21.8) is indeed the power set of mirror-based stigmatic optical systems. This is because all the possible configurations of mirrors work as a single team to be stigmatic. In consequence, the shape or orientation of the previous mirrors does not matter, the set of systems designed with equation (21.8) contains all the possible mirror-based stigmatic optical systems even if they are non-symmetric. The last statement is true because the derivation of equation (21.8) reveals that there is a uniqueness theorem related to equation (21.8). Since there is only one solution to the problem which is equation (21.8). This should not surprise us since we have seen in most of this treatise that there is a unique theorem related to stigmatic optical systems.

The study of equation (21.8) ends only in rigorous but abstract results. In fact equation, (21.8) can be used as a potential candidate to design mirror optical systems from scratch. In the next section, we show an illustrative example of how equation (21.8) can be used as a starting point.

## 21.3 Illustrative example

An example of a mirror-base system design has been made with equation (21.8). Just like the last chapter, the ray-tracing analysis is done with Quadoa optical CAD. The results obtained with equation (21.8) are exported to Quadoa optical CAD directly as a point cloud and then converted as $xy$ polynomial with an internal tool of the software.

In figure 21.2 can be seen the ray tracing of the illustrative non-symmetric example. This particular system is composed of four mirrors, $N = 3$. The last mirror $N + 1$ is obtained directly with equation (21.8). For the example, the wavelength chosen is 550 nm. Equation (21.8) is such that despite the system being non-symmetric, the system is stigmatic. The result is confirmed in the spot diagram of figure 21.3 and it is as expected.

Now it is important to remark that equation (21.8) is not just useful to get stigmatic systems for a single field. It can be used as a good starting point for a system with more off-axis objects. The idea is first to get the initial candidate directly with equation (21.8). Then the optimization process can be implemented. The optimization process follows the standard paradigm in lens design and it really depends on the constraint of the software used to implement it. The details of the optimization process are out of the scope of this book. When the optimization begins the analytical lens design ends.

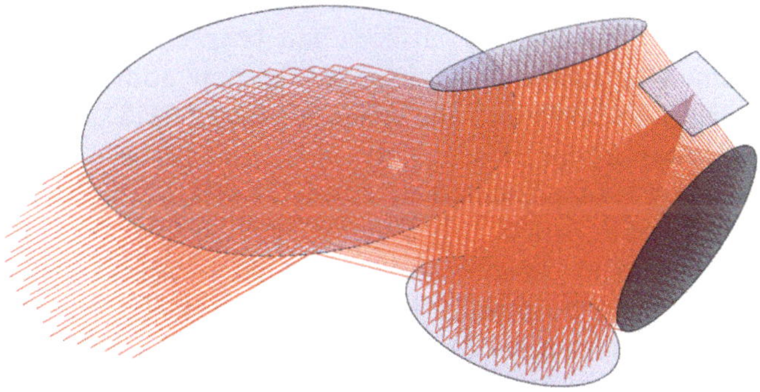

**Figure 21.2.** Ray tracing illustrative example before optimization.

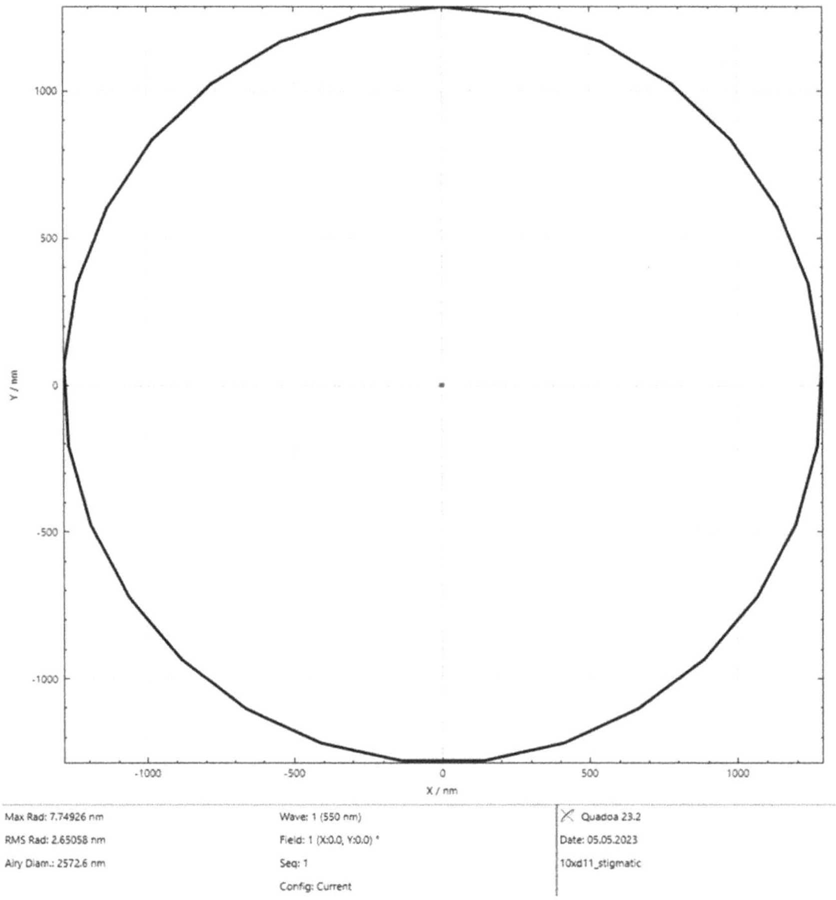


**Figure 21.3.** Spot diagram of the illustrative example before optimization.

In the example presented the constraints are a field of view of $8.6^\circ \times 6.5^\circ$ and an image plane of $3 \times 4$ mm. After the optimization of the design presented in figure 21.2 the results can be seen in figure 21.4. When the optimization of the analytical lens design ends. All the surfaces had been changed during the optimization. The system is diffracted limited for all the new additional fields. The Airy disk is the gray disk that surrounds the spots, see figure 21.5. The modulus transfer function is presented in figure 21.6.

Even though the design presented in figure 21.4 is obtained particularly with analytical tools and optimization, it has remarkable performance. It may be due to the fact that the optimization process starts with a good candidate given by equation (21.8). Note that during the optimization the decanters and the tilts had not been modified. That gives equation (21.8) potential use as a starting point generator for mirror-base optical systems.

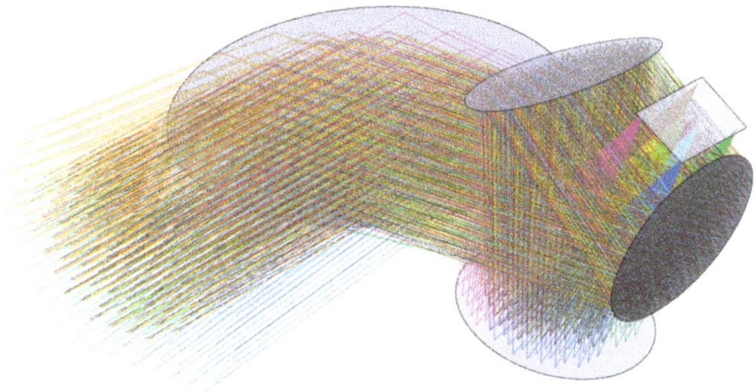

**Figure 21.4.** Ray tracing of the illustrative example after optimization.

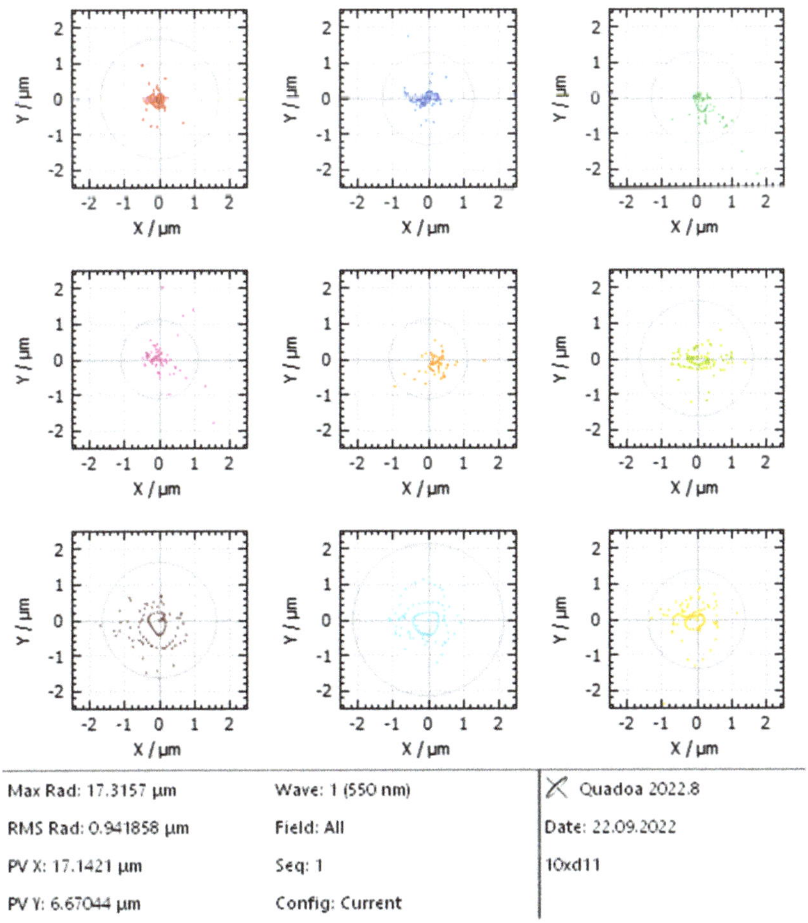


**Figure 21.5.** Spot diagrams of the illustrative example after optimization.

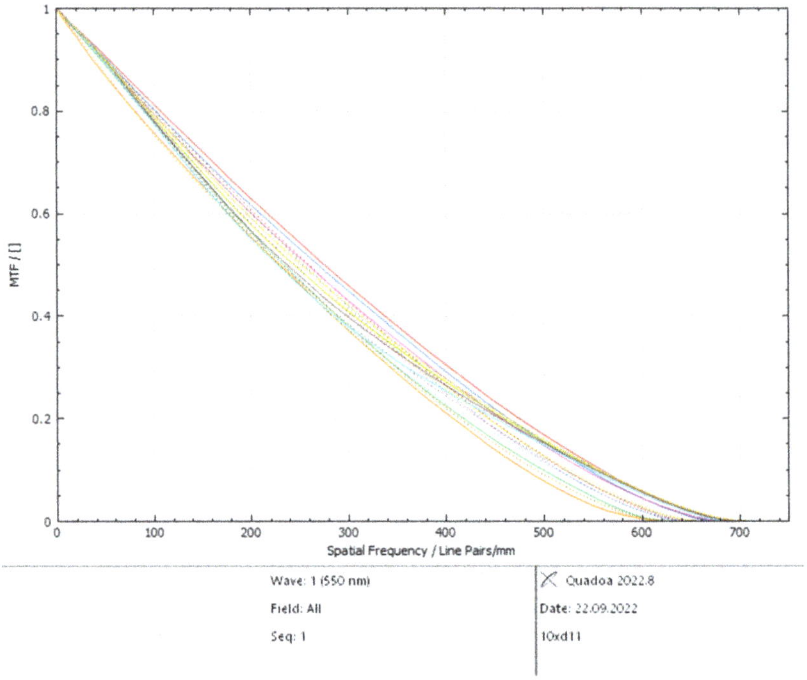


**Figure 21.6.** Modulus transfer function of the illustrative example after optimization.

Finally, it is important to remark that equation (21.8) is very versatile. It is not limited to plane-symmetric systems or confocal systems since it expresses the power set of mirror-base stigmatic optical systems.

## 21.4 Conclusions

In this chapter, the power set of the stigmatic mirror-based optical system is presented. This set can be described by equation (21.8).

The deduction of equation (21.8) is totally analytical and it has a uniqueness theorem related. Thus equation (21.8) expresses all the possible mirror-based systems that are stigmatic. An illustrative example of equation (21.8) has been also shown and the results are as expected by the theory. Equation (21.8) can be successfully used for designing optical systems from scratch.

## References

Alvarado-Martínez J J, Granados-Agustín F S, Montiel S V Y, Vázquez-Villa A and Cornejo-Rodríguez A A 2022 Design of a compact off-axis four-mirror objective system for a thermal camera *Appl. Opt.* **61** A43–9

Bauer A and Rolland J P 2021 Roadmap for the unobscured three-mirror freeform design space *Opt. Express* **29** 26736–44

Bauer A, Schiesser E M and Rolland J P 2018 Starting geometry creation and design method for freeform optics *Nat. Commun.* **9** 1–11

Caron J and Bäumer S 2021 Progress in freeform mirror design for space applications *Int. Conf. on Space Optics—ICSO 2020* vol 11852 (Bellingham, WA: International Society for Optics and Photonics)

Duerr F and Thienpont H 2021 Freeform imaging systems: Fermat's principle unlocks 'first time right' design *Light Sci. Appl.* **10** 1–12

González-Acuña R G 2021 Two-mirror system for tunable apodization *Appl. Opt.* **60** 10756–60

González-Acuña R G 2022a Analytical equations for a nonconfocal stigmatic three-freeform-mirror system *Appl. Opt.* **61** 8317–22

González-Acuña R G 2022b Design of a pair of aplanatic mirrors *Appl. Opt.* **61** 1982–6

González-Acuña R G 2022c Set of all possible stigmatic pairs of mirrors *Appl. Opt.* **61** 2513–7

González-Acuña R G 2023 Power set of mirror-based non-symmetric stigmatic optical systems *Appl. Opt.* **62** 536–40

González-Acuña R G, Chaparro-Romo H A and Gutiérrez-Vega J C 2020 *Analytical Lens Design* (Bristol: IOP Publishing)

González-Acuña R G and Chaparro-Romo H A 2020 *Stigmatic Optics* (Bristol: IOP Publishing)

Kelley J L 2017 *General Topology* (New York: Courier Dover Publications)

Kingslake R and Johnson R B 2009 *Lens Design Fundamentals* (New York: Academic)

Liu Q, Zhou Z, Jin Y and Shen W 2016 Optical design of free-form surface two-mirror telescopic objective with ultrawide field of view *8th Int. Symp. on Advanced Optical Manufacturing and Testing Technologies: Large Mirrors and Telescopes* vol 9682 (Bellingham, WA: SPIE) pp 133–40

Meng Q, Wang H, Wang K, Wang Y, Ji Z and Wang D 2016 Off-axis three-mirror freeform telescope with a large linear field of view based on an integration mirror *Appl. Opt.* **55** 8962–70

Meng Q, Wang H, Liang W, Yan Z and Wang B 2019 Design of off-axis three-mirror systems with ultrawide field of view based on an expansion process of surface freeform and field of view *Appl. Opt.* **58** 609–15

Narasimhan B, Benitez P, Miñano J C, Chaves J, Grabovickic D, Nikolic M and Infante J 2015 Design of three freeform mirror aplanat *Novel Optical Systems Design and Optimization XVIII* vol 9579 (Bellingham, WA: SPIE) pp 125–35

Nie Y, Shafer D R, Ottevaere H, Thienpont H and Duerr F 2021 Automated freeform imaging system design with generalized ray tracing and simultaneous multi-surface analytic calculation *Opt. Express* **29** 17227–45

Optic Studio 2019 *Optic Studio User Manual 19.4* (Bellevue, WA: Zemax LLC)

Papa J C, Howard J M and Rolland J P 2018 Three-mirror freeform imagers *Optical Design and Engineering VII* vol 10690 (Bellingham, WA: International Society for Optics and Photonics) p 106901D

Papa J C, Howard J M and Rolland J P 2019 Automatic solution space exploration for freeform optical design *Freeform Optics* (Washington, DC: Optical Society of America), p FM4B–1

Quadoa Optical CAD 2022 *Quadoa Optical CAD Software Manual V* (Berlin: Quadoa Optical Systems) https://www.quadoa.com

Reshidko D and Sasian J 2018 Method for the design of nonaxially symmetric optical systems using free-form surfaces *Opt. Eng.* **57** 101704

Rolland J P, Davies M A, Suleski T J, Evans C, Bauer A, Lambropoulos J C and Falaggis K 2021 Freeform optics for imaging *Optica* **8** 161–76

Sasián J 2019 Method of confocal mirror design *Opt. Eng.* **58** 015101

Silva-Lora A and Torres R 2020 Aplanatism in stigmatic optical systems *Opt. Lett.* **45** 6390–3

Silva-Lora A and Torres R 2022 Achromatic stigmatism: achromatic Cartesian ovoid *J. Opt. Soc. Am.* A **39** 1524–32

Singer W, Totzeck M and Gross H 2005 *Handbook of Optical Systems, Vol 2: Physical Image Formation* (New York: Wiley)

Wu W, Jin G and Zhu J 2019 Optical design of the freeform reflective imaging system with wide rectangular FOV and low F-number *Results Phys.* **15** 102688

Xie Y, Mao X, Li J, Wang F, Wang P, Gao R, Li X, Ren S, Xu Z and Dong R 2020 Optical design and fabrication of an all-aluminum unobscured two-mirror freeform imaging telescope *Appl. Opt.* **59** 833–40

Yang T, Zhu J, Wu X and Jin G 2015 Direct design of freeform surfaces and freeform imaging systems with a point-by-point three-dimensional construction-iteration method *Opt. Express* **23** 10233–46

Zhang B, Jin G and Zhu J 2021 Towards automatic freeform optics design: coarse and fine search of the three-mirror solution space *Light Sci. Appl.* **10** 1–11

**IOP** Publishing

Analytical Lens Design (Second Edition)

**Rafael G González-Acuña, Héctor A Chaparro-Romo and Julio C Gutiérrez-Vega**

# Algorithms

Some of the most used algorithms in this treatise are shown in Wolfram language.

**On-axis stigmatic collector singlet lens**

```
(* parameters *)
za[ra_]:=-ra^2/(4*50);

f[ra_]:=-ta+n t+tb+Sign[ta] Sqrt[ra^2+(ta-za[ra])^2];

\[Tau][ra_]:=za[ra]-t-tb;

\[CurlyRho]z[ra_]:=((za^\[Prime])[ra] (ra+(-ta+za[ra]) (za^\[Prime])[ra]))/(-Sign[ta]n Sqrt[ra
    ^2+(ta-za[ra])^2] (1+((za^\[Prime])[ra])^2))+Sqrt[1-(ra+(-ta+za[ra]) (za^\[Prime])[ra])
    ^2/(n^2 (ra^2+(ta-za[ra])^2) (1+((za^\[Prime])[ra])^2))]/Sqrt[1+((za^\[Prime])[ra])^2];

\[CurlyRho]r[ra_]:=(ra+(-ta+za[ra]) (za^\[Prime])[ra])/(-Sign[ta]n Sqrt[ra^2+(ta-za[ra])^2]
    (1+((za^\[Prime])[ra])^2))-((za^\[Prime])[ra] Sqrt[1-(ra+(-ta+za[ra]) (za^\[Prime])[ra])^2/(
    n^2 (ra^2+(ta-za[ra])^2) (1+((za^\[Prime])[ra])^2))])/Sqrt[1+((za^\[Prime])[ra])^2];

\[CurlyTheta][ra_]:=(-n f[ra]-ra
\[CurlyRho]r[ra]-\[CurlyRho]z[ra] \[Tau][ra]+Sign[n]Sign[tb]Sqrt[(n f[ra]+ra \[CurlyRho]r[ra]+\[
    CurlyRho]z[ra] \[Tau][ra])^2-(1-n^2) (ra^2-f[ra]^2+\[Tau][ra]^2)])/(1-n^2)

zb[ra_]:=za[ra]+\[CurlyTheta][ra]\[CurlyRho]z[ra];
rb[ra_]:=ra+\[CurlyTheta][ra]\[CurlyRho]r[ra];

(* Define system configuration 's *)
n = 1.5;
t = 5;
ta = -60;
tb = 60;
Rmax = 12.5;
zb[0]

(* System plot *)
g1 = Plot[{za[ra]}, {ra, -Rmax, Rmax}, PlotStyle -> {Black},
   AspectRatio -> Automatic, AxesLabel -> {r, z}];
g2 = ParametricPlot[{rb[ra], zb[ra]}, {ra, -Rmax, Rmax},
   AspectRatio -> 1/1, AxesOrigin -> {0, 0}, PlotStyle -> {Black},
   Mesh -> None];
Show[g1, g2, PlotRange -> All]
```

 

## On-axis stigmatic collimator singlet lens

```
(* parameters *)
za[ra_]:=-ra^2/(4*50);

\[CurlyRho]z[ra_]:=((za^\[Prime])[ra] (ra+(-ta+za[ra]) (za^\[Prime])[ra]))/(-Sign[ta]n Sqrt[ra
    ^2+(ta-za[ra])^2] (1+((za^\[Prime])[ra])^2))+Sqrt[1-(ra+(-ta+za[ra]) (za^\[Prime])[ra])
    ^2/(n^2 (ra^2+(ta-za[ra])^2) (1+((za^\[Prime])[ra])^2))]/Sqrt[1+((za^\[Prime])[ra])^2];

\[CurlyRho]r[ra_]:=(ra+(-ta+za[ra]) (za^\[Prime])[ra])/(-Sign[ta]n Sqrt[ra^2+(ta-za[ra])^2]
    (1+((za^\[Prime])[ra])^2))-((za^\[Prime])[ra] Sqrt[1-(ra+(-ta+za[ra]) (za^\[Prime])[ra])^2/(
    n^2 (ra^2+(ta-za[ra])^2) (1+((za^\[Prime])[ra])^2))])/Sqrt[1+((za^\[Prime])[ra])^2];

\[CurlyTheta][ra_]:=((-1+n)t-ta+za[ra]+Sign[ta]Sqrt[ra^2+ (za[ra]-ta)^2])/(n-\[CurlyRho]z[ra]);
zb[ra_]:=za[ra]+\[CurlyTheta][ra]\[CurlyRho]z[ra];
rb[ra_]:=ra+\[CurlyTheta][ra]\[CurlyRho]r[ra];

(* Define system configuration 's *)
n = 1.5;
t = 5;
ta = -60;
Rmax = 15;
zb[0]

(* System plot *)
g1 = Plot[{za[ra]}, {ra, -Rmax, Rmax}, PlotStyle -> {Black},
   AspectRatio -> Automatic, AxesLabel -> {r, z}];
g2 = ParametricPlot[{rb[ra], zb[ra]}, {ra, -Rmax, Rmax},
   AspectRatio -> 1/1, AxesOrigin -> {0, 0}, PlotStyle -> {Black},
   Mesh -> None];
Show[g1, g2, PlotRange -> All]
```

## On-axis stigmatic singlet lens infinite object finite image

```
(* parameters *)
za[ra_]:=Module[{h},h=-ra^2/(4*50)];

f[ra_]:=Module[{h},h=n t+tb-za[ra]];

\[Tau][ra_]:=Module[{h},h=za[ra]-t-tb];

\[CurlyRho]z[ra_]:=Module[{h},h=(za^\[Prime])[ra]^2/(n+n (za^\[Prime])[ra]^2)+Sqrt
    [1+(-1+1/(1+(za^\[Prime])[ra]^2))/n^2]/Sqrt[1+(za^\[Prime])[ra]^2]];

\[CurlyRho]r[ra_]:=Module[{h},h=(za^\[Prime])[ra] (1/(n+n (za^\[Prime])[ra]^2)-Sqrt
    [1+(-1+1/(1+(za^\[Prime])[ra]^2))/n^2]/Sqrt[1+(za^\[Prime])[ra]^2])];

\[CurlyTheta][ra_]:=Module[{h},h=(-n f[ra]-ra \[CurlyRho]r[ra]-\[CurlyRho]z[ra] \[Tau][ra]+Sign[n
    ]Sign[tb]Sqrt[(n f[ra]+ra \[CurlyRho]r[ra]+\[CurlyRho]z[ra] \[Tau][ra])^2-(1-n^2) (ra^2-f[ra
    ]^2+\[Tau][ra]^2)])/(1-n^2)]

zb[ra_]:=Module[{h},h=za[ra]+\[CurlyTheta][ra]\[CurlyRho]z[ra]];

rb[ra_]:=Module[{h},h=ra+\[CurlyTheta][ra]\[CurlyRho]r[ra]];

(* Define system configuration 's *)
n = 1.5;
t = 5;
tb = 60;
Rmax = 18.5;
zb[0]

(* System plot *)
g1 = Plot[{za[ra]}, {ra, -Rmax, Rmax}, PlotStyle -> {Black},
   AspectRatio -> Automatic, AxesLabel -> {r, z}];
g2 = ParametricPlot[{rb[ra], zb[ra]}, {ra, -Rmax, Rmax},
   AspectRatio -> 1/1, AxesOrigin -> {0, 0}, PlotStyle -> {Black},
   Mesh -> None];
Show[g1, g2, PlotRange -> All]
```

## Single-lens telescope

```
(* parameters *)
za[ra_] := ra^2/80;


\[CurlyRho]z[ra_] :=
  Derivative[1][za][ra]^2/(n + n Derivative[1][za][ra]^2) + Sqrt[(
   n^2 + (-1 + n^2) Derivative[1][za][ra]^2)/(
   n^2 (1 + Derivative[1][za][ra]^2))]/Sqrt[
   1 + Derivative[1][za][ra]^2];


\[CurlyRho]r[ra_] := -((
   Derivative[1][za][
     ra] (-1 +
      n Sqrt[1 + Derivative[1][za][ra]^2] Sqrt[(
       n^2 + (-1 + n^2) Derivative[1][za][ra]^2)/(
       n^2 (1 + Derivative[1][za][ra]^2))]))/(
   n (1 + Derivative[1][za][ra]^2)));

\[CurlyTheta][ra_] := ((-1 + n) t)/(n - \[CurlyRho]z[ra])

zb[ra_] := za[ra] + \[CurlyTheta][ra] \[CurlyRho]z[ra];

rb[ra_] := ra + \[CurlyTheta][ra] \[CurlyRho]r[ra];

(* Define system configuration 's *)
n = 1.5;
t = 15;
Rmax = 20;
zb[0]

(* System plot *)
g1 = Plot[{za[ra]}, {ra, -Rmax, Rmax}, PlotStyle -> {Black},
   AspectRatio -> Automatic, AxesLabel -> {r, z}];
g2 = ParametricPlot[{rb[ra], zb[ra]}, {ra, -Rmax, Rmax},
   AspectRatio -> 1/1, AxesOrigin -> {0, 0}, PlotStyle -> {Black},
   Mesh -> None];
Show[g1, g2, PlotRange -> All]
```

## Gaxicon

```
(* parameters *)
ra[r]:=Module[{h},h=Abs[r]]

za[ra_]:=Module[{h},h=-ra^2/(4fa)]
\[Rho]z[ra_]:=Module[{h},h=(za^\[Prime])[ra]^2/(n+n (za^\[Prime])[ra]^2)+Sqrt[(n^2+(-1+n^2) (za^\[Prime])[ra]^2)/(n^2 (1+(za^\[Prime])[ra]^2))]/Sqrt[1+(za^\[Prime])[ra]^2]];

\[Rho]r[ra_]:=Module[{h},h=-(((za^\[Prime])[ra] (-1+n Sqrt[1+(za^\[Prime])[ra]^2] Sqrt[(n^2+(-1+n^2) (za^\[Prime])[ra]^2)/(n^2 (1+(za^\[Prime])[ra]^2))]))/(n (1+(za^\[Prime])[ra]^2)))];

\[Koppa][ra_]:=Module[{h},h=(t (-1+n Sqrt[1+w^2])+w ra-(-1+Sqrt[1+w^2]) za[ra])/(n Sqrt[1+w^2]-w \[Rho]r[ra]-\[Rho]z[ra])];

zb[ra_]:=Module[{h},h=za[ra]+\[Koppa][ra]\[Rho]z[ra]]

rb[ra_]:=Module[{h},h=ra+\[Koppa][ra]\[Rho]r[ra]]

(* Define system configuration 's *)
s1 = 1;
s2 = 1;
w = -0.1;
n = 1.5;
fa = 20;
t = 4;
basez = -5;
topez = 35;
rmax = 25;
zb[0.01]
rb[0.01]

(* System plot *)
g1 = Plot[{za[ra], basez, topez}, {ra, -rmax, rmax},
   PlotStyle -> {Black, Transparent, Transparent},
   AspectRatio -> Automatic, AxesLabel -> {r, z}];

g2 = ParametricPlot[{rb[ra], zb[ra]}, {ra, 0.01, rmax},
   AspectRatio -> 1/1, AxesOrigin -> {0, 0}, PlotStyle -> {Black},
   Mesh -> None];

g3 = ParametricPlot[{-rb[ra], zb[ra]}, {ra, 0.01, rmax},
   AspectRatio -> 1/1, AxesOrigin -> {0, 0}, PlotStyle -> {Black},
   Mesh -> None];
Show[g1, g2, g3]
```

## Off-axis stigmatic singlet lens

```
(* parameters *)
za[ra_]:=Module[{h},h=ra^2/100];

\[CurlyRho]r0[ra_]:=Module[{h},h=-ha/(n Sqrt[(ha)^2+(ta)^2])]

\[CurlyRho]z0[ra_]:=Module[{h},h=Sqrt[1-(-ha)^2/(n^2 ((ha)^2+(ta)^2))]]

f[ra_]:=Module[{h},h=Sqrt[(ta)^2+(ha)^2]+n t+Sqrt[(\[CurlyRho]r0[ra]t-hb)^2+(tb)^2]-Sqrt[(ra-ha)^2+(za[ra]-ta)^2]];

\[Tau][ra_]:=Module[{h},h=za[ra]-t \[CurlyRho]z0[ra]-tb];

\[CurlyRho]r[ra_]:=Module[{h},h=(-ha+ra+(-ta+za[ra]) (za^\[Prime])[ra])/(n Sqrt[(ha-ra)^2+(ta-za[ra])^2] (1+((za^\[Prime])[ra])^2))-((za^\[Prime])[ra] Sqrt[1-(-ha+ra+(-ta+za[ra]) (za^\[Prime])[ra])^2/(n^2 ((ha-ra)^2+(ta-za[ra])^2) (1+((za^\[Prime])[ra])^2))])/Sqrt[1+((za^\[Prime])[ra])^2]];

\[CurlyRho]z[ra_]:=Module[{h},h=((za^\[Prime]) [ra](-ha+ra+(-ta+za[ra]) (za^\[Prime])[ra]))/(n Sqrt[(ha-ra)^2+(ta-za[ra])^2] (1+((za^\[Prime])[ra])^2))+ Sqrt[1-(-ha+ra+(-ta+za[ra]) (za^\[Prime])[ra])^2/(n^2 ((ha-ra)^2+(ta-za[ra])^2) (1+((za^\[Prime])[ra])^2))]/Sqrt[1+((za^\[Prime])[ra])^2]];

\[CurlyTheta][ra_]:=Module[{h},h=(-n f[ra]-(ra-hb) \[CurlyRho]r[ra]-\[CurlyRho]z[ra] \[Tau][ra]+s1 Sqrt[(n f[ra]+(ra-hb)\[CurlyRho]r[ra]+\[CurlyRho]z[ra] \[Tau][ra])^2-(1-n^2) ((ra-hb)^2-f[ra]^2+\[Tau][ra]^2)])/(1-n^2)]

zb[ra_]:=Module[{h},h=za[ra]+\[CurlyTheta][ra]\[CurlyRho]z[ra]];

rb[ra_]:=Module[{h},h=ra+\[CurlyTheta][ra]\[CurlyRho]r[ra]];

(* Define system configuration 's *)
s1=1;
n=1.5;
t=10;
ta=-20;
tb=25;
ha=5;
hb=-3;
Rmax=14;
Sqrt[(zb[0]-za[0])^2+(rb[0]-0)^2]

(* System plot *)
g1=Plot[{za[ra]},{ra,-Rmax,Rmax},PlotStyle->{Black},AspectRatio->Automatic,AxesLabel->{r,z}];

g2=ParametricPlot[{rb[ra],zb[ra]},{ra,-Rmax,Rmax},AspectRatio->1/1, AxesOrigin->{0,0}, PlotStyle->{Black}, Mesh ->None];
Show[g1,g2,PlotRange->All]
```

## On-axis stigmatic triplet lens

```
<< MaTeX`

n0=1;
n1=1.5;
n2=1;
n3=1.5;
n4=1;

f1=60;
f2=80;
f3=−70;

t0=−50;
t1=8;
t2=11;
t3=9;
t4=90;
t=t1+t2+t3+t4;

step=0.05; (*0.05*)
rmax=17.5;
density=1;
d=22;

Z1[R1_]:=R1^2/(4f1);

na[ra_]:=Sqrt[Z1'[ra]^2+1];

Zz2[R2_]:=t1−R2^2/(4f2);

Zz3[R3_]:=t1+t2−(R3)^2/(4f3 ) ;

z1=Table[Z1[R1],{R1,step,Floor[rmax],step}];

dz1=Table[Z1'[R1],{R1,step,Floor[rmax],step}];

r1=Table[R1,{R1,step,Floor[rmax],step}];

zz2=Table[Zz2[R2],{R2,step,Floor[rmax],step}];

zz3=Table[Zz3[R3],{R3,step,Floor[rmax],step}];



L1=Join[Reverse[Table[{z1[[i]], r1 [[ i ]]},{ i ,1, Length[z1]}]], List [{0,0}], Table[{z1[[ i ]],−r1 [[ i ]]},{ i
    ,1, Length[z1]}]];

\[CurlyRho]z1=(n0(dz1(r1+(−t0+z1) dz1)))/(n1 Sqrt[r1^2+(t0−z1)^2] (1+(dz1)^2))+Sqrt[1−(n0^2
    (r1+(−t0+z1) dz1)^2)/(n1^2 (r1^2+(t0−z1)^2) ((1+(dz1)^2)))]/Sqrt[(1+(dz1)^2)];

\[CurlyRho]r1=(n0(r1+(−t0+z1) dz1))/(n1 Sqrt[r1^2+(t0−z1)^2] (1+(dz1)^2))−(dz1 Sqrt[1−(n0^2
    (r1+(−t0+z1) dz1)^2)/(n1^2 (r1^2+(t0−z1)^2) ((1+(dz1)^2)))])/Sqrt[(1+(dz1)^2)];

```

```
\[CurlyTheta]1=1/\[CurlyRho]r1^2 (−r1 \[CurlyRho]r1−2 f2 \[CurlyRho]z1+2 Sqrt[f2] Sqrt[\[
    CurlyRho]r1^2 (t1−z1)+r1 \[CurlyRho]r1 \[CurlyRho]z1+f2 \[CurlyRho]z1^2]);

r2=r1+\[CurlyTheta]1 \[CurlyRho]r1;

z2=z1+\[CurlyTheta]1 \[CurlyRho]z1;

L2=Join[Reverse[Table[{Re[z2[[i]]], Re[r2[[ i ]]]},{ i ,1, Length[z2]}]], List [{t1 ,0}], Table[{Re[z2[[i
    ]]],− Re[r2 [[ i ]]]},{ i ,1, Length[z2]}]];

dr21=ListConvolve[{1,−1},r2]/ListConvolve[{1,−1},r1];

dr2=Join[dr21,{Last[dr21]}];

dz21=ListConvolve[{1,−1},z2]/ListConvolve[{1,−1},r1];

dz2=Join[dz21,{Last[dz21]}];

\[CurlyRho]r2=−((n1 dr2((r1−r2) dr2+(z1−z2)dz2))/(n2 Sqrt[(r1−r2)^2+(z1−z2)^2] (dr2^2+dz2^2)
    ))−(dz2 Sqrt[1−(n1^2 ((r1−r2) dr2+(z1−z2)dz2)^2)/(n2^2 ((r1−r2)^2+(z1−z2)^2) (dr2^2+
    dz2^2))])/Sqrt[(dr2^2+dz2^2)];

\[CurlyRho]z2=−((n1 dz2 ((r1−r2) dr2+(z1−z2)dz2))/(n2 Sqrt[(r1−r2)^2+(z1−z2)^2] (dr2^2+dz2
    ^2)))+(dr2 Sqrt[1−(n1^2 ((r1−r2) dr2+(z1−z2)dz2)^2)/(n2^2 ((r1−r2)^2+(z1−z2)^2) (dr2
    ^2+dz2^2))])/Sqrt[(dr2^2+dz2^2)];

\[CurlyTheta]2=1/\[CurlyRho]r2^2 (−r2 \[CurlyRho]r2−2 f3 \[CurlyRho]z2+2 Sqrt[f3] \[Sqrt]((t1+t2
    −z2) \[CurlyRho]r2^2+r2 \[CurlyRho]r2 \[CurlyRho]z2+f3 \[CurlyRho]z2^2));

r3=r2+\[CurlyTheta]2 \[CurlyRho]r2;

z3=z2+\[CurlyTheta]2 \[CurlyRho]z2;

L3=Join[Reverse[Table[{Re[z3[[i]]], Re[r3 [[ i ]]]},{ i ,1, Length[z3]}]], List [{t1+t2,0}], Table[{Re[z3[[i
    ]]],− Re[r3 [[ i ]]]},{ i ,1, Length[z3]}]];

dr31=ListConvolve[{1,−1},r3]/ListConvolve[{1,−1},r1];

dr3=Join[dr31,{Last[dr31]}];

dz31=ListConvolve[{1,−1},z3]/ListConvolve[{1,−1},r1];

dz3=Join[dz31,{Last[dz31]}];

\[CurlyRho]r3=−((n2 dr3((r2−r3) dr3+(z2−z3) dz3))/(n3 Sqrt[(r2−r3)^2+(z2−z3)^2] ((dr3)^2+(dz3
    )^2)))−(dz3 Sqrt[1−(n2^2 ((r2−r3) dr3+(z2−z3) dz3)^2)/(n3^2 ((r2−r3)^2+(z2−z3)^2)((dr3)
    ^2+(dz3)^2))])/Sqrt[((dr3)^2+(dz3)^2)];

\[CurlyRho]z3=−((n2 dz3 ((r2−r3) dr3+(z2−z3) dz3))/(n3 Sqrt[(r2−r3)^2+(z2−z3)^2] ((dr3)^2+(
    dz3)^2)))+(dr3 Sqrt[1−(n2^2 ((r2−r3) dr3+(z2−z3) dz3)^2)/(n3^2 ((r2−r3)^2+(z2−z3)^2)((
    dr3)^2+(dz3)^2))])/Sqrt[((dr3)^2+(dz3)^2)];

F=−n0 t0+n1 t1+n2 t2+n3 t3+n4 t4−(n0 Sqrt[r1^2+(−t0+z1)^2]+n1 Sqrt[(−r1+r2)^2+(−z1+z2)
    ^2]+n2 Sqrt[(−r2+r3)^2+(−z2+z3)^2]);
```

```
\[CurlyTheta]3=1/(n3^2-n4^2) (F n3+n4^2 (r3 \[CurlyRho]r3+(-t+z3) \[CurlyRho]z3)-1/2 \[Sqrt
    ](-4 (-n3^2+n4^2) (-F^2+n4^2 (r3^2+(t-z3)^2))+4 (F n3+n4^2 (r3 \[CurlyRho]r3+(-t+z3)
    \[CurlyRho]z3))^2));

r4=r3+\[CurlyTheta]3 \[CurlyRho]r3;

z4=z3+\[CurlyTheta]3 \[CurlyRho]z3;

L4=Join[Reverse[Table[{Re[z4[[i]]], Re[r4[[ i ]]]},{ i ,1, Length[z4]}]], List [{t1+t2+t3,0}],Table[{Re[
    z4[[i]]],- Re[r4[[ i ]]]},{ i ,1, Length[z4]}]];

dr41=ListConvolve[{1,-1},r4]/ListConvolve[{1,-1},r1];
dr4=Join[dr41,{Last[dr41]}];

dz41=ListConvolve[{1,-1},z4]/ListConvolve[{1,-1},r1];
dz4=Join[dz41,{Last[dz41]}];

\[CurlyRho]r4=-((n3 dr4((r3-r4) dr4+(z3-z4) dz4))/(n4 Sqrt[(r3-r4)^2+(z3-z4)^2] ((dr4)^2+(dz4
    )^2)))-(dz4 Sqrt[1-(n3^2 ((r3-r4) dr4+(z3-z4) dz4)^2)/(n4^2 ((r3-r4)^2+(z3-z4)^2)((dr4)
    ^2+(dz4)^2))])/Sqrt[((dr4)^2+(dz4)^2)];

\[CurlyRho]z4=-((n3 dz4 ((r3-r4) dr4+(z3-z4) dz4))/(n4 Sqrt[(r3-r4)^2+(z3-z4)^2] ((dr4)^2+(
    dz4)^2)))+(dr4 Sqrt[1-(n3^2 ((r3-r4) dr4+(z3-z4) dz4)^2)/(n4^2 ((r3-r4)^2+(z3-z4)^2)((
    dr4)^2+(dz4)^2))])/Sqrt[((dr4)^2+(dz4)^2)];

Surface := {ParametricPlot[{ra^2/(4 f1), ra}, {ra, -rmax, rmax},
    PlotStyle -> {Black}, Axes -> False,
    PlotRange -> {{-60, 130}, {-1.5 rmax, 1.5 rmax}}],
  ParametricPlot[{t1 - ra^2/(4 f2), ra}, {ra, L2[[1, 2]],
    L2[[Length[L2], 2]]}, PlotStyle -> {Black}],
  ParametricPlot[{t1 + t2 - ra^2/(4 f3), ra}, {ra, L3[[1, 2]],
    L3[[Length[L3], 2]]}, PlotStyle -> {Black}], ListLinePlot[{L4}],
  Graphics[{Black,
    Line[{{rmax^2/(4 f1), rmax}, {rmax^2/(4 f2), rmax}}]}]}

Show[Surface, ImageSize -> {1280, Automatic}]
```

www.ingramcontent.com/pod-product-compliance
Ingram Content Group UK Ltd.
Pitfield, Milton Keynes, MK11 3LW, UK
UKHW051904100926
8556IPUK00005B/762

9 780750 357722